SOLAR ENERGY

Renewable Energy and the Environment

ENERGY AND THE ENVIRONMENT

SERIES EDITOR
Abbas Ghassemi
New Mexico State University

PUBLISHED TITLES

Solar Energy: Renewable Energy and the Environment
Robert Foster, Majid Ghassemi, Alma Cota

Wind Energy: Renewable Energy and the Environment
Vaughn Nelson

SOLAR ENERGY

Renewable Energy and the Environment

Robert Foster
Majid Ghassemi
Alma Cota

CRC Press
Taylor & Francis Group
Boca Raton London New York

CRC Press is an imprint of the
Taylor & Francis Group, an **informa** business

CRC Press
Taylor & Francis Group
6000 Broken Sound Parkway NW, Suite 300
Boca Raton, FL 33487-2742

© 2010 by Taylor and Francis Group, LLC
CRC Press is an imprint of Taylor & Francis Group, an Informa business

International Standard Book Number: 978-1-4200-7566-3 (Hardback)

Library of Congress Cataloging-in-Publication Data

Solar energy : renewable energy and the environment / Robert Foster ... [et al.].
    p. cm. -- (Energy and the environment)
  Includes bibliographical references and index.
  ISBN 978-1-4200-7566-3 (hardcover : alk. paper)
   1. Solar energy. 2. Renewable energy sources--Environmental aspects. I. Foster, Robert, 1962 Apr. 25- II. Title. III. Series.

TJ810.S4897 2009
621.47--dc22   2009014967

Visit the Taylor & Francis Web site at
http://www.taylorandfrancis.com

and the CRC Press Web site at
http://www.crcpress.com

Contents

Series Preface

By 2050 the demand for energy could double or even triple as the global population grows and developing countries expand their economies. All life on Earth depends on energy and the cycling of carbon. Energy is essential for economic and social development and also poses an environmental challenge. We must explore all aspects of energy production and consumption, including energy efficiency, clean energy, the global carbon cycle, carbon sources, and sinks and biomass, as well as their relationship to climate and natural resource issues. Knowledge of energy has allowed humans to flourish in numbers unimaginable to our ancestors.

The world's dependence on fossil fuels began approximately 200 years ago. Are we running out of oil? No, but we are certainly running out of the affordable oil that has powered the world economy since the 1950s. We know how to recover fossil fuels and harvest their energy for operating power plants, planes, trains, and automobiles; this leads to modifying the carbon cycle and additional greenhouse gas emissions. The result has been the debate on availability of fossil energy resources; peak oil era and timing for anticipated end of the fossil fuel era; price and environmental impact versus various renewable resources and use; carbon footprint; and emissions and control, including cap and trade and emergence of "green power."

Our current consumption has largely relied on oil for mobile applications and coal, natural gas, and nuclear or water power for stationary applications. In order to address the energy issues in a comprehensive manner, it is vital to consider the complexity of energy. Any energy resource, including oil, coal, wind, and biomass, is an element of a complex supply chain and must be considered in its entirety as a system from production through consumption. All of the elements of the system are interrelated and interdependent. Oil, for example, requires consideration for interlinking of all of the elements, including exploration, drilling, production, water, transportation, refining, refinery products and byproducts, waste, environmental impact, distribution, consumption/application, and, finally, emissions.

Inefficiencies in any part of the system have an impact on the overall system, and disruption in one of these elements causes major interruption in consumption. As we have experienced in the past, interrupted exploration will result in disruption in production, restricted refining and distribution, and consumption shortages. Therefore, any proposed energy solution requires careful evaluation and, as such, may be one of the key barriers to implementing the proposed use of hydrogen as a mobile fuel.

Even though an admirable level of effort has gone into improving the efficiency of fuel sources for delivery of energy, we are faced with severe challenges on many fronts. These include population growth, emerging economies, new and expanded usage, and limited natural resources. All energy solutions include some level of risk, including technology snafus, changes in market demand, and economic drivers. This is particularly true when proposing an energy solution involving implementation of untested alternative energy technologies.

There are concerns that emissions from fossil fuels will lead to changing climate with possibly disastrous consequences. Over the past five decades, the world's collective greenhouse gas emissions have increased significantly—even as increasing efficiency has resulted in extending energy benefits to more of the population. Many propose that we improve the efficiency of energy use and conserve resources to lessen greenhouse gas emissions and avoid a climate catastrophe. Using fossil fuels more efficiently has not reduced overall greenhouse gas emissions for various reasons, and it is unlikely that such initiatives will have a perceptible effect on atmospheric greenhouse gas content. Although the correlation between energy use and greenhouse gas emissions is debatable, there are effective means to produce energy, even from fossil fuels, while controlling emissions. Emerging technologies and engineered alternatives will also manage the makeup of the atmosphere, but will require significant understanding and careful use of energy.

We need to step back and reconsider our role in and knowledge of energy use. The traditional approach of micromanagement of greenhouse gas emissions is not feasible or functional over a long period of time. More assertive methods to influence the carbon cycle are needed and will be emerging in the coming years. Modifications to the cycle mean that we must look at all options in managing atmospheric greenhouse gases, including various ways to produce, consume, and deal with energy. We need to be willing to face reality and search in earnest for alternative energy solutions. Some technologies appear to be able to assist; however, all may not be viable. The proposed solutions must not be in terms of a "quick approach," but rather as a more comprehensive, long-term (10, 25, and 50+ years) approach based on science and utilizing aggressive research and development. The proposed solutions must be capable of being retrofitted into our existing energy chain. In the meantime, we must continually seek to increase the efficiency of converting energy into heat and power.

One of the best ways to define sustainable development is through long-term, affordable availability of resources, including energy. There are many potential constraints to sustainable development. Foremost of these is the competition for water use in energy production, manufacturing, and farming versus a shortage of fresh water for consumption and development. Sustainable development is also dependent on the Earth's limited amount of soil; in the not too distant future, we will have to restore and build soil as a part of sustainable development. Hence, possible solutions must be comprehensive and based on integrating our energy use with nature's management of carbon, water, and life on Earth as represented by the carbon and hydrogeological cycles.

Obviously, the challenges presented by the need to control atmospheric greenhouse gases are enormous and require "out of the box" thinking, innovative approaches, imagination, and bold engineering initiatives in order to achieve sustainable development. We will need to exploit energy even more ingeniously and integrate its use with control of atmospheric greenhouse gases. The continued development and application of energy is essential to the development of human society in a sustainable manner through the coming centuries.

All alternative energy technologies are not equal; they have various risks and drawbacks. When evaluating our energy options, we must consider all aspects, including performance against known criteria, basic economics and benefits, efficiency, processing and utilization requirements, infrastructure requirements, subsidies and credits, and waste and the ecosystem, as well as unintended consequences such as impacts on natural resources and the environment. Additionally, we must include the overall changes and the emerging energy picture based on current and future efforts to modify fossil fuels and evaluate the energy return for the investment of funds and other natural resources such as water.

A significant driver in creating this book series focused on alternative energy and the environment and was initiated as a consequence of lecturing around the country and in the classroom on the subject of energy, environment, and natural resources such as water. Water is a precious commodity in the West in general and the Southwest in particular and has a significant impact on energy production, including alternative sources, due to the nexus between energy and water and the major correlation with the environment and sustainability-related issues. The correlation among these elements, how they relate to each other, and the impact of one on the other are understood; however, integration and utilization of alternative energy resources into the energy matrix has not been significantly debated.

Also, as renewable technology implementation grows by various states nationally and internationally, the need for informed and trained human resources continues to be a significant driver in future employment. This has resulted in universities, community colleges, and trade schools offering minors, certificate programs, and, in some cases, majors in renewable energy and sustainability. As the field grows, the demand increases for trained operators, engineers, designers, and architects able to incorporate these technologies into their daily activity. Additionally, we receive daily deluges of flyers, e-mails, and texts on various short courses available for parties interested in solar, wind, geothermal, biomass, and other types of energy. These are under the umbrella of retooling an individual's career and providing the trained resources needed to interact with financial, governmental, and industrial organizations.

In all my interactions in this field throughout the years, I have conducted significant searches for integrated textbooks that explain alternative energy resources in a suitable manner that would complement a syllabus for a potential course to be taught at the university and provide good reference material for parties getting involved in this field. I have been able to locate a number of books on the subject matter related to energy; energy systems; and resources such as fossil nuclear, renewable energy, and energy conversion, as well as specific books on the subjects of natural resource availability, use, and impact as related to energy and environment. However, books that are correlated and present the various subjects in detail are few and far between.

We have therefore started a series in which each text addresses specific technology fields in the renewable energy arena. As a part of this series, there are textbooks on wind, solar, geothermal, biomass, hydro, and other energy forms yet to be developed. Our texts are intended for upper level undergraduate and graduate students and informed readers who have a solid fundamental understanding of science and mathematics. Individuals and organizations that are involved with design development of the renewable energy field entities and interested in having reference material available to their scientists and engineers, consulting organizations, and reference libraries will also be interested in these texts. Each book presents fundamentals as well as a series of numerical and conceptual problems designed to stimulate creative thinking and problem solving.

I wish to express my deep gratitude to my wife, Maryam, who has served as a motivator and intellectual companion and too often has been the victim of this effort. Her support, encouragement, patience, and involvement have been essential to the completion of this series.

Abbas Ghassemi, PhD

The Series Editor

Dr. Abbas Ghassemi is the director of Institute for Energy and Environment (IE&E) and professor of chemical engineering at New Mexico State University. In addition to teaching and research, he oversees the operations of WERC: A Consortium for Environmental Education and Technology Development, the Southwest Technology Development Institute (SWTDI), and the Carlsbad Environmental Monitoring and Research Center (CEMRC) and has been involved in energy, water, risk assessment, process control, pollution prevention, and waste minimization areas for a number of industries throughout the United States for the past 20 years. He has also successfully led and managed a number of peer-reviewed scientific evaluations of environmental, water, and energy programs for the U.S. Department of Energy, the U.S. Environmental Protection Agency, national laboratories, and industry. Dr. Ghassemi has over 30 years of industrial and academic experience in risk assessment and decision theory; renewable energy; water quality and quantity; pollution control technology and prevention; energy efficiency; process control, management, and modification; waste management; and environmental restoration. He has authored and edited several textbooks and many publications and papers in the areas of energy, water, waste management, process control, sensors, thermodynamics, transport phenomena, education management, and innovative teaching methods. Dr. Ghassemi serves on a number of public and private boards, editorial boards, and peer-review panels and holds MS and PhD degrees in chemical engineering, with minors in statistics and mathematics, from New Mexico State University, and a BS degree in chemical engineering, with a minor in mathematics, from the University of Oklahoma.

Preface

The twenty-first century is rapidly becoming the "perfect energy storm"; modern society is faced with volatile energy prices and growing environmental concerns, as well as energy supply and security issues. Today's society was founded on hydrocarbon fuel—a finite resource that already is one of the main catalysts for international conflicts, which is likely to intensify in the future. The global energy appetite is enormous, representing over $6 trillion per year, or about 13% of global gross domestic product (GDP). Unfortunately, the vast majority of this energy is not efficiently utilized for buildings, vehicles, or industry. This is especially true in the United States, which has about double the per-capita and GDP energy usage rates as compared to the European Union and Japan. The inefficient use of energy strongly exacerbates the global energy crisis. It is time to shed the outdated "burn, baby, burn" hydrocarbon energy thinking with a new energy vision; the time for clean energy solutions is here. Only through energy efficiency and renewable energy technologies can modern civilization extricate itself from the gathering perfect energy storm.

The United States is addicted to the consumption of fossil fuels. The country obtains about two-fifths of its energy from petroleum, about one-fourth from coal, and another quarter from natural gas. Two-thirds of oil in the United States is imported; if business continues as usual, by 2020, the country will import three-fourths of its oil. In 2006, the United States spent $384 billion on imported oil. By 2030, carbon fuels will still account for 86% of U.S. energy use with a business-as-usual approach. The United States uses about 100 quadrillion BTUs (29,000 TWh) annually. From this, 39% is energy for buildings, 33% for industry, and 28% for transportation. On average, the country uses 1.4 times more energy than the European Union and Japan in industry, 2.5 times more energy in buildings, and 1.8 times more in transportation. Like the United States, these countries are very much dependent on oil imports. However, in comparison to the United States, Japan uses only 53% energy per capita and 52% energy per GDP, while the European Union uses only 48% and 64%, respectively.

The new global energy realities have brought the highest energy prices in history. Sustained price volatility will continue, with large spikes and drops of energy prices tracking global economic trends. Peak oil is predicted by many within the next decade. The North American energy infrastructure and workforce are aging. China and India are now new global energy customers causing major impacts on primary fuel prices. By 2030, China is projected to import as much oil as the United States does now. Trigger events such as blackouts, hurricanes, floods, and fires further increase volatility due to tight supplies. Food, metal, and transportation prices are rising as a result of increased energy demand.

In addition to costs and availability of fossil fuels, a worse panorama results from counting the increase of the millions of tons per year of carbon dioxide emissions—the main gas precursor of the greenhouse effect. Future CO_2 emission increments will be originated mainly in developing countries as population and industry grow. The current CO_2 average concentration in the atmosphere is about 400 parts per million (ppm)—the highest ever experienced by the Earth. Maintaining as much reliance on fossil fuels as today, by 2050, such concentration may exceed 700 or 800 ppm.

At higher concentration, the few degrees gained in Earth's average temperature exert several grave impacts on food safety, water, the ecosystem, and the environment. Currently, only half a Celsius degree increase has been enough for catastrophic natural disasters to occur. To limit sea level rise to only 1 m and species loss to 20% by the end of this century, additional warming must be limited to 1°C. This means stabilizing atmospheric CO_2 at about 450–500 ppm. The United States is the second largest emitter of CO_2 emissions after China. The United States currently emits 23% of global CO_2 and needs to reduce CO_2 by 60 to 80% by midcentury. If the Greenland ice sheet melted, global sea level would rise 7 m; if East and West Antarctica ice sheets melt, sea levels would rise an additional 70 m. Through the widespread burning of fossil fuels, humanity is creating the largest ecological disaster since the disappearance of the dinosaurs.

However, all is not bad news. There are options to slow the detriment of the natural media; appropriate use of resources is the key. During the last decade, a great level of consciousness of climate change and energy was achieved around the world and, most importantly, among governments. Countries need a safe, clean, secure, and affordable energy future. Reduced reliance on oil and a switchover to clean technologies will create new local jobs. Millions of under- or unemployed people in Africa, Asia, the Middle East, etc., could find gainful employment in this new sector. To start switching, policies must be created to move toward clean and sustainable energy solutions. Requiring significant energy production from renewable energy sources and increasing energy efficiency are two basic steps toward a more secure and clean energy future that we can take now.

The United States can increase energy production from clean energy sources like the Sun and wind. States such as California, New Mexico, and Texas have already begun to lead the way with renewable energy portfolio standards. The Obama administration has proceeded to set national standards that require an increasing amount of electricity to come from renewable energy resources like solar, wind, and geothermal energy. Execution of the president's plan of 10% renewable energy generation by 2012 and 25% by 2025 is greatly needed. Wind energy development is already booming in the United States due to state portfolio requirements and the federal production tax credit. The United States now has over 25,000 MW of wind power, producing 1% of the nation's electricity, with another 8,500 MW under construction. The goals of the Department of Energy are that 20% of the nation's electricity must be generated from wind power; this requires about 300,000 MW of wind, which is an achievable goal with plenty of wind availability in the Midwest.

Despite three decades of heavy investment in electrification projects by less developed nations—often at huge environmental and social costs—about 2 billion people in developing regions still lack electricity for basic needs and economic growth. Hundreds of millions of households around the globe rely solely on kerosene lamps for lighting, disposable batteries for radios, and, in some cases, car batteries recharged weekly for television. These people have no access to good health care, education, or reliable income. For most of them, there is little likelihood of receiving electricity from conventional grid sources in the foreseeable future. Renewable energy sources can provide local jobs while improving their standard of living.

The cost of bringing utility power via transmission and distribution lines to nonelectrified villages is high, especially considering the typically small household electrical loads and the fact that many villages are located at great distances, over difficult terrain, from the existing grid. Stand-alone solar and wind energy systems can cost-effectively provide modest levels of power for lighting, communication, fans, refrigerators, water pumping, etc. Using a least-cost model, some governments and national utilities, such as in Brazil, China, Central America, South Africa, Mexico, and elsewhere, have used photovoltaic (PV) and wind systems, in an integrated development tool for electrification planning, as either centralized or distributed solutions.

Solar and wind energy are now providing the lowest cost options for economic and community development in rural regions around the globe, while supplying electricity, creating local jobs, and promoting economic development with clean energy resources. Rural regions in the Americas will greatly benefit from solar and wind electrification in the coming years. PV technology provides power for remote water pumping and for disinfection of community water supplies. For larger load requirements, the combination of PV and wind technologies, with a diesel generator and battery storage, into hybrid configurations provides higher system reliability at a more reasonable cost than with any one technology alone.

Large-scale wind systems are becoming economically attractive at $0.06–0.08 kWh for bulk utility electric power generation—large-scale solar thermal systems cost is approximately double this value. Although not as economically attractive as wind and solar thermal power for bulk power generation, PV has an even more important role to play in rural regions as a power source for remote and distributed applications due to its reliability and inherent modularity. PV energy costs have declined from about \$60/kWh in 1970 to \$1/kWh in 1980 to under \$0.25/kWh for grid-tied installations today. Module efficiencies have increased with commercially available modules that are 15–22% efficient, and research laboratory cells demonstrate efficiencies above 40%. Commercial PV module reliability has improved to last 30 years or longer.

This book intends to provide field engineers and engineering students with detailed knowledge for converting solar radiation into a suitable energy supply. Within this book, solar energy technical fundamentals are presented to give a clear understanding on how solar energy can be captured for later use. Such energy can be collected by two types of devices: thermosolar collectors, which transform solar energy into heat, and PV modules, which directly convert the energy intrinsic within light into electricity. Other important types of solar receivers use mirrors or lenses to redirect solar radiation toward a solar collector; the purpose is to focus as much energy as possible into a particular point or volume.

The authors have a century of solar energy experience among them and have conducted extensive solar research and project implementation around the globe, much of which is cited in this book. Although great technical advances in solar technology have been made, many solar energy system installations have failed—often due to simple causes; the lessons learned are also discussed in this book. For this reason, special emphasis has been placed on the practical aspects of solar technology implementation. Economics, politics, capacity building, technical capabilities, market building, and replication are the main supporting actors to develop a solar energy future that provides local jobs in troubled regions, supplies clean energy, and reduces global warming emissions. As the worldwide perfect energy storm approaches, solar energy will be one of the keys to lessening its potentially harmful impacts. The authors hope that the students and readers who use this book will be inspired to pursue a clean energy future and will choose the solar path.

Acknowledgments

As one of the authors, I would like to acknowledge the help and support of scores of dedicated people and renewable energy development program colleagues I have worked with over the years. Special thanks goes to the past and present staff, students, and contractors at New Mexico State University, including Omar Carrillo, Luis Estrada, Martín Gomez, Gabriela Cisneros, Abraham Ellis, Soumen Ghosh, Lisa Büttner, Ronald Donaghe, Steven Durand, Cary Lane, Marty Lopez, Sherry Mills, Laura Orta, Ron Polka, Vern Risser, Martín Romero, Rudi Schoenmackers, Therese Shakra, Sorn Stoll, Kinney Stevens, Anita Tafoya, Gloria Vásquez, John Wiles, and Walter Zachritz, all of whose mentorships are reflected within these pages. Thanks to my talented brother James Foster for helping with some of the drawings used in the book.

I also want to thank the past USAID/DOE Mexico Renewable Energy Program (MREP) team core and especially to Charles Hanley, Vipin Gupta, Warren Cox, Max Harcourt, Elizabeth Richards, Ron Pate, Gray Lowrey, and John Strachan at Sandia National Laboratories. Thanks to the USAID staffers who "got it" and understood the power of renewables as a tool within development programs, especially to Art Danart, Patricia Flanagan, Jorge Landa, John Naar, Odalis Perez, Ross Pumfrey, Frank Zadroga. Credit also goes to Bud Annan formerly with DOE who made MREP possible. Likewise kudos to Richard Hansen and Eric Johnson of Enersol/GTC, Michael Cormier. Steve Cook, and Sharon Eby Cornet of EPSEA, Mike Ewert of NASA, David Corbus, Ian Baring-Gould, Larry Flowers, and David Renee of NREL, Alberto Rodriguez of Peace Corps DR, David Bergeron and Billy Amos of SunDanzer Refrigeration, Lloyd Hoffstatter and Dave Panico of SunWize, Chris Rovero and Bikash Pandey of Winrock International, Ken Starcher of WTAMU, Ernesto Terrado of the World Bank, Mike Bergey, Windy Dankoff, Shannon Graham, Ron Kenedi, Andy Kruse, Ivonne Maldonado, Larry Mills, Rob Muhn, Ken Olsen, Ron Orozco, Terry Schuyler, and Pete Smith.

The project content would not have been possible without the hard work of our global counterparts in the field, especially Marcela Ascencio, Arnoldo Bautista, Marco Borja, Rafael Cabanillas, José Luis Esparza, Claudio Estrada, Carlos Flores, Rodolfo Martínez, Octavio Montufar, Victor Meraz, Lilia Ojinaga Ray, Jesús Parada, Arturo Romero, Aarón Sánchez, and Adolfo Tres Palacios of Mexico; Jorge Lima of Brazil; Pablo Espinoza and Raul Sapiain of Chile; Danilo Carranza, Janeybi Faringthon, Héctor Luis Mercedes of the Dominican Republic; Hugo Arriaza, Ivan Azurdia, Carolina Palma, and Saul Santos of Guatemala; Christiam Aguilar, Loyda Alonso, Ethel Enamorado Davis, Leonardo Matute, and Diana Solis of Honduras; Izumi Kaizuka of Japan; Susan Kinney, Elieneth Lara, and Herminia Martínez of Nicaragua; Deon Raubenheimer of South Africa; and my deep appreciation to the many, many other solar pioneers too numerous to mention here that I have had the pleasure to work and journey with over the years.

Robert E. Foster

The Authors

Robert Foster has a quarter century of experience applying solar and wind energy technologies and has implemented hundreds of solar and wind projects in over 30 countries. He has worked since 1989 at New Mexico State University (NMSU) as a program manager for the College of Engineering at the Southwest Region Solar Experiment Station and the Institute for Energy and the Environment. He is presently on assignment for NMSU in Kabul as the deputy chief of party for the U.S. Agency for International Development (USAID), Afghanistan Water, Agriculture, and Technology Transfer Program. He has assisted with numerous renewable energy programs for the U.S. Department of Energy, National Renewable Energy Laboratory, Sandia National Laboratories, USAID, National Aeronautics and Space Administration, National Science Foundation, Winrock International, World Bank, Institute of International Education, industry, utilities, and foreign governments. He was the technical manager for Sandia National Laboratories under the USAID/DOE Mexico Renewable Energy Program from 1992 to 2005, as well as technical advisor for Winrock International for the USAID Electrical Sector Restructuring Project in the Dominican Republic from 1997 to 1999.

Mr. Foster is a returned Peace Corps Volunteer from the Dominican Republic (1985–1988), where he built community water supply projects and worked with pioneering the use of rural PV systems for developing countries with Enersol Associates. Prior to that, he worked at Cole Solar Systems in Austin, Texas fabricating and installing solar hot-water systems. He holds patents on solar distillation and cofounded SolAqua, Inc., which fabricates solar water purification systems in Texas. He received the governor's award for renewable energy development in the state of Chihuahua, Mexico, and was also honored with the Guatemalan Renewable Energy Award by the Fundación Solar. Mr. Foster holds a BS degree in mechanical engineering from the University of Texas at Austin and an MBA from NMSU, where he completed his thesis on the Mexican PV market. He is past chairman and board member of the Texas Solar Energy Society and the El Paso Solar Energy Association. He has published over 120 papers and articles and 90 technical reports on solar energy, wind, energy, evaporative cooling, waste heat, and geothermal energy. He has taught more than 160 technical workshops on renewable energy technologies for thousands of engineers and technicians around the globe.

Majid Ghassemi is a research associate professor at the New Mexico Institute of Mining and Technology (NM Tech) in the Institute for Engineering Research and Applications, where he is currently conducting research on energy-efficient wall panels for the U.S. Department of Energy (DOE). He is also a co-principal investigator for DOE on atmospheric waste reduction through energy-efficient, PV-powered building construction. He arrived at NM Tech in 2002 as an associate professor and has worked in various programs, including the areas of sustainable energy and energy efficiency at the Magdalena Ridge Observatory. He has worked with MIT and General Electric researchers on wind energy in New Mexico and has also researched hydrogen production by solar energy for fuel cell use. Dr. Ghassemi has assisted the Institute for Engineering Research Applications with energy conservation projects and microelectromagnetic pumps and liquid metal heat pipes for space applications. He has taught courses in thermodynamics, heat transfer, and thermal fluid systems' design. In 2002, he served as an associate visiting professor with the University of Texas at El Paso, where he worked on fuel cells and solar water purification systems. He is currently an asssociate professor with K. N. Toosi University in Tehran, Iran, where he is responsible for teaching undergraduate as well as graduate courses in the area of thermal science, including advanced conduction heat transfer, convection and heat transfer, fundamentals of heat transfer, and thermodynamics.

Dr. Ghassemi has supervised several undergraduate, masters, and PhD students. He was a visiting associate professor in aerospace engineering at Sharif University in Tehran, where he taught heat transfer. From 1997 to 2002, he was director of the thermal division of AERC in Tehran, where he was responsible for thermal design and fabrication of small satellites and space applications. He also helped design the national energy laboratory in Iran from 1996 to 1997. From 1995 to 1996, he was a professor in mechanical engineering at the University of New Mexico. He served as a senior scientist at Mission Research Corporation in Albuquerque, New Mexico from 1993 to 1994, where he worked in the thermal and environmental sciences.

Dr. Ghassemi received his PhD degree in mechanical engineering from Iowa State University in 1993. He received his MS and BS degrees in mechanical engineering from the University of Mississippi. He has coauthored 5 published books on heat transfer and thermal design and has published 21 journal papers and more than 30 conference papers.

Alma Cota is a research professor at the Autonomous University of Ciudad Juárez in Mexico, where she lectures on chemistry, energy, and environmental topics for the chemistry department. Dr. Cota has a Ph.D. in chemical engineering from New Mexico State University, where she also completed her postdoctoral work on photovoltaic power systems. She holds a BS degree in chemical engineering from the University of Sonora and a MS degree in solar energy from the National Autonomous University of Mexico – Center for Energy Research. Dr. Cota has extensive experience with a wide variety of solar energy systems including solar drying and disinfection of sludge wastes and water disinfection. She worked for 6 years at the Southwest Region Solar Experiment Station on photovoltaic systems where she assisted with the DOE/USAID Mexico Renewable Energy Program managed by Sandia National Laboratories from 1998-2004.

The Contributors

Jeannette M. Moore is a research assistant and electrical engineer at Sandia National Laboratories (SNLA) in Albuquerque, New Mexico. She is currently involved in the U.S. Department of Energy (DOE) "Solar America Cities" program, providing technical and project management assistance for various U.S. cities.

Vaughn Nelson is a renewable energy pioneer who has been active since the early 1970s. He is professor emeritus with the Alternative Energy Institute at West Texas A&M University (WTA&M). Dr. Nelson's primary work has been on wind resource assessment, education and training, applied R&D, and rural applications of wind energy with the U.S. Department of Agriculture (USDA). He has more than 30 years' experience in solar and wind research.

1 Introduction to Solar Energy

1.1 THE TWENTY-FIRST CENTURY'S PERFECT ENERGY STORM

The twenty-first century is forming into the perfect energy storm. Rising energy prices, diminishing energy availability and security, and growing environmental concerns are quickly changing the global energy panorama. Energy and water are the keys to modern life and provide the basis necessary for sustained economic development. Industrialized societies have become increasingly dependent on fossil fuels for myriad uses. Modern conveniences, mechanized agriculture, and global population growth have only been made possible through the exploitation of inexpensive fossil fuels. Securing sustainable and future energy supplies will be the greatest challenge faced by all societies in this century.

Due to a growing world population and increasing modernization, global energy demand is projected to more than double during the first half of the twenty-first century and to more than triple by the end of the century. Presently, the world's population is nearly 7 billion, and projections are for a global population approaching 10 billion by midcentury. Future energy demands can only be met by introducing an increasing percentage of alternative fuels. Incremental improvements in existing energy networks will be inadequate to meet this growing energy demand. Due to dwindling reserves and ever-growing concerns over the impact of burning carbon fuels on global climate change, fossil fuel sources cannot be exploited as in the past.

Finding sufficient supplies of clean and sustainable energy for the future is the global society's most daunting challenge for the twenty-first century. The future will be a mix of energy technologies with renewable sources such as solar, wind, and biomass playing an increasingly important role in the new global energy economy. The key question is: How long it will take for this sustainable energy changeover to occur? And how much environmental, political, and economic damage is acceptable in the meantime? If the twenty-first century sustainable energy challenge is not met quickly, many less-developed countries will suffer major famines and social instability from rising energy prices. Ultimately, the world's economic order is at stake.

Approximately one-third of the world's population lives in rural regions without access to the electric grid, and about half of these same people live without access to safe and clean water. Solar energy is unique in that it can easily provide electricity and purified water for these people today with minimal infrastructure requirements by using local energy resources that promote local economic development.

Unfortunately, traditional fossil fuel energy use has had serious and growing negative environmental impacts, such as CO_2 emissions, global warming, air pollution, deforestation, and overall global environmental degradation. Additionally, fossil fuel reserves are not infinite or renewable; the supply is limited. Without a doubt, there will be significant changes in our society's modern energy infrastructure by the end of the twenty-first century. A future mix that includes sustainable energy sources will contribute to our prosperity and health. Our future energy needs must be met by a mix of sustainable technologies that have minimal environmental impacts. Potentially, many of these technologies will use solar energy in all its forms, permitting gradual evolution into a hydrogen-based economy. A renewable energy revolution is our hope for a sustainable future. Clearly, the future belongs to clean energy sources and to those who prepare for it now.

1.2 RENEWABLE ENERGY FOR RURAL DEVELOPMENT

Given that the need for power grows much faster for less developed nations than for those that are already industrialized, this changing energy panorama will significantly impact how power is supplied to developing regions. Industrialized countries need to clean up their own energy production acts, while encouraging developing countries not to follow in their footsteps, but rather to leapfrog to clean energy technologies directly.

Despite three decades of major investments by less developed nations and multilaterals on electrification projects (often at huge environmental and social costs), nearly 2 billion people in developing regions around the globe still lack electricity. Over 1 billion people are also without access to safe drinking water. Millions of households rely solely on kerosene lamps for lighting and disposable batteries for radios. For most of these people, there is little likelihood of ever receiving electricity from conventional grid sources. However, there is growing momentum in supplying electricity to developing regions using solar and wind energy sources. Both solar and wind energy technologies offer energy independence and sustainable development by using indigenous renewable energy resources and by creating long-term local jobs and industries.

The cost of bringing utility power via transmission and distribution lines to nonelectrified villages is great. This is largely due to small household electrical loads and the fact that many villages are located at great distances over difficult terrain from the existing grid. Stand-alone solar and wind energy systems can provide cost-effective, modest levels of power for lighting, communication, fans, refrigerators, water pumping, etc. Using a least-cost model, some governments and national utilities, such as those in Brazil, India, Central America, South Africa, Mexico and elsewhere, have used PV and wind systems as an integrated development tool for electrification planning as either centralized or distributed solutions.

Two decades ago, PV technology was relatively unknown. The Dominican Republic was one of the early proving grounds for developing rural PV electrification efforts. The nonprofit group Enersol Associates began work in 1984, offering technical assistance and training to Dominican businesses. Nonprofit organizations also worked to develop a market for rural PV technology. Enersol began to work closely with the Peace Corps using seed funding from the U.S. Agency for International Development (USAID) to help set up a revolving fund offering rural farmers low-interest loans to purchase small PV systems.

The work of this nongovernment organization (NGO) later evolved into private enterprise as companies such as Soluz formed in the Dominican Republic and Honduras. Gradually throughout the developing world, small solar companies began to form as PV module manufacturers began to establish distributor networks to serve remote, nonelectrified areas. The model of rural off-grid PV systems (Figure 1.1) has spread globally with over 5 million systems installed. More total kiloWatts of grid-tie PV systems are installed each year; however, numerically more small, off-gird systems are installed annually.

Over time, the focus of PV projects has changed. Installation of PV systems solely for remote sites has expanded to include the promotion of rural economic development through PV. PV provide power for remote water pumping, refrigeration, and water treatment of community water supplies. Solar distillation can meet individual household potable water needs from even the most contaminated and brackish water sources. For larger load requirements, the combination of PV and wind technologies with diesel generators and battery storage has proved that hybrid configurations provide higher system reliability at a more reasonable cost than with any one technology alone.

Solar thermal energy represents the most competitive but often overlooked solar technology option. Domestic solar hot water heating systems typically have cost paybacks from 5 to 7 years—much better than grid-tied PV systems, where payback may take decades, if ever. Additionally, large-scale solar thermal concentrating solar power (CSP) plants have better economies of scale than PV for utility power generation at almost half the kiloWatt-hour cost.

FIGURE 1.1 Remote PV-powered school for satellite-assisted education in the Lempira Province of Honduras.

Solar and wind energy often provide least-cost options for economic and community development in rural regions around the globe, while supplying electricity, creating local jobs, and promoting economic development with clean energy resources. PV projects in developing nations have provided positive change in the lives of the rural people. Yet there is still much to do to educate, institutionalize, and integrate renewable technologies for maximum benefit for all. One of the greatest challenges is to work on reforming energy policies and legal frameworks to create a context that permits the sustainable development of renewable energy technologies.

1.3 RENEWABLE ENERGY SOLUTIONS

There are many different types of energy. Kinetic energy is energy available in the motion of particles—wind energy is one example of this. Potential energy is the energy available because of the position between particles—for example, water stored in a dam, the energy in a coiled spring, and energy stored in molecules (gasoline). There are many examples of energy: mechanical, electrical, thermal, chemical, magnetic, nuclear, biological, tidal, geothermal, and so on. Renewable energy denotes a clean, nontoxic energy source that cannot be exhausted.

The primary renewable energy sources are the Sun, wind, biomass, tides, waves, and the Earth's heat (geothermal). Solar energy is referred to as renewable and/or sustainable energy because it will be available as long as the Sun continues to shine. Estimates for the life of the main stage of the Sun are another 4 to 5 billion years. Wind energy is derived from the uneven heating of the Earth's surface due to more heat input at the equator with the accompanying transfer of water by evaporation and rain. In this sense, rivers and dams for hydroenergy are stored solar energy. Another aspect of solar energy is the conversion of sunlight into biomass by photosynthesis. Animal products such as whale oil and biogas from manure are derived from this form of solar energy. Tidal energy is primarily due to the gravitational interaction of the Earth and the moon. Another renewable energy is geothermal, due to heat from the Earth generated by decay of radioactive particles from when the solar system formed. Volcanoes are fiery examples of geothermal energy reaching the surface of the Earth from the hot and molten interior.

Overall, about 14% of the world's energy comes from biomass—primarily wood and charcoal, but also crop residue and even animal dung for cooking and some heating. This contributes to deforestation and the loss of topsoil in developing countries.

Fossil fuels are stored solar energy from past geological ages (i.e., ancient sunlight). Even though the quantities of oil, natural gas, and coal are large, they are finite and resources are sufficient to power the industrialized world anywhere from a few more decades to a few more centuries, depending on the resource. There are also large environmental costs associated with fossil fuel exploitation—from habitat loss and destruction due to strip mining and oil spills to global warming of the atmosphere largely caused by the combustion by-product of carbon dioxide.

The advantages of renewable energy are many: sustainability (cannot be depleted), ubiquity (found everywhere across the world in contrast to fossil fuels and minerals), and essentially non-polluting and carbon free. The disadvantages of renewable energy are: variability, low density, and generally higher initial cost for conversion hardware. For different forms of renewable energy, other disadvantages or perceived problems are: visual pollution, odor from biomass, perceived avian issues with wind plants, large land requirements for solar conversion, and brine from many geothermal sources.

1.4   GLOBAL SOLAR RESOURCE

Solar energy is the energy force that sustains life on Earth for all plants, animals, and people. It provides a compelling solution for all societies to meet their needs for clean, abundant sources of energy in the future. The source of solar energy is the nuclear interactions at the core of the Sun, where the energy comes from the conversion of hydrogen into helium. Sunlight is readily available, secure from geopolitical tensions, and poses no threat to our environment and our global climate systems from pollution emissions.

Solar energy is primarily transmitted to the Earth by electromagnetic waves, which can also be represented by particles (photons). The Earth is essentially a huge solar energy collector receiving large quantities of solar energy that manifest in various forms, such as direct sunlight used for plant photosynthesis, heated air masses causing wind, and evaporation of the oceans resulting as rain, which forms rivers and provides hydropower.

Solar energy can be tapped directly (e.g., PV); indirectly as with wind, biomass, and hydropower; or as fossil biomass fuels such as coal, natural gas, and oil. Sunlight is by far the largest carbon-free energy source on the planet. More energy from sunlight strikes the Earth in 1 hour (4.3×10^{20} J) than all the energy consumed on the planet in a year (4.1×10^{20} J). Although the Earth receives about 10 times as much energy from sunlight each year as that contained in all the known reserves of coal, oil, natural gas, and uranium combined, renewable energy has been given a dismally low priority by most political and business leaders.

We are now witnessing the beginning of a global paradigm shift toward clean energy in response to the twenty-first century perfect energy storm that is forming. As conventional energy prices rise, new and cleaner alternatives will begin to emerge and become economically more competitive. Energy solutions for the future depend on local, national, and world policies. Solutions also depend on individual choices and the policies that we implement as a society. This does not mean that we have to live in caves to negate our energy inputs, but we do have to make wise energy choices and conserve by methods such as driving fuel-efficient vehicles and insulating our homes, to name a few. To overcome the twenty-first century perfect energy storm, we will all have to work together cooperatively while doing our individual parts.

PROBLEMS

1.1.   Describe how the global economy depends on fossil fuels today.
1.2.   What do you think will be the key energy solutions to meeting the twenty-first century's global energy challenge?
1.3.   Do you believe that it is a greater priority for wealthier, industrialized countries to install grid-tie PV systems or for poorer, less developed countries to adopt off-grid PV systems?
1.4.   Describe three things that you can do practically in your life today to reduce your energy footprint.

2 Solar Resource

2.1 INTRODUCTION

Our planet faces significant challenges in the twenty-first century because energy consumption is expected to double globally during the first half of this century. Faced with increasingly constrained oil supplies, humanity must look to other sources of energy, such as solar, to help us meet the growing energy demand. A useful measure of the level of a country's development is through its energy consumption and efficiency. Excessive fossil fuel energy use not only has caused severe and growing damage to the environment from greenhouse gas emissions and oil spills, but also has brought political crises to countries in the form of global resource conflicts and food shortages. Solar and other forms of renewable energy offer a practical, clean, and viable solution to meet our planet's growing environmental and energy challenges.

Solar radiation is the most important natural energy resource because it drives all environmental processes acting at the surface of the Earth. The Sun provides the Earth with an enormous amount of energy. The energy stored by the oceans helps maintain the temperature of the Earth at an equilibrium level that allows for stability for a broad diversity of life.

Naturally, the Sun has always held the attention of humanity and been the subject of worship by many cultures over the millennia, such as the Egyptians, Incans, Greeks, and Mayans, among many others. The potential of solar energy to produce heat and electricity to be supplied for our modern economies in a variety of productive activities has been widely demonstrated but not yet widely adopted around the globe due to relatively cheap fossil fuels. Although the solar energy source is inexhaustible and free, it is not the most convenient energy source because it is not constant during the day and not readily dispatched. In contrast, modern lifestyles demand a continuous and reliable supply of energy. However, there are ways to overcome these shortfalls.

In order to understand solar energy, this chapter discusses the resources, including energy irradiated from the Sun, the geometrical relationship between the Sun and the Earth, and orientation of energy receivers, as well as the importance of acquiring reliable solar information for engineering design, operation, and management of solar technologies.

2.2 SUN–EARTH GEOMETRIC RELATIONSHIP

The amount and intensity of solar radiation reaching the Earth's surface depends on the geometric relationship of the Earth with respect to the Sun. Figure 2.1 shows this geometric relationship and its effects for different seasons in both hemispheres. The position of the Sun, at any moment at any place on Earth, can be estimated by two types of calculations: first, by simple equations where the inputs are the day of the year, time, latitude, and longitude, and, secondly, by calculations through complex algorithms providing the exact position of the Sun. Mostly, such algorithms are valid for a limited period varying from 15 to 100 years; the best uncertainties achieved are greater than ±0.01 (Blanco-Muriel et al. 2001; Michalsky 1988). Ibrahim and Afshin (2004) summarized a step-by-step procedure for implementing an algorithm developed by Meeus (1998) to calculate the solar angles in the period from the years 2000 B.C. to 6000 A.D. for which uncertainties of ±0.0003 were accomplished. This chapter includes only calculations from geometry in order to understand the nature of the variant incoming solar radiation.

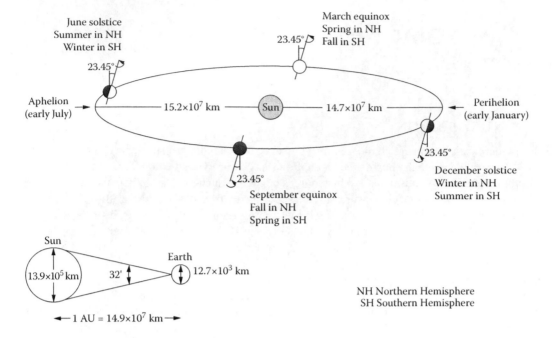

FIGURE 2.1 Earth–Sun geometric relationships.

2.2.1 EARTH–SUN DISTANCE

The Earth has a diameter of 12.7×10^3 km, which is approximately 110 times less than the Sun's. The Earth orbits approximately once around the Sun every 365 days. The Earth's orbit's eccentricity is very small, about 0.0167, which causes the elliptical path to be nearly circular. The elliptical path of the Earth varies from 14.7×10^7 km in early January—the closest distance to the Sun, called *perihelion*—to 15.2×10^7 km in early July—the farthest distance, called *aphelion*. The average Earth–Sun distance of 14.9×10^7 km is defined as the *astronomical unit* (AU), which is used for calculating distances within the solar system. However, the Earth is about 4% closer to the Sun at the perihelion than the aphelion. The Sun subtends an angle of $32'$ on the Earth at a 1 AU distance. Equation 2.1 (derived by Spencer, 1971, in terms of Fourier series) gives the Earth–Sun distance (E_0) in astronomical units with a maximum error of ±0.0001.

$$E_0 = \left(\frac{r_0}{r}\right)^2 = 1.000110 + 0.03422\cos\Gamma + 0.00128\sin\Gamma +$$
$$+ 0.000719\cos 2\Gamma + 0.000077\sin 2\Gamma \tag{2.1}$$

where r_0 is equal to 1 AU, r is the Earth–Sun distance, Γ is the daily angle in radians given as

$$\Gamma = 2\pi\frac{n-1}{365}, \tag{2.2}$$

and n is the day of the year ($1 \leq n \leq 365$) and can be calculated from Table 2.1. A less complex expression for E_0 was proposed by Duffie and Beckman (1991). Slight differences are found between both equations; for simplicity, calculations within this text use Equation 2.3.

TABLE 2.1
Declination and Earth–Sun Distance of the Representative Averaged Days for Months

*i*th day of the month	Month	*n* for *i*th day of the month	Julian Day of the year *n*	Declination δ in degrees	Earth–Sun distance E_0 in AU
17	January	*i*	17	−20.92	1.03
16	February	31 + *i*	47	−12.95	1.02
16	March	59 + *i*	75	−2.42	1.01
15	April	90 + *i*	105	9.41	0.99
15	May	120 + *i*	135	18.79	0.98
11	June	151 + *i*	162	23.09	0.97
17	July	181 + *i*	198	21.18	0.97
16	August	212 + *i*	228	13.45	0.98
15	September	243 + *i*	258	2.22	0.99
15	October	273 + *i*	288	−9.60	1.01
14	November	304 + *i*	318	−18.91	1.02
10	December	334 + *i*	344	−23.05	1.03

Source: Adapted from Duffie, J. A., and W. A. Beckman. 1991. *Solar Engineering of Thermal Processes,* 2nd ed., 919. New York: John Wiley & Sons.

$$E_0 = 1 + 0.033 \cos\left(\frac{360n}{365}\right) \tag{2.3}$$

2.2.2 Apparent Path of the Sun

The Earth rotates at an approximately constant rate on its axis once in about 24 hours. Such rotation in the eastward direction gives the sense that the Sun moves in the opposite direction. The so-called ecliptic is the apparent path that the Sun traces out in the sky while it goes from east to west during the day. The plane of the ecliptic is the geometric plane containing the mean orbit of the Earth around the Sun. Due to the overall interacting forces among the planets, the Sun is not always exactly in such a plane, but rather, may be some arc seconds out of it.

The rotation axis of the Earth is tilted 23.45° from being perpendicular to the ecliptic plane and remains constant as the Earth orbits the Sun as pointed out in Figure 2.1. As a result, the angle between the Sun and a point on the surface of the Earth varies throughout the year and, with this, the length of day also changes. The length of a solar day for a specific location may differ by as much as 15 minutes throughout the year, with an average of 24 hours. Seasons are also caused by the constant tilt of Earth with respect to the ecliptic plane; when the northern axis is pointing to the direction of the Sun, it is summer in the Northern Hemisphere and winter in the Southern Hemisphere. Both hemispheres receive the same amount of light, but the Southern receives it at a more glancing angle; hence, it is less concentrated and does not warm up as much as the Northern Hemisphere. The reverse holds true when the Earth's southern axis is pointing toward the Sun. The Earth is also about 4% further from the Sun during the Southern Hemisphere winter as compared to the Northern Hemisphere winter; thus, Southern winters are colder than Northern.

Day length is determined by the length of time when the Sun is above the horizon and varies throughout the year as the Earth–Sun geometric relationships change. Such geometrical changes are clearly perceived by the apparent movement of the Sun in the sky during the year. Again, the Earth's tilt has a great effect on what an observer sees, depending on whether he or she is in the Northern or Southern Hemisphere, as shown in Figure 2.2.

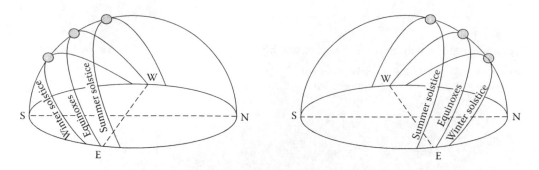

FIGURE 2.2 Apparent daily path of the Sun in the sky throughout the year for an observer in the Northern (left) and Southern Hemispheres (right).

In wintertime, for the Northern Hemisphere, days are short and the Sun is at a low angle in the sky, rising not exactly in the east, but instead just south of east and setting south of west. The shortest day of the year occurs on December 21, the winter solstice, when the Sun is the lowest in the southern sky. Each day after the winter solstice, the Sun begins to rise closer to the east and set closer to the west until it rises exactly in the east and sets exactly in the west. This day, about March 21, is called the vernal or spring *equinox* and it lasts for 12 hours.

After the spring equinox, the Sun still continues to follow a higher path through the sky, with the days growing longer, until it reaches the highest point in the northern sky on the summer *solstice;* this occurs on June 21. This day is the longest because the Sun traces the highest path through the sky and is directly over the Tropic of Cancer when the Northern Hemisphere is tilted toward the Sun at its maximum extent. Because this day is so long, the Sun does not rise exactly from the east, but rather to the north of east and sets to the north of west, allowing it to be above the horizon longer than 12 hours. After the summer solstice, the Sun follows a lower path through the sky each day until it reaches the point where it is again in the sky for exactly 12 hours. This is the fall equinox. Just like the spring equinox, the Sun will rise exactly east and set exactly west. After the fall equinox, the Sun will continue to follow a lower path through the sky and the days will grow shorter until it reaches its lowest path at the winter solstice.

The same cycle occurs for the Southern Hemisphere during the year. The shortest day occurs about June 21, the winter solstice. The Sun continues to increase its altitude in the sky and on about September 21 the Southern Hemisphere spring equinox is reached. Every place on Earth experiences a 12-hour day twice a year on the spring and fall equinoxes. Then, around December 21, the highest point in the sky occurs, the longest day of the year for the Southern Hemisphere when the Sun lies directly over the Tropic of Capricorn. Later, the 12-hour day occurs again around March 21. After this, the Sun continues to follow a lower path through the sky until it closes the cycle for the Southern winter solstice.

2.2.3 Earth and Celestial Coordinate Systems

Any location on Earth is described by two angles, *latitude* (ϕ) and *longitude* (λ). Figure 2.3 sketches the Earth coordinate system indicating the latitude and longitude constant lines. The latitude corresponds to the elevation angle between a hypothetical line from the center of Earth to any point on the surface and its projection on the equator plane. Latitude values fall between $90° < \phi < -90°$; latitude is zero at the equator, $90°$ at the northern pole, and $-90°$ at the southern pole. As for the longitude angle, imaginary lines extended from pole to pole are called meridians; these lines are at constant longitude. For each meridian crossing the equator's circle, there is an angle assigned. The meridian passing through the old Royal Astronomical Observatory in Greenwich, England, is the one chosen as zero longitude and known as the Prime Meridian. Longitudes are measured from 0 to

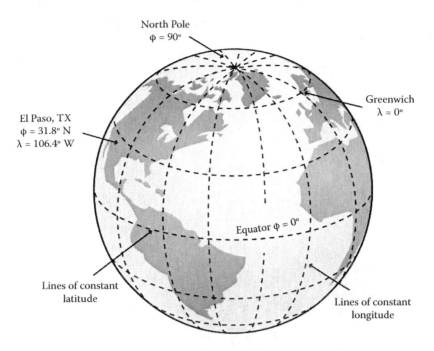

FIGURE 2.3   Earth coordinate system.

180° east of the Prime Meridian and 180° west (or –180°). For a particular location, the imaginary line that divides the sky in two and passes directly overhead is then the location's *meridian*. The abbreviations a.m. and p.m. come from the terms ante meridian and post meridian, respectively.

To determine the amount of solar energy received on any point of the Earth's surface, more than latitude and longitude angles are needed. When the Earth coordinate system is extended to the celestial sphere, as in Figure 2.4, it is possible to calculate the exact position of the Sun with respect to a horizontal surface at any point on Earth.

The *celestial sphere* is a hypothetical sphere of infinite radius whose center is the Earth and on which the stars are projected. This concept is used to measure the position of stars in terms of angles, independently of their distances. The north and south celestial poles of the celestial sphere are aligned with the northern and southern poles of the Earth. The celestial equator lies in the same plane as the Earth's equator does. Analogous to the longitude on Earth, the *right ascension* angle (χ) of an object on the celestial sphere is measured eastward along the celestial equator; lines of constant right ascension run from one celestial pole to the other, defining $\chi = 0°$ for the March equinox—the place where the Sun is positioned directly over Earth's equator.

Similarly to the latitude concept on Earth, the declination δ on the celestial sphere is measured northward or southward from the celestial equator plane. Lines of constant declination run parallel to the celestial equator and run in numerical values from +90° to –90°. Because of the Earth's yearly orbital motion, the Sun appears to circle the ecliptic up to an inclination of 23.45° to the celestial equator, $-23.45° < \delta < 23.45°$ with $\delta = 0°$ at the equator for the equinoxes, –23.45° on the December solstice, and +23.45° on the June solstice.

Several expressions to calculate declination in degrees have been reported. One of the most cited is Equation 2.4, which was derived by Spencer (1971) as function of the daily angle given by Equation 2.2. Some other simpler equations used in solar applications are Equation 2.5 by Perrin de Brichambaut (1975) and Equation 2.6 by Cooper (1969). Although slight differences exist among them, for great accuracy Spencer's equation is the best with a maximum error of 0.0006 radian (Iqbal 1983). However, for simplicity, Cooper's equation is used throughout this text:

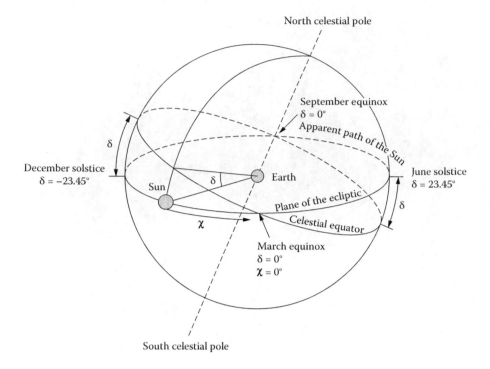

FIGURE 2.4 Celestial coordinate system.

$$\delta = 0.006918 - 0.399912\cos\Gamma + 0.070257\sin\Gamma$$
$$-0.006758\cos 2\Gamma + 0.000907\sin 2\Gamma \qquad (2.4)$$
$$-0.002697\cos 3\Gamma + 0.00148\sin 3\Gamma$$

$$\delta = \arcsin\left\{0.4\sin\left(\frac{360}{365}(n-80)\right)\right\} \qquad (2.5)$$

$$\delta = 23.45\sin\left(\frac{360}{365}(n+284)\right) \qquad (2.6)$$

where n is the day of the year.

2.2.4 POSITION OF THE SUN WITH RESPECT TO A HORIZONTAL SURFACE

In addition to the fixed celestial coordinate systems on the sky, to describe the Sun's position with respect to a horizontal surface on Earth at any time, other angles based on the Earth's coordinates need to be understood: *solar altitude* (α_s), *zenith* (θ_z), *solar azimuth* (γ_s), and *hour* (ω) *angles*. Figure 2.5 presents the geometric relationships among these angles to determine the position of the Sun in the sky at any time. The solar altitude is measured in degrees from the horizon of the projection of the radiation beam to the position of the Sun. When the Sun is over the horizon, $\alpha_s = 0°$ and when it is directly overhead, $\alpha_s = 90°$. In most latitudes, the Sun will never be directly overhead; that only happens within the tropics. Because the zenith is the point directly overhead and 90° away

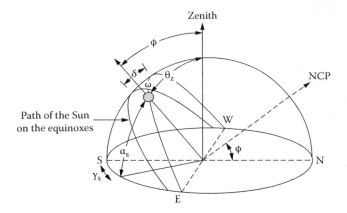

FIGURE 2.5 Position of the Sun in the sky relative to the solar angles.

from the horizon, the angle of the Sun relative to a line perpendicular to the Earth's surface is called the zenith angle, θ_z, so that

$$\alpha_s + \theta_z = 90^o \tag{2.7}$$

and the zenith angle is given by

$$\cos\theta_z = \sin\phi\sin\delta + \cos\delta\cos\phi\cos\omega \tag{2.8}$$

Also, there is a strong relationship between the solar azimuth and hour angles. The solar azimuth is the angle on the horizontal plane between the projection of the beam radiation and the north–south direction line. Positive values of γ_s indicate the Sun is west of south and negative values indicate when the Sun is east of south. The hour angle ω is the angular distance between the Sun's position at a particular time and its highest position for that day when crossing the local meridian at the solar noon. Because the Earth rotates approximately once every 24 hours, the hour angle changes by 15° per hour and moves through 360° over the course of the day. The hour angle is defined to be zero at solar noon, a negative value before crossing the meridian, and a positive after crossing.

As mentioned before, the length of the day varies for all latitudes during the year and, with this, the solar altitude α_s also changes hourly and daily. This angle can be calculated in terms of declination δ, latitude ϕ, and hour ω angles by using the next equation:

$$\sin\alpha_s = \sin\phi\sin\delta + \cos\phi\cos\delta\cos\omega \tag{2.9}$$

To make certain that Equation 2.9 does not fail at any point because the arcsine of a negative number does not exist, it is better to implement Equations 2.10 through 2.12:

$$\sin^2\alpha_s = (\sin\alpha_s)(\sin\alpha_s) \tag{2.10}$$

Because $\sin^2\alpha_s + \cos^2\alpha_s = 1$, then,

$$\cos\alpha_s = (1 - \sin^2\alpha_s)^{1/2} \tag{2.11}$$

$$\alpha_s = \text{atan}\left(\frac{\sin\alpha_s}{\cos\alpha_s}\right) \qquad (2.12)$$

Example 2.1

Calculate the zenith and solar altitude angles for a latitude of 32.34° north at (a) 10:30 a.m. and (b) 3:15 p.m. solar time on April 17.

Solution:

(a)   On April 17, $n = 107$ calculated from Table 2.1; then, the declination gives

$$\delta = 23.45\sin\left(\frac{360}{365}(107+284)\right) = 10.14^\circ$$

Daily, the Sun moves through the sky 15° each hour; at solar noon (local meridian), the value of the hour angle is zero and takes negative values during mornings and positive values in afternoons. At 10:30 a.m., $\omega = -22.5°$. From Equation 2.8,

$$\cos\theta_z = \sin\left(32.34^\circ\right)\sin\left(10.14^\circ\right)+\cos\left(10.14^\circ\right)\cos\left(32.34^\circ\right)\cos\left(-22.5^\circ\right)$$

$$\theta_z = 30.4^\circ$$

and, from Equation 2.9,

$$\sin\alpha_s = \sin\phi\sin\delta+\cos\phi\cos\delta\cos\omega$$

$$\alpha_s = 59.6^\circ$$

(b)   At 3:15 p.m., $\omega = 48.75°$, $\theta_z = 50°$, and $\alpha_s = 40°$.

Figure 2.6 shows the direction of the solar radiation beam for three particular declinations. When $\delta = 0°$, during the equinoxes, the equators of the Sun and the Earth fall in the same plane (i.e., both rotation axes are parallel); for $\delta = -23.45°$, on the December solstice, the North Pole of the Earth points 23.45° away from being parallel to the Sun's rotation axis, making the South Pole more exposed to the solar radiation. When $\delta = +23.45°$, on the June solstice, the North Pole is closer 23.45° to the Sun and the South Pole is farther by the same angular distance. When $\delta = 0°$, the behavior of the solar altitude α_s as a function of the hour angle or solar time, according to Equation 2.9, is symmetrical for both hemispheres.

Figures 2.7, 2.8, and 2.9 plot solar altitude along the day for several latitudes. When $\delta = 0°$, it can be seen that all latitudes on Earth experience a 12-hour solar day. The maximum solar altitude of 90° is achieved on the equator; at noon the Sun is right on the zenith and high α_s are experienced by locations near the equator. The farther from the equator a location is, the less solar altitude is observed. At the poles, the path of the Sun has almost zero values for α_s, just as if the whole day were sunset. For $\delta \neq 0°$, the behavior of the solar altitude during the day is no longer symmetrical. When $\delta = -23.45°$ (Figure 2.8), the Northern Hemisphere locations with $\phi = 70...90°$ are not illuminated at all during the day and only negative values from Equation 2.12 are obtained; in contrast, the South Pole is fully illuminated. For $\phi = -90°$, the solar altitude remains constant at 23.45° during the 24 hours. The locations with $\phi = -40...0°$ experience the greatest solar altitude during the day.

The opposite occurs during the June solstice, $\delta = +23.45°$. The Southern Hemisphere locations with $\phi = -70...-90°$ are not illuminated at all during the day and the North Pole is fully illuminated (Figure 2.9). The locations with $\phi = 0...40°$ experience the greatest solar altitude; for $\phi = +90°$, the solar altitude remains constant at 23.45° during the 24 hours.

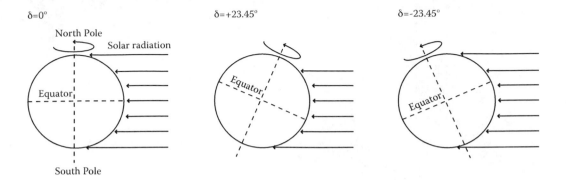

FIGURE 2.6 Direction of incoming solar radiation beam into Earth during the equinoxes with $\delta = 0°$, on the June solstice at $\delta = +23.45°$, and on the December solstice at $\delta = -23.45°$.

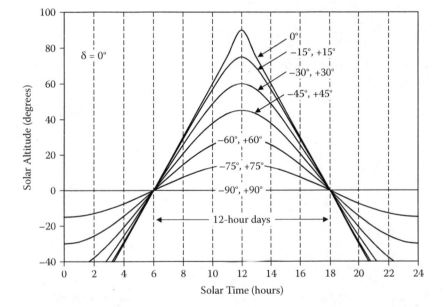

FIGURE 2.7 Solar altitude during the day for different latitudes during the equinoxes when $\delta = 0°$.

The solar azimuth angle, γ_s, can be calculated in terms of declination δ, latitude ϕ, and hour ω angles following Braun and Mitchell's (1983) formulation. Equation 2.13 for azimuth angle depends on a pseudo solar azimuth angle, γ'_s, and three constants, C_1, C_2, and C_3, which are used to find out which quadrant the Sun is in at any moment, for any day, and at any location:

$$\gamma_s = C_1 C_2 \gamma'_s + C_3 \left(\frac{1 - C_1 C_2}{2} \right) 180 \tag{2.13}$$

where γ'_s is a pseudo solar azimuth angle, γ_s, for the first or fourth quadrant

$$\sin \gamma'_s = \frac{\sin \omega \cos \delta}{\sin \theta_z} \tag{2.14}$$

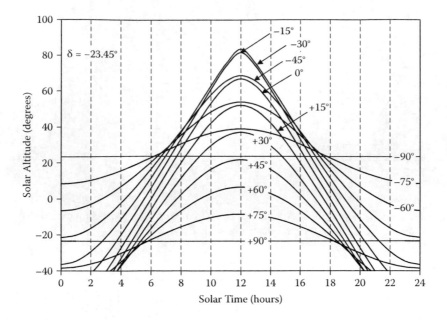

FIGURE 2.8 Solar altitude during the day for different latitudes during the December solstice when $\delta = -23.45°$.

The calculation of C_1 determines whether or not the Sun is within the first or fourth quadrants and above the horizon:

$$C_1 = 1 \begin{cases} \text{if } |\omega| < \omega_{WE} \\ \text{or} \\ \text{if } \left|\dfrac{\tan \delta}{\tan \phi}\right| > 1 \end{cases} , \ \text{-1 otherwise}$$

(2.15)

where ω_{WE} is the hour angle when the Sun is due east or west and can be obtained as

$$\cos \omega_{WE} = \frac{\tan \delta}{\tan \phi}$$

(2.16)

The constant C_2 includes the variables of latitude and declination. C_2 will take the value of 1 when $\phi = 0°$, $\phi = \delta$, or $|\phi| > |\delta|$ and will become -1 when $\phi \neq 0°$ and $|\phi| < |\delta|$.

$$C_2 = 1 \ \text{if } \phi(\phi\text{-}\delta) \geq 0, \ \text{-1 otherwise}$$

(2.17)

Calculation of C_3 defines whether or not the Sun has passed the local meridian (i.e., identifies whether it is morning or afternoon):

$$C_3 = 1 \ \text{if } \omega \geq 0, \ \text{-1 otherwise}$$

(2.18)

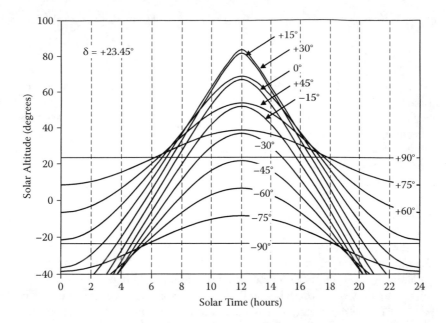

FIGURE 2.9 Solar altitude during the day for different latitudes during the June solstice when $\delta = +23.45°$.

Example 2.2

Determine the solar azimuth angle on May 1 for a latitude of 45° at 11:15 a.m.

Solution:

On May 1, from Table 2.1, $n = 121$, from Equation 2.6, $\delta = 14.9°$ and at 11:15 a.m., $\omega = -11.25°$
To solve Equation 2.13 for γ_s, γ_s', C_1, C_2, and C_3, must be calculated. From Equation 2.8,

$$\cos\theta_z = \sin(45°)\sin(14.9°) + \cos(45)\cos(14.9°)\cos(-11.25°)$$

$$\theta_z = 31.6°$$

Substituting ϕ, δ, and θ_z into Equation 2.14,

$$\sin\gamma_s' = \frac{\sin(45°)\cos(14.9°)}{\sin(31.6°)}$$

$$\gamma_s' = -21.1°$$

For the constants,

$$\cos\omega_{WE} = \frac{\tan(14.9°)}{\tan(45°)}$$

$$\omega_{WE} = 74.6°$$

According to Equations 2.15, 2.17, and 2.18, $C_1 = 1$, $C_2 = 1$, and $C_3 = -1$; then,

$$\gamma_s = \gamma_s' = -21.1°$$

To locate the position of the Sun in the sky at any time, for any day, and for any location, a plot of the solar altitude α_s versus azimuth γ_s at different times throughout the year is commonly used. This diagram is called a *sun chart* and it is built for any particular latitude. A sun chart consists of several curves, each of which represents the Sun's path for a particular day of each month; each curve works for 2 days of the year. Also, equivalent times for the specific day-plotted paths are also shown in sun charts by the lines connecting the curves. A sun chart for the 45° latitude in the Northern Hemisphere is presented in Figure 2.10. This exhibits the longest day of the year during the summer solstice, with a 23.45° declination, reaching the maximum α_s value of 68.45° at solar noon.

The shortest path or shortest day in such a sun chart occurs on December 21, with a maximum α_s of 21.5°. During the equinoxes (March 21 and September 21), the Sun rises exactly in the east and sets exactly in the west. This agrees with the sun chart because the Sun rises at an azimuth angle of –90° at 6 a.m. and sets at 90° at 6 p.m., predicting the 12-hour days (as also seen in Figure 2.7, where α_s is plotted along the day at different latitudes). Figure 2.11 shows sun charts for several latitudes for the representative days of each month. Each plot works for negative and positive latitudes; however, the order of the representative day curves change according to the table included in the same figure.

When the solar incident radiation on a horizontal-solar collector is calculated, two new angles should be defined. The *slope-surface angle* (β) indicates how inclined the collector is from the horizontal; on a horizontal collector, $\beta = 0°$. The allowed range for β goes from 0 to 180°. The other relevant angle for calculations corresponds to the *surface-azimuth angle* (γ), which indicates how far the solar collector deviates from the north–south axis. This angle is measured between the horizontal projection of the surface normal and the north–south direction line, with 0 due south and negative values to the east of such an axis; $-180° \leq \gamma \leq -180°$.

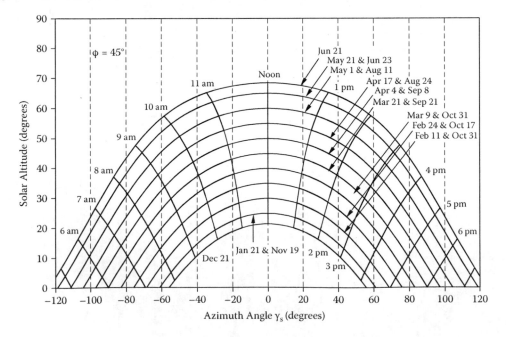

FIGURE 2.10 Sun chart for latitude 45° north.

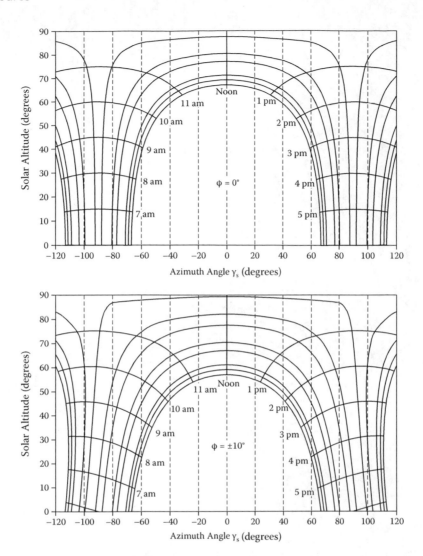

Assignation of representative day to curves from outside to inside or from top to bottom			
$\phi \geq 0°$		$\phi < 0°$	
Representative day of the month	δ	Representative day of the month	δ
June 11	23.09	December 10	−23.05
July 17	21.18	January 17	−20.92
May 15	18.79	November 14	−18.91
August 16	13.45	February 16	−12.95
April 15	9.41	October 15	−9.60
September 15	2.22	March 16	−2.42
March 16	−2.42	September 15	2.22
October 15	−9.60	April 15	9.41
February 16	−12.95	August 16	13.45
November 14	−18.91	May 15	18.79
January 17	−20.92	July 17	21.18
December 10	−23.05	June 11	23.09

FIGURE 2.11A Sun charts at latitudes 0° and ±10°.

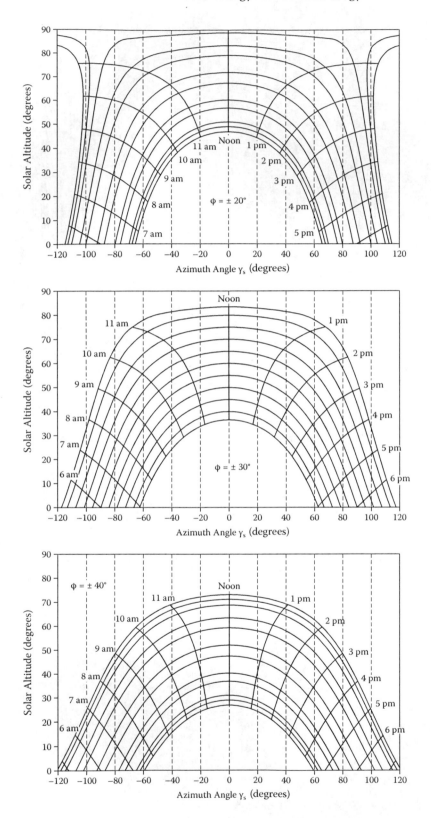

FIGURE 2.11B Sun charts at latitudes ±20°, ±30°, and ±40°.

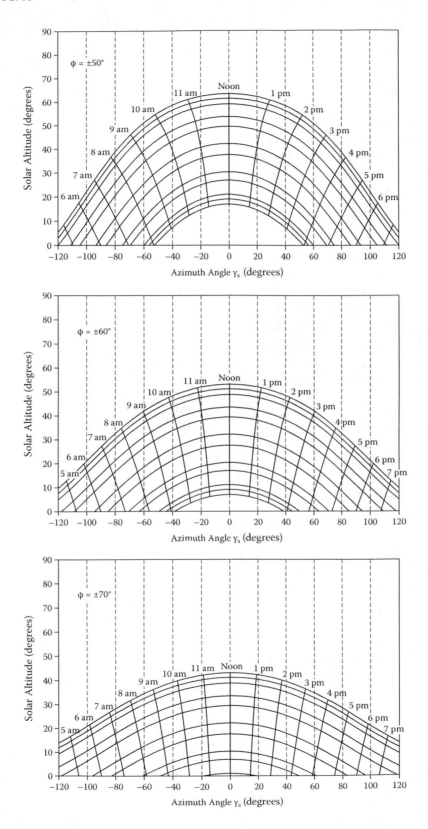

FIGURE 2.11C Sun charts at latitudes ±50°, ±60°, and ±70°.

2.2.5 POSITION OF THE SUN WITH RESPECT TO A TILTED SURFACE

The maximum solar energy collection is achieved when the Sun's rays are perpendicular to the collecting area (i.e., parallel to the surface normal). This can be achieved only when solar tracking systems are used to modify the slope or the surface azimuth or both angles during the collector's operation. However, these systems are more expensive than the fixed ones due to their moving components.

The fixed-β collectors are the most practical receivers and the most widely installed throughout the world. In order that the fixed-β collectors capture most of the annual incoming solar radiation, the surfaces must always be tilted facing the equator. As demonstrated in Figure 2.11, the maximum solar altitude for each day is reached around noon when the solar azimuth angle is around zero (i.e., around the north–south line). For dates when the Sun is at low maximum solar altitudes, it is convenient to install the collectors with greater β to minimize the angle between the Sun's rays and the surface normal. For periods when the Sun follows higher paths through the sky, β must be small. Several criteria might be used to select β, such as maximum collection for the greatest energy demand period or optimization during the whole year. Another option could be having several positions in the systems so that the collector could be manually fixed at several β values over the year.

The last angle to be defined, which completely relates the solar radiation to a surface, is the *solar incidence angle* (θ). This is the angle between the solar radiation beam incident on a surface and the imaginary line normal to such a surface. At $\theta = 0°$, the Sun's rays are perpendicular to the surface and, when $\theta = 90°$, the Sun's rays are parallel to the surface. Maximum solar gain for any solar intensity is achieved when the incidence angle is zero because the cross section of light is not spread out and also because surfaces reflect more light when the light rays are not perpendicular to the surface. Figure 2.12 presents the geometric relationship between the solar angles in a horizontal surface and in one tilted by a β slope. The angle of incidence can be calculated by any of the following equations:

$$
\begin{aligned}
\cos \theta = {}& \sin \delta \sin \phi \cos \beta \\
& - \sin \delta \cos \phi \sin \beta \cos \gamma \\
& + \cos \delta \cos \phi \cos \beta \cos \omega \\
& + \cos \delta \sin \phi \sin \beta \cos \gamma \cos \omega \\
& + \cos \delta \sin \beta \sin \gamma \sin \omega
\end{aligned}
$$

(2.19)

$$
\cos \theta = \cos \theta_z \cos \beta + \sin \theta_z \sin \beta \cos (\gamma_s - \gamma)
$$

(2.20)

For horizontal surfaces $\beta = 0°$, the angle of incidence becomes the zenith angle $\theta = \theta_z$. For this particular case, Equation 2.19 is reduced to Equation 2.8; then, the *sunset hour angle* (ω_{sunset}) can be derived when $\theta_z = 90°$:

$$
\cos \omega_{sunset} = -\tan \phi \tan \delta
$$

(2.21)

Because 1 hour equals 15° of the Sun traveling through the sky, the number of daylight hours (*N*) can be determined by solving Equation 2.21 for ω_{sunset} and converting the resultant degrees into hours:

$$
N = \frac{2}{15} \cos^{-1} (-\tan \phi \tan \delta)
$$

(2.22)

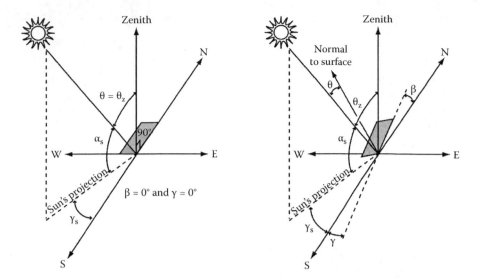

FIGURE 2.12 Solar angles for a horizontal solar surface facing south (left) and for a tilted surface facing south with an arbitrary surface azimuth angle.

For vertical surfaces with $\beta = 90°$, Equation 2.19 becomes

$$\cos \theta = -\sin \delta \cos \phi \cos \gamma$$
$$+ \cos \delta \sin \phi \cos \gamma \cos \omega$$
$$+ \cos \delta \sin \gamma \sin \omega \tag{2.23}$$

Equation 2.22 could be useful in the calculation of energy gain in building through windows. For tilted surfaces, other than $\beta = 0°$ or $\beta = 90°$, toward exactly south or north with $\gamma = 0°$ or $\gamma = 180°$, respectively, the last term of Equation 2.19 is zero.

$$\cos \theta = \sin \delta \sin \phi \cos \beta$$
$$- \sin \delta \cos \phi \sin \beta$$
$$+ \cos \delta \cos \phi \cos \beta \cos \omega$$
$$+ \cos \delta \sin \phi \sin \beta \cos \omega \tag{2.24}$$

When a solar collector is installed, if there is not a physical obstruction, such as buildings or any other object that cannot be removed, the collector must be aligned on the true north–south axis in order to capture effectively the solar energy during the day. The south- or north-pointing direction of the surface will depend on the difference between latitude and declination.

$$\text{if} \left(\phi - \delta\right) > 0, \gamma = 0°$$
$$\text{if} \left(\phi - \delta\right) < 0, \gamma = 180° \tag{2.25}$$

The amount of solar energy incoming in collectors depends strongly on the β values. The different declinations, experienced during the year, affect the optimum slope for surfaces. Figure 2.13 shows the geometrical analysis to select the best surface slope along the year for both hemispheres. For collectors with such slopes, the solar incidence angle θ is zero at solar noon because the Sun's rays are normal to the surface. The slopes for maximizing energy capture for Northern Hemisphere latitudes when $(\phi - \delta) > 0$ are as follows:

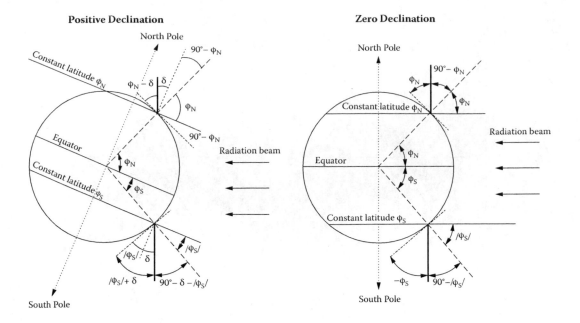

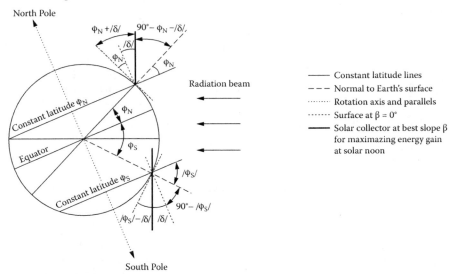

FIGURE 2.13 Geometric relationship for solar collectors perpendicular to the solar radiation beam at solar noon when $\delta = +23.45°$ (upper left), $0°$ (upper right), and $-23.45°$.

$$\beta\left(\gamma_s = 0°\right) = \begin{cases} \phi - \delta & \text{for } \delta > 0 \\ \phi & \text{for } \delta = 0 \\ \phi + |\delta| & \text{for } \delta < 0 \end{cases}$$

(2.26)

For the Southern Hemisphere latitudes, when $(\phi - \delta) < 0$, the surface must be oriented toward the north and the best slope is the following:

$$\beta\left(\gamma_s = 180^\circ\right) = \begin{cases} |\phi| + \delta & \text{for } \delta > 0 \\ |\phi| & \text{for } \delta = 0 \\ |\phi| - |\delta| & \text{for } \delta < 0 \end{cases}$$

(2.27)

Figure 2.14 demonstrates the same angular relationship between the incidence angle θ of the radiation beam incoming on a β-fixed surface, regardless of whether it is facing south or north, at an arbitrary latitude ϕ, and the incidence angle to a horizontal surface at a latitude $\phi^* = \phi - \beta$. For the Northern Hemisphere, Equation 2.19 can be simplified as

$$\cos\theta = \sin\left(\phi - \beta\right)\sin\delta + \cos\left(\phi - \beta\right)\cos\delta\cos\omega$$

(2.28)

and for the Southern Hemisphere as

$$\cos\theta = \sin\left(\phi + \beta\right)\sin\delta + \cos\left(\phi + \beta\right)\cos\delta\cos\omega$$

(2.29)

At solar noon, for the south-facing tilted surfaces in the Northern Hemisphere,

$$\theta_{noon} = \left|\phi - \delta - \beta\right|$$

(2.30)

and for the Southern Hemisphere,

$$\theta_{noon} = \left|-\phi + \delta - \beta\right|$$

(2.31)

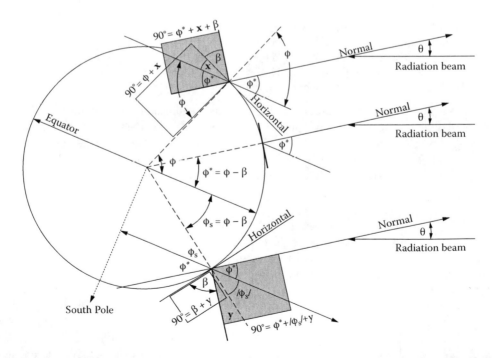

FIGURE 2.14 Angular relationship between the incidence angle θ of the radiation beam incoming in a β-fixed surface at any latitude ϕ and the incidence angle to a horizontal surface at a latitude $\phi^* = \phi - \beta$.

When $\beta = 0$, the angle of incidence is the zenith angle, and Equations 2.30 and 2.31 for the Northern and Southern Hemispheres, respectively, become Equations 2.32 and 2.33:

$$\theta_{z,noon} = |\phi - \delta| \tag{2.32}$$

$$\theta_{z,noon} = |-\phi + \delta| \tag{2.33}$$

Example 2.3

Calculate the solar incidence and zenith angles on a solar collector located at El Paso, Texas (31.8° north; 106.4° west), at 11:30 a.m. on March 3, if the surface is (a) 30° tilted from the horizontal and pointed 10° west south, (b) $\beta = 40°$ and $\gamma = 10°$, (c) $\beta = 30°$ and $\gamma = 0°$, (d) $\beta = 40°$ and $\gamma = 0°$, (e) $\beta = \phi - |\delta|$ and $\gamma = 0°$, and (f) $\beta = \phi - |\delta|$ and $\gamma = 0°$ at solar noon.

Solution:

(a)　On March 3, $n = 62$ and, from Equation 2.6, $\delta = -7.5°$. At 11:30 a.m., $\omega = -7.5°$. More known data are $\phi = 31.8°$, $\gamma = 10°$, and $\beta = 30°$:

$$\begin{aligned}
\cos\theta = {} & \sin\left(-7.5°\right)\sin\left(31.8°\right)\cos\left(30°\right) \\
& - \sin\left(-7.5°\right)\cos\left(31.8°\right)\sin\left(30°\right)\cos\left(10°\right) \\
& + \cos\left(-7.5°\right)\cos\left(31.8°\right)\cos\left(30°\right)\cos\left(-7.5°\right) \\
& + \cos\left(-7.5°\right)\sin\left(31.8°\right)\sin\left(30°\right)\cos\left(10°\right)\cos\left(-7.5°\right) \\
& + \cos\left(-7.5°\right)\sin\left(30°\right)\sin\left(10°\right)\sin\left(-7.5°\right)
\end{aligned}$$

$$\theta = 15.7°$$

(b)　$\gamma = 10°$ and $\beta = 40°$ gives $\theta = 13.8°$ and $\theta_z = 39.9°$.
(c)　$\gamma = 0°$ and $\beta = 30°$ gives $\theta = 11.9°$ and $\theta_z = 39.9°$.
(d)　$\gamma = 0°$ and $\beta = 40°$ gives $\theta = 7.46°$ and $\theta_z = 39.9°$.
(e)　$\gamma = 0°$ and the optimal $\beta = \phi - |\delta| = 31.8 - 7.5 = 24.3°$ gives $\theta = 16.8°$ and $\theta_z = 39.9°$.
(f)　$\gamma = 0°$, $\beta = \phi - |\delta| = 24.3°$, and $\omega = 0°$ gives $\theta = 15.1°$ and $\theta_z = 39.3°$.

From this exercise, it can be demonstrated that surfaces facing south gain the most possible solar energy incoming because the incidence angle is minimized. On the other side, by modifying the β slope and getting closer to its optimal value of $\beta = \phi - |\delta|$ for north latitudes when experiencing a negative declination (as shown in Figure 2.13 and Equation 2.26), the solar incidence angle takes the zero value—the best possible for a β-fixed surface for that specific day.

2.3　EQUATION OF TIME

All points at constant longitude experience noon and any other hour at the same time. Local time (LT), also known as solar time, is a measure of the position of the Sun relative to a locality. At noon local time, the Sun goes through its highest position in the sky. Figure 2.16 graphically shows the equation of time as a function of the Julian day and declination. The universal time (UT) can be defined as the local time at the zero meridian. To avoid confusion due to infinite local times, time zones were introduced under the concept of standard time. Standard time (SDT) was proposed by Sandford Fleming in 1879; this consisted of dividing the world into 24 time zones, each one covering exactly 15° because the Earth rotates 15° per hour. Political consider-ations have now increased the number of standard time zones to 39 (shown in Figure 2.15). Local standard time (LST) is the same time in the entire time zone. In addition, the clock is generally

WORLD MAP OF TIME ZONES

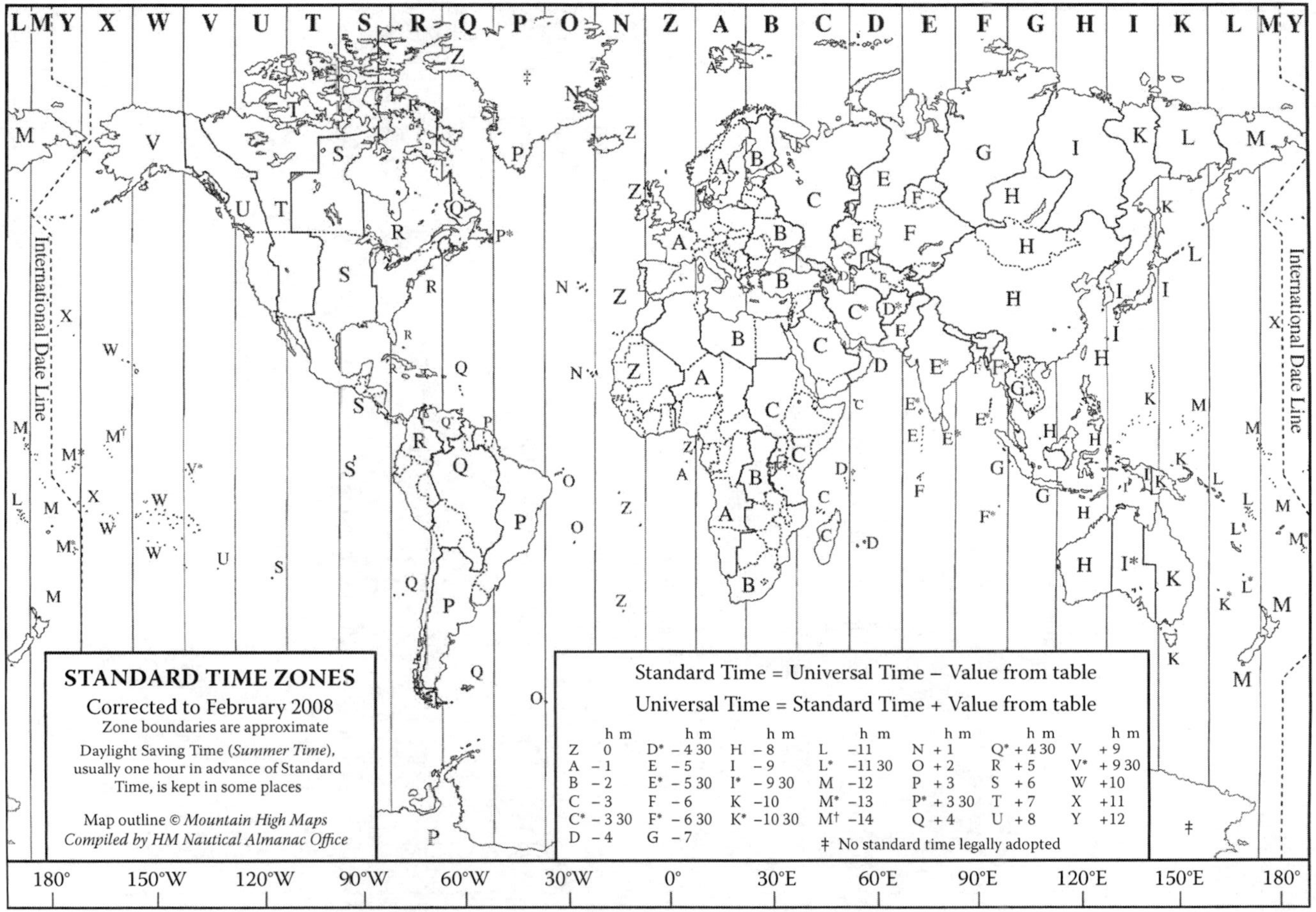

STANDARD TIME ZONES

Corrected to February 2008
Zone boundaries are approximate
Daylight Saving Time (*Summer Time*),
usually one hour in advance of Standard
Time, is kept in some places

Map outline © *Mountain High Maps*
Compiled by HM Nautical Almanac Office

Standard Time = Universal Time − Value from table
Universal Time = Standard Time + Value from table

	h m		h m		h m		h m		h m
Z	0	D*	−4 30	L	−11	N	+1	Q*	+4 30	V	+9
A	−1	E	−5	L*	−11 30	O	+2	R	+5	V*	+9 30
B	−2	E*	−5 30	M	−12	P	+3	S	+6	W	+10
C	−3	F	−6	M*	−13	P*	+3 30	T	+7	X	+11
C*	−3 30	F*	−6 30	M†	−14	Q	+4	U	+8	Y	+12
D	−4	G	−7								

‡ No standard time legally adopted

FIGURE. 2.15 World time zones. This map is a copyrighted production of H. M. Nautical Almanac Office. (With permission.)

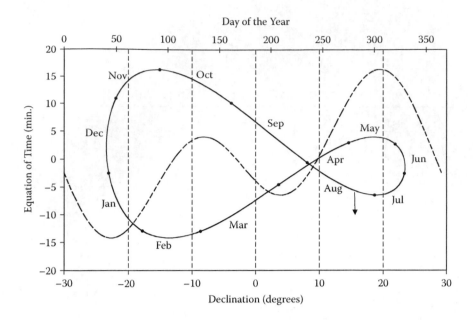

FIGURE 2.16   Equation of time as a function of the day of the year and declination.

shifted 1 hour forward between April and October to make better use of sunlight, purportedly to save energy.

The relationship between solar time and standard time must be known to describe the position of the Sun. For most places where standard zones advance by hour, the adjustment of solar time for longitude can be done by the subtraction of the observer's longitude (λ_{local}) from the standard meridian longitude (λ_{STD}) for the observer's time zone and multiplying it by the 4 minutes the Sun takes to move 1° through the sky. Equation 2.34 estimates the time difference in minutes between solar time and standard time plus a correction due to the irregularity of the natural length of a day. Such irregularity is caused by the noncircular orbit of the Earth spinning around the Sun and the inclination of the north–south axis relative to the Sun:

$$LT - SDT = 4\left(\lambda_{STD} - \lambda_{local}\right) + E_t \tag{2.34}$$

where E_t is known as the equation of time as function of the daily angle Γ given by Equation 2.2:

$$E_t = (0.000075 + 0.001868\cos\Gamma - 0.032077\sin\Gamma - \\ -0.014615\cos 2\Gamma - 0.04089\sin 2\Gamma)(229.18) \tag{2.35}$$

Example 2.4

What is the solar time in El Paso, Texas (31.8° north; 106.4° west), at 11 a.m. mountain time on March 3?

Solution:

On March 3, $n = 62$:

$$\Gamma = 2\pi\frac{62-1}{365} = 1.05 \text{ rad}$$

From Equation 2.33,

$$LT = STD + 4\left(105° - 106.4°\right) + E_t$$

and, from Equation 2.34, the time equation gives

$$E_t = (0.000075 + 0.001868\cos(1.05) - 0.032077\sin(1.05) -$$
$$-0.014615\cos(2 \times 1.05) - 0.04089\sin(2 \times 1.05))(229.18)$$
$$= -12.54\,\text{min}$$

The solar time or local time is

$$LT = 11 \text{ h} + (4\left(105° - 106.4°\right) - 12.54)\,\text{min}$$
$$= 10:43 \text{ h}$$

2.4   STRUCTURE OF THE SUN

The Sun is a typical middle-aged star with a diameter of 1.39×10^6 km, a mass of 2×10^{30} kg, and a luminosity of 4×10^{26} W (Tayler 1997). The Sun is a plasma, primarily composed of 70% hydrogen and 28% helium. This changes over time as hydrogen is converted to helium in its core by thermonuclear reactions. Every second, 700 million tons of hydrogen is converted into helium.

The Sun is composed of the core, the radiation and the convection zones, and its atmosphere. The conditions of the Sun vary greatly along its radius. The core, with a radius of $0.2R,$ is the source of all the Sun's energy and it contains half of the Sun's mass. The temperature and pressure in this zone are extreme: 1.5×10^7 K and 250×10^9 atm, with a density of 150 g/cm^3—13 times greater than that of solid lead. The combination of high temperature and high density creates the correct environment for the thermonuclear reaction to take place; two atoms of hydrogen come together to produce one heavier atom of helium, releasing a great amount of energy.

Once energy is produced in the core, it travels from the center to the outer regions. The region immediate to the core is identified as the radiation zone because energy is transported by radiation and it extends to $0.7R$. It takes thousands of years for the energy released by the core to exit this zone. The temperature in the radiation zone is about 5×10^6 K. Once the energy has left this zone and its temperature has dropped down to 2×10^6 K, rolling turbulent motions of gases arise; this is known as the convection zone. It takes around a week for the hot material to bring its energy to the top of the convection zone. This layer extends from $0.7R$ to R.

The solar atmosphere, the exterior of the Sun, is composed of the photosphere, chromosphere, and the corona. The photosphere corresponds to the lowest and densest part of the atmosphere; in the interior of the Sun, the gas becomes much denser so that is not possible to see through it. Because the Sun is completely made of gas and there is no hard surface, the photosphere is usually referred to as the Sun's surface. The photosphere's temperature is about 5×10^3 K. Above the photosphere is a layer of gas, approximately 2×10^3 km thick, known as the chromosphere. In this layer, energy continues to be transported by radiation but it also presents convective patterns, with the presence of reddish flames extending several thousands of kilometers and then falling again. The outermost layer is called the corona. The shape of this is mostly determined by the magnetic field of the Sun, forming dynamic loops and arches. The corona emits energy of many different wavelengths that emerge from the interior of the Sun, from long wavelength radio waves to short wavelength x-rays.

The outermost layers of the Sun exhibit differential rotation—that is, each latitude rotates at slightly different speeds due to the fact that the Sun is not a solid body like the Earth. The surface rotates faster at the equator than at the areas by the poles. It rotates once every 25 days at the equator and 36 days near the poles.

2.5 ELECTROMAGNETIC RADIATION

Electromagnetic radiation is self-propagated in wave form through space with electric and magnetic components as seen in Figure 2.17. These components oscillate at right angles to each other and to the direction of propagation and are in phase with each other. An electromagnetic wave is characterized by its wavelength (λ) and frequency (f). Because a wave consists of successive troughs or crests, the wavelength is the distance between two identical adjacent points in the repeating cycles of the propagating wave, and the frequency is defined as the number of cycles per unit of time. The electromagnetic wave spectrum covers energy having wavelengths from thousands of meters, such as the very long radio waves, to fractions of the size of an atom, such as the very short gamma ray waves. The units for wavelength vary from picometers (pm) to megameters (Mm); for the frequency, the most common unit is the hertz (Hz), which is the inverse of time (1/seconds). Frequency is inversely proportional to wavelength according to

$$f = \frac{v}{\lambda}$$

$$(2.36)$$

where v is the speed of the wave; in vacuum $v = c = 299{,}792{,}458$ m/s—the speed of the light is less in other media.

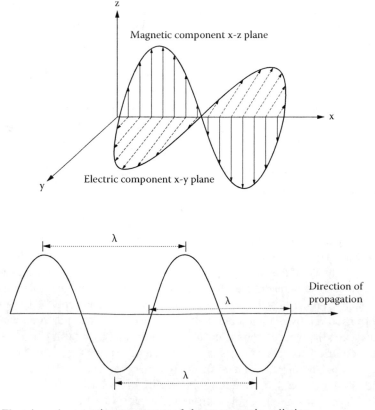

FIGURE 2.17   Electric and magnetic components of electromagnetic radiation.

As waves cross boundaries between different media, their speed and wavelength change but their frequencies remain constant. The high-frequency electromagnetic waves have a short wavelength and high energy; low-frequency waves have a long wavelength and low energy.

Because the energy of an electromagnetic wave is quantized as well, a wave consists of discrete packets of energy called photons. Its energy (E) depends on the frequency (f) of the electromagnetic radiation according to Planck's equation:

$$E = h f = \frac{h\nu}{\lambda}$$

(2.37)

where h is the constant of Planck ($h \approx 6.626069 \times 10^{-34}$ J-s or 4.13527 μeV/GHz).

Electromagnetic radiation is classified by wavelength or frequency ranges into electrical energy, radio, microwave, infrared, the visible region we perceive as light, ultraviolet, x-rays, and gamma rays; their limits on wavelength and frequency are listed in Table 2.2. There is no fixed division between regions; in reality, often some overlap exists between neighboring types of electromagnetic energy.

All objects at temperatures greater than 0 K emit energy as electromagnetic radiation due to the movement of the electrons. To study the mechanisms of interchange of energy between radiation and mass, the concept of *blackbody* was defined. A blackbody is an ideal concept and refers to a perfect absorbing body of thermal radiation, with no reflection and transmission involved. Because no light is reflected or transmitted, the object appears black when it is cold. If the blackbody is hot, these properties make it also an ideal source of thermal radiation. For a blackbody, the spectral absorption factor (α_λ) is equal to the emissivity (ε_λ); this relation is known as *Kirchhoff's law* of thermal radiation. Then, for all wavelengths, the next equation applies:

$$\alpha_\lambda = \varepsilon_\lambda = 1$$

(2.38)

The emissivity of a material, other than a blackbody, is the ratio of the energy radiated by the material to the energy radiated by a blackbody at the same temperature. It is a measure of a material's ability to absorb and radiate energy. Any real object would have $\varepsilon_\lambda < 1$.

The spectral radiation intensity emitted by a blackbody at all wavelengths (I_λ^b) at a temperature T is given by Planck's law:

$$I_\lambda^b = \frac{C_1}{\lambda^5} \frac{1}{\exp(C_2/\lambda T) - 1}$$

(2.39)

where $C_1 = 3.746 \times 10^{-16}$ Wm^2 and $C_2 = 0.014384$ mK are the Planck's first and second radiation constants, respectively, and T is the absolute temperature in Kelvin.

TABLE 2.2
Limits in the Spectrum of Electromagnetic Radiation

Region	Wavelength range (nm)	Frequency range (Hz)
Gamma rays	$1 \times 10^{-5} - 1 \times 10^{-1}$	$3 \times 10^{22} - 3 \times 10^{18}$
x-rays	$1 \times 10^{-1} - 10$	$3 \times 10^{18} - 3 \times 10^{16}$
Ultraviolet	$10 - 400$	$3 \times 10^{16} - 7.5 \times 10^{14}$
Visible light	$400 - 800$	$7.5 \times 10^{14} - 3.75 \times 10^{14}$
Infrared	$800 - 1 \times 10^{6}$	$3.75 \times 10^{14} - 3 \times 10^{11}$
Microwave	$1 \times 10^{6} - 1 \times 10^{9}$	$3 \times 10^{11} - 3 \times 10^{8}$
Radio waves	$1 \times 10^{9} - 1 \times 10^{13}$	$3 \times 10^{8} - 3 \times 10^{4}$

The integration of Planck's law over the whole electromagnetic spectrum gives the total energy radiated per unit surface area of a blackbody per unit of time—also called irradiance. The Stefan–Boltzmann law states that the total irradiance is directly proportional to the fourth power of the blackbody absolute temperature:

$$I^b = \int_0^\infty I_\lambda^b d\lambda = \int_0^\infty \frac{C_1}{\lambda^5} \frac{1}{\exp(C_2/\lambda T) - 1} d\lambda = \sigma T^4$$

(2.40)

where σ is the constant of Stefan–Boltzmann ($\sigma = 5.67 \times 10^{-8}$ W/(m^2 K^4)). The hotter an object is, the shorter is the wavelength range at which it will emit most of its radiation and the higher is the frequency for maximal radiation power. Wien's displacement law states that there is an inverse relationship between the peak wavelength of the blackbody's emission and its temperature:

$$\lambda_{max} T = 2.897\ 768 \times 10^6\ \text{nmK}$$

(2.41)

where λ_{max} is the wavelength in nanometers at which the maximum radiation emission occurs and T is the blackbody temperature in Kelvin.

Figure 2.18 shows that as the temperature of a blackbody increases, the spectral distribution and power of light emission change. The hotter an object is, the greater is energy emission at every wavelength and the shorter is the wavelength for the maximum emission. In this figure, the blackbody spectral irradiances at four temperatures are compared. At a low temperature of 500 K, a blackbody emitter has essentially no power emitted in the visible and near infrared portions of the spectrum; it will emit low-power radiation at wavelengths predominantly greater than 1,000 nm. When the blackbody is heated to 1,000–2,000 K, it will glow red because the spectrum of emitted light shifts to higher energies and into the visible spectrum (400–800 nm). If the temperature of

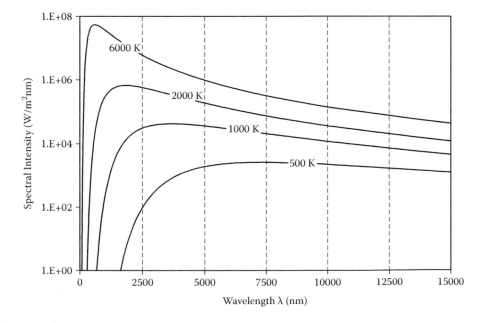

FIGURE 2.18   Spectral intensity distribution of blackbody radiation.

the blackbody is further increased to 6,000 K, radiation is emitted at wavelengths across the visible spectrum from red to violet and the light appears white.

Figure 2.19 is a useful representation to compare radiation emission from different bodies at different temperatures. In this figure, the energy output is normalized and then the wavelengths at which maximum intensity occurs are found.

Example 2.5

What is the wavelength at which the maximum monochromatic emission occurs for a star behaving as a blackbody at 8,000 K?

Solution:

According to the Wien's displacement law,

$$\lambda_{max} = \frac{2.897\ 768\ \times 10^6\ nmK}{8000\ K} = 362\ nm$$

2.6 SOLAR SPECTRAL DISTRIBUTION

The enormous amount of energy radiated from the Sun derives from the extremely high temperatures within its different layers. The Sun radiates throughout the entire electromagnetic spectrum from the shortest x-rays to long-wavelength radio waves. By far the greatest amount of the radiation falls in the visible range, and the shape of the solar spectrum is quite similar to a blackbody spectrum for an effective temperature near 5,800 K, peaking near 480 nm. Figure 2.20 illustrates the solar spectrum from 200 nm in the ultraviolet to 2,000 nm in the near infrared. The integral of this part of the spectrum accounts for almost 94% of the radiant energy from the Sun. The smooth curve overlying the solar spectrum corresponds to that of a blackbody with a temperature of 5,800 K. Table 2.3 presents the fraction of the solar irradiance from 200 to 10,000 nm.

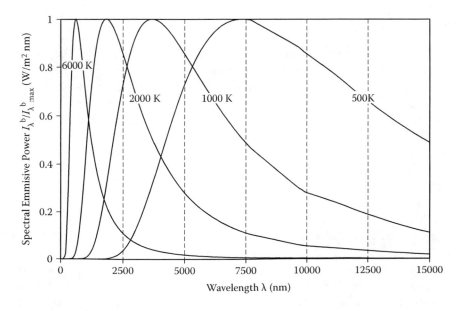

FIGURE 2.19 Normalized spectral intensity distribution of blackbody radiation.

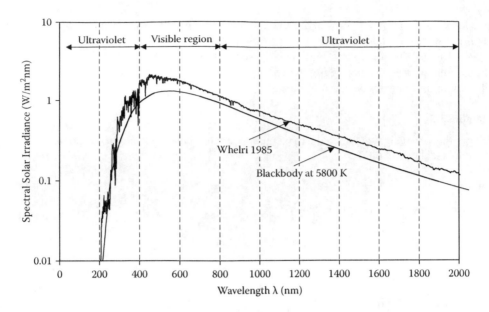

FIGURE 2.20   Solar spectrum and blackbody radiation at 5,800 K.

Example 2.6

What is the fraction of the power emitted by the Sun in the visible region of the electromagnetic spectrum solar?

Solution:

The visible ranges from 400 to 800 nm, so the fraction corresponds to

$$f(\lambda_2 - \lambda_1) = f(800 - 400) = 0.54963 - 0.07858 = 0.47105$$

2.7   SOLAR CONSTANT

Solar radiation is a general term for the electromagnetic radiation emitted by the Sun. Given the amount of energy radiated by the Sun and the geometrical relationship between the Earth and the Sun, the amount of radiation intercepted by the outer limits of the Earth's atmosphere is nearly constant. The varying solar energy output should be referred to as the total solar irradiance (TSI),[*] whereas the long-term average of TSI is commonly known as the solar constant (I_{SC}). The solar constant can be defined as the TSI integrated over the whole electromagnetic spectrum incoming to a hypothetical surface perpendicular to the Sun's rays and located outside the atmosphere at 1-AU distance, per unit of time and per unit of area.

TSI was first monitored from space with the launch of the *Nimbus 7* spacecraft in 1978 (Hickey et al. 1980). Afterward, different space experiments (HF on *Nimbus 7,* ACRIM I on *SMM,* ACRIM II on *UARS,* VIRGO on *SOHO,* and TIM on *SORCE*) have monitored TSI (Fröhlich 2006). According to the daily irradiance measurements from different instruments (Fröhlich 2006; Willson and Mordvinov 2003; Dewitte et al. 2004), TSI is not constant over time (Figure 2.21). Currently, the most successful models relate the irradiance variability to the evolution of the solar surface magnetic field (Foukal and Lean 1986, 1988; Chapman, Cookson, and Dobias 1996; Fligge and Solanki 1998; Fligge, Solanki, and Unruh 2000; Ermolli, Berrilli, and Florio 2003; Wenzler, Solanki, and

---

[*] Irradiance is referred to as an instantaneous amount of power (i.e., radiation per unit time).

TABLE 2.3
Fraction of Solar Irradiance from the Ultraviolet to the Infrared Region

λ (nm)	$f_{0-\lambda}$	λ (nm)	$f_{0-\lambda}$	λ (nm)	$f_{0-\lambda}$
200	0.012	1150	0.74963	2100	0.93041
250	0.00149	1200	0.76828	2137	0.93328
300	0.01112	1250	0.78622	2200	0.93615
350	0.03982	1300	0.802	2250	0.93902
400	0.07854	1350	0.81706	2302	0.94188
450	0.14117	1400	0.82998	2342	0.94332
500	0.2124	1450	0.84217	2402	0.94619
550	0.27962	1500	0.85293	2442	0.94762
600	0.34511	1550	0.86369	2517	0.95049
650	0.40709	1600	0.87302	3025	0.96269
700	0.46068	1650	0.88163	3575	0.96986
750	0.50659	1700	0.88952	4085	0.97345
800	0.54963	1750	0.89669	5085	0.97704
850	0.58815	1800	0.90315	5925	0.97847
900	0.62259	1850	0.90889	7785	0.9799
950	0.65265	1900	0.91391	10075	0.98062
1000	0.68084	1950	0.91821	∞	1.00000
1050	0.7063	2000	0.92323		
1100	0.72883	2050	0.92682		

Note: Data calculated from Wehrli, C. 1985. Extraterrestrial solar spectrum. Publication no. 615, Physikalisch-Meteorologisches Observatorium + World Radiation Center (PMO/WRC) Davos Dorf, Switzerland, July.

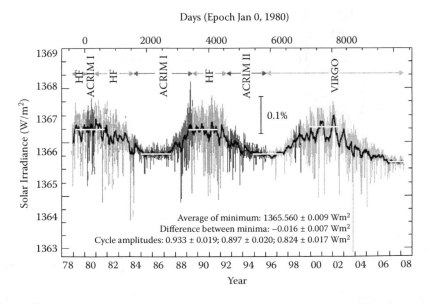

FIGURE 2.21   Total solar irradiance monitored from spacecraft experiments (Frölich 2006).

Krivova 2005; Wenzler et al. 2004, 2006). They reproduce irradiance changes with high accuracy on scales from days to solar rotation. The absolute minimum and maximum daily TSI for the experimental data set obtained from November 1978 to January 2003 (24.2 years) were 1363 and 1368 W/m², respectively. The processing of these numbers resulted in 1365.0 and 1367.2 W/m², respectively, yielding a mean value of 1366.1 W/m² and a half-amplitude of 1.1 W/m² (i.e., ±0.08% of the mean).

This value of the mean TSI confirms the solar constant value, which has been standardized (ASTM 2000). It is also only 0.9 W/m² less than the value of 1367 W/m² recommended by the World Meteorological Organization (WMO) in 1981 with an uncertainty of 1%. The difference between the two values is not significant. Nevertheless, the latest determination of I_{SC} (1366.1 W/m²) is used in this book.

2.8 EXTRATERRESTRIAL SOLAR RADIATION

Some variations in the extraterrestrial radiation above the atmosphere are not due to solar changes but rather to the Earth–Sun distance throughout the year as stated in Equation 2.3:

$$I_o = I_{SC} \left(1 + 0.033 \cos \frac{360\,n}{365}\right) \tag{2.42}$$

where I_o is the extraterrestrial radiation, I_{SC} is the solar constant, and n is the day of the year. The units are Joules per second per square meter (J/s-m²).

Also of interest is the amount of beam energy received by a horizontal surface outside the atmosphere at any time. This value corresponds to the maximum possible if there were no atmosphere:

$$H_o = I_{SC} \left(1 + 0.033 \cos \frac{360\,n}{365}\right) \sin \alpha_s \tag{2.43}$$

where H_o is the extraterrestrial solar radiation on a horizontal surface and α_s is the solar altitude in Equation 2.9. The integration of Equation 2.42 from sunshine to sunrise gives the extraterrestrial daily insolation on a horizontal surface:

$$\overline{H}_o = I_{SC} \frac{24}{\pi} \left(1 + 0.033 \cos \frac{360\,n}{365}\right) \left(\frac{\pi \omega_{\text{sunset}}}{180} \sin \phi \sin \delta + \cos \phi \cos \delta \cos \omega_{\text{sunset}}\right) \tag{2.44}$$

Figure 2.22 and Table 2.4 present the monthly average daily extraterrestrial insolation on a horizontal surface for both hemispheres. The calculation was based on $I_{sc} = 1366.1$ W/m².

Example 2.7

Determine the monthly average solar radiation on a horizontal surface outside the atmosphere at latitude 31.8° north on March 3.

Solution:

From Equation 2.21,

$$\cos \omega_{\text{sunset}} = -\tan \phi \tan \delta$$
$$= -\tan\left(31.8°\right) \tan\left(-7.5°\right)$$
$$\omega_{\text{sunset}} = 85.3°$$

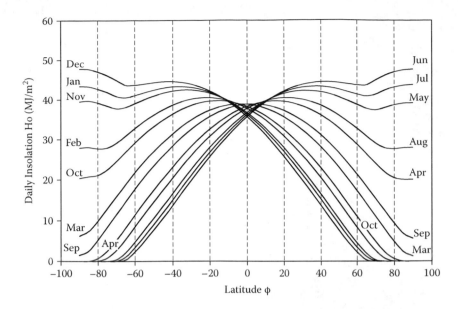

FIGURE 2.22 Monthly average daily extraterrestrial insolation on a horizontal surface.

$$\overline{H}_o = (1366.1)\frac{24 \times 3600}{\pi}\left(1 + 0.033\cos\frac{360 \times 62}{365}\right)$$

$$\left(\frac{\pi(85.3)}{180}\sin(31.8)\sin(-7.5) + \cos(31.8)\cos(-7.5)\cos(85.3)\right)$$

$$= 28.1 \text{ MJ/m}^2$$

An approximate value can be obtained from interpolation of data presented in Table 2.4 or using Figure 2.22.

2.9 TERRESTRIAL SOLAR RADIATION

In space, solar radiation is practically constant; on Earth, it varies with the day of the year, time of the day, the latitude, and the state of the atmosphere. In solar engineering, the surfaces that capture or redirect solar radiation are known as solar collectors. The amount of solar radiation striking solar collectors depends also on the position of the surface and on the local landscape.

Solar radiation can be converted into useful forms of energy such as heat and electricity using a variety of thermal and photovoltaic (PV) technologies, respectively. The thermal systems are used to generate heat for hot water, cooking, heating, drying, melting, and steam engines, among others. Photovoltaics are used to generate electricity for grid-tied or stand-alone off-grid systems. There are also applications where ultraviolet solar energy is used in chemical reactions.

When electromagnetic waves are absorbed by an object, the energy of the waves is typically converted to heat. This is a very familiar effect because sunlight warms surfaces that it irradiates. Often this phenomenon is associated particularly with infrared radiation, but any kind of electromagnetic radiation will warm object that absorbs it. Electromagnetic waves can also be reflected or scattered, in which case their energy is redirected or redistributed as well.

The *total solar radiation* incident on either a horizontal (*H*) or tilted plane (*I*) consists of three components: beam, diffuse, and reflected radiation. As sunlight passes through the atmosphere, some of it is absorbed, scattered, and reflected by air molecules, water vapor, clouds, dust, and

TABLE 2.4
Monthly Average Daily Extraterrestrial Insolation on a Horizontal Surface[a]

Latitude	Jan	Feb	Mar	Apr	May	Jun	Jul	Aug	Sep	Oct	Nov	Dec
−90	43.31	27.82	6.20	0.00	0.00	0.00	0.00	0.00	1.38	20.36	39.41	47.76
−85	43.21	27.96	7.35	0.01	0.00	0.00	0.00	0.00	2.51	20.75	39.62	47.64
−80	42.63	27.64	9.85	0.67	0.00	0.00	0.00	0.04	5.18	21.09	39.10	46.98
−75	41.81	27.82	12.86	2.49	0.00	0.00	0.00	0.80	8.32	22.41	38.35	46.08
−70	40.74	29.07	15.90	5.22	0.42	0.00	0.04	2.76	11.50	24.54	37.66	44.83
−65	40.43	30.84	18.87	8.23	2.16	0.37	1.10	5.47	14.63	26.82	38.02	43.60
−60	40.97	32.67	21.71	11.32	4.70	2.29	3.34	8.42	17.67	29.06	38.93	43.59
−55	41.66	34.41	24.40	14.40	7.54	4.82	6.04	11.47	20.58	31.16	39.90	43.89
−50	42.29	35.98	26.92	17.42	10.54	7.66	8.96	14.53	23.35	33.09	40.78	44.19
−45	42.78	37.35	29.23	20.34	13.59	10.65	12.00	17.55	25.94	34.79	41.50	44.37
−40	43.07	38.48	31.33	23.14	16.65	13.72	15.07	20.48	28.35	36.26	42.00	44.38
−35	43.12	39.36	33.20	25.78	19.65	16.81	18.12	23.29	30.54	37.48	42.26	44.16
−30	42.92	39.97	34.81	28.24	22.56	19.86	21.12	25.96	32.50	38.43	42.26	43.71
−25	42.46	40.30	36.17	30.50	25.36	22.84	24.02	28.46	34.21	39.11	42.00	42.99
−20	41.73	40.35	37.25	32.54	28.01	25.72	26.79	30.77	35.67	39.49	41.46	42.02
−15	40.73	40.11	38.04	34.35	30.48	28.46	29.41	32.86	36.85	39.59	40.64	40.80
−10	39.47	39.59	38.55	35.90	32.76	31.03	31.85	34.72	37.76	39.40	39.56	39.31
−5	37.95	38.78	38.77	37.19	34.82	33.43	34.09	36.33	38.38	38.92	38.21	37.58
0	36.18	37.69	38.70	38.20	36.65	35.61	36.12	37.69	38.71	38.16	36.60	35.61
5	34.18	36.33	38.33	38.94	38.24	37.58	37.91	38.77	38.75	37.11	34.75	33.42
10	31.95	34.72	37.67	39.38	39.57	39.31	39.46	39.58	38.50	35.79	32.66	31.03
15	29.53	32.85	36.73	39.54	40.63	40.79	40.74	40.10	37.95	34.21	30.37	28.45
20	26.93	30.76	35.51	39.40	41.42	42.02	41.77	40.34	37.11	32.38	27.88	25.71
25	24.17	28.45	34.02	38.98	41.93	42.99	42.52	40.29	36.00	30.31	25.22	22.83
30	21.28	25.95	32.28	38.27	42.17	43.70	43.01	39.95	34.61	28.03	22.41	19.85
35	18.29	23.28	30.29	37.29	42.14	44.15	43.24	39.34	32.96	25.55	19.49	16.80
40	15.24	20.46	28.07	36.04	41.85	44.37	43.22	38.46	31.07	22.89	16.48	13.71
45	12.17	17.52	25.65	34.53	41.32	44.36	42.96	37.33	28.94	20.08	13.42	10.64
50	9.14	14.50	23.03	32.79	40.58	44.17	42.51	35.96	26.59	17.14	10.37	7.64
55	6.20	11.44	20.25	30.83	39.66	43.87	41.91	34.39	24.05	14.12	7.38	4.81
60	3.48	8.39	17.32	28.69	38.65	43.57	41.26	32.65	21.33	11.04	4.54	2.28
65	1.20	5.43	14.27	26.42	37.70	43.57	40.78	30.82	18.46	7.96	2.03	0.36
70	0.06	2.71	11.14	24.08	37.27	44.80	41.19	29.05	15.47	4.98	0.35	0.00
75	0.00	0.73	7.97	21.87	37.87	46.05	42.30	27.83	12.41	2.31	0.00	0.00
80	0.00	0.02	4.86	20.43	38.61	46.95	43.12	27.60	9.35	0.60	0.00	0.00
85	0.00	0.00	2.29	20.01	39.06	47.49	43.62	27.87	6.73	0.01	0.00	0.00
90	0.00	0.00	1.25	20.07	39.21	47.67	43.79	27.98	5.66	0.00	0.00	0.00

[a] MJ/m^2.

pollutants. The *diffuse solar radiation* is the portion scattering downward from the atmosphere that arrives at the Earth's surface and the energy reflected on the surface from the surroundings. For a horizontal surface, this is expressed as H_d and for a tilted one as I_d. The solar radiation that reaches the Earth's surface without being modified in the atmosphere is called *direct beam solar radiation*; H_b for a horizontal and I_b for a tilted surface. Atmospheric conditions can reduce direct beam radiation by 10% on clear, dry days and by nearly 100% during dark, cloudy days. Measurements of solar energy are typically expressed as total solar radiation on a horizontal or tilted surface and calculated from the relationship

$$I = I_b + I_d \tag{2.45}$$

$$H = H_b + H_d \tag{2.46}$$

In designing and sizing solar energy systems, the quantification of the amount of solar energy incoming to solar collectors can be represented as irradiance and insolation. *Irradiance* is the instantaneous radiant power incident on a surface, per unit area. Usually, it is expressed in Watts per square meter. The integration of the irradiance over a specified period of time corresponds to the insolation. Typically, the integration represents hourly, daily, monthly, and yearly data.

Another useful definition of amount of energy corresponds to the *peak sun hours* (PSH). This definition equals the power received by a 1 m^2 horizontal surface during total daylight hours with the corresponding hypothetical number of hours for which irradiance would have been constant at one kW/m^2. Figure 2.23 is a representation of the PSH received on a clear day. The PSH is a useful value for comparison of the energy differences received daily, monthly, seasonally, and yearly for one site, and also to evaluate different locations. It is common to find a solar resource map with annually or average PSH values (Figure 2.24).

Realistically, the disadvantage of solar-powered systems is that energy supply is not continuous and constant during the day and also varies from day to day throughout the year. PSH is the energy parameter use when sizing PV systems; the criteria vary from (1) the month with the maximum demand of energy, (2) the month with the lowest PSH, or (3) the yearly average PSH. Design decisions reflect the investment, backup, cogeneration, and storage systems selected.

The *air mass* (*m*) is an indication of the length of the path that solar radiation travels through the atmosphere. At sea level, *m* = 1 means that the Sun is directly overhead at the zenith and the radiation travels through the thickness of 1 atm (i.e., solar noon). For zenith angles θ_z from 0 to 70° at sea level, Equation 2.47 is a close approximation to calculate the air mass.

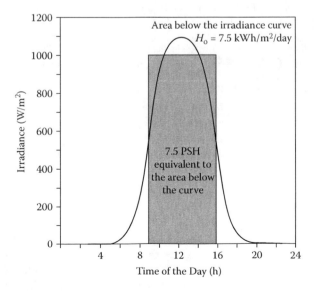

FIGURE 2.23 Peak Sun hour representation.

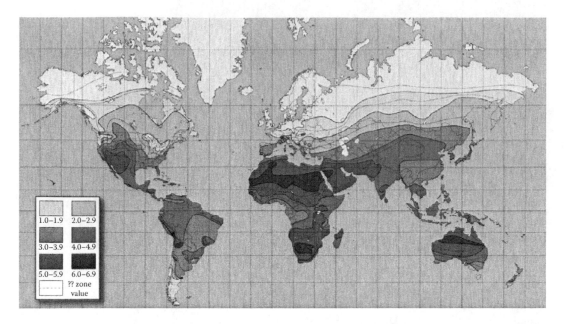

FIGURE 2.24 Global horizontal insolation map for April in kWh/m²/day (*Source*: NASA).

$$m = \frac{1}{\cos\theta_z}$$

(2.47)

For higher zenith angles, the effect of the Earth's curvature becomes significant and must be taken into account. The Earth's atmospheric gases scatter blue light more than red at one air mass. For an observer on the Earth at sunrise or sunset, when sunlight's path is longest through the atmosphere, the orange and red colors dominate because most of the violet, blue, green, and yellow light is scattered. This color change is produced because the Sun's rays must pass through much more atmosphere. Refraction as the Sun sets can sometimes even be seen as a "green flash" during the last seconds just before the Sun goes below the horizon (e.g., over water in tropical regions).

2.10 MEASUREMENT OF TERRESTRIAL SOLAR RADIATION

Solar radiation data are required for resource assessment, model development, system design, and collector testing—among other activities in solar engineering and research. The basic solar radiation measurements are the beam, diffuse, and global radiation components. The expense of radiometric stations and high maintenance make impossible the spatially continuous mapping of solar radiation. Due to the scarcity of real data, the use of representative sites where irradiance data are measured or modeled has been a common practice for engineering calculations. In the United States, the National Solar Radiation Database (NSRDB 1994) includes data for 239 locations that can be used to simulate systems throughout the country. There is also a global world meteorological organization network.

However, whereas this practice may be acceptable for standard energy calculations, nearby site extrapolation may prove widely inaccurate when site- or time-specific data are needed; this is particularly true for concentrating solar power (CSP) applications where direct normal solar radiation is required. The International Energy Agency Solar Heating and Cooling Program (IEA-SHCP) developed and evaluated techniques for estimating solar radiation at locations between network sites, using both measured and modeled data (Zelenka et al. 1992). In addition to classical statistical techniques, new methods such as satellite-based techniques have been investigated. Although they are less accurate than ground-based measurements, they may be more suitable to generate site- or time-specific data at arbitrary locations and times.

The most commonly used instruments to measure solar radiation today are based on either the thermoelectric or the photoelectric effects. The thermoelectric effect is achieved using a thermopile that comprises collections of thermocouples, which consist of dissimilar metals mechanically joined together. They produce a small current proportional to their temperature. When thermopiles are appropriately arranged and coated with a dull black finish, they serve as nearly perfect blackbody detectors that absorb energy across the entire range of the solar spectrum. The hot junction is attached to one side of a thin metallic plate. The other side of the plate is blackened to be highly absorptive when exposed to the Sun's radiation. The cold junction is exposed to a cold cavity within the instrument. The output is compensated electrically for the cavity temperature. The amount of insolation is related to the elevated temperature achieved by the hot junction and the electromagnetic force generated. The response is linearized and calibrated so that the output voltage can be readily converted to the radiative flux. The PV sensors are simpler and have instantaneous response and good overall stability. The PV effect occurs when solar radiation strikes a light-sensitive detector; atoms in the detector absorb some of the photons' energy. In this excited state, which may be produced only by light in a specific range of wavelengths, the atoms release electrons, which can flow through a conductor to produce an electrical current. The current is proportional to the intensity of the radiation striking the detector. The major disadvantage of these sensors is that their spectral response is not uniform in the solar band.

Instruments used to measure the transmission of sunlight through Earth's atmosphere fall into two general categories: instruments that measure radiation from the entire sky and instruments that measure only direct solar radiation. Within each of these categories, instruments can be further subdivided into those that measure radiation over a broad range of wavelengths and those that measure only specific wavelengths. The full-sky instruments need an unobstructed 360° view of the horizon, without significant obstacles. Full-sky instruments are called radiometers or, in the case of solar monitors, pyranometers (Figure 2.25). Good quality ones are typically about 15 cm in diameter. The sensor is under one or two hemispherical glass domes. The glass is specially formulated to

FIGURE 2.25 Pyranometer Eppley Model PSP, first-class reference instrument, as defined by the World Meteorological Organization. (Courtesy of CIE-UNAM.)

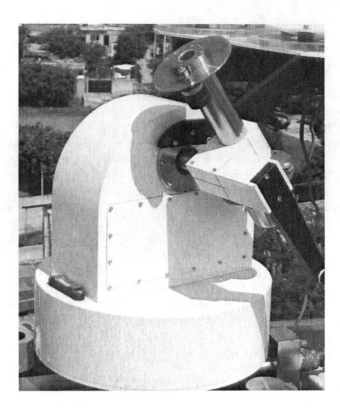

FIGURE 2.26 Pyrheliometer. (Courtesy of CIE-UNAM.)

transmit solar radiation over a wide range of wavelengths and is isolated thermally from the sensor. The pyranometer is intended for use in the permanently mounted horizontal position for which it is calibrated.

The absolute calibration coefficients for pyranometers in units of microvolts per Watt/square meter should be traceable to an internationally accepted reference, such as that maintained at the World Radiation Center (WRC).

Although broadband detectors are required for measuring total solar radiation, an inexpensive alternative is to use PV detectors such as silicon-based solar cells. Their major disadvantage is that their spectral response is different from the solar spectrum. Typically, they respond to sunlight in the range from 400 to 1,100 nm, with a peak response in the near-infrared, around 900 nm. Under normal outdoor sunlight conditions, this introduces a potential error of a few percent. Commercial pyranometers that use silicon-based sensors are much less expensive than thermopile-based pyranometers.

The direct sunlight radiation is measured with pyrheliometers (Figure 2.26). These are designed to view only light coming directly from the Sun. The radiation incident on the detector is restricted to a narrow cone of the sky to avoid scattered light. The sensor is located at the base of a tube fitted with annular diaphragms where only nearly normal incident radiation reaches. The tubes housing the detector at the bottom are about 50 cm long. This instrument automatically tracks the Sun under computer control; the solar disk subtends about 0.5°.

2.11 TERRESTRIAL INSOLATION ON TILTED COLLECTORS

When designing solar energy systems or conducting performance monitoring, it is necessary to account for the availability of solar data in order to calculate the amount of solar radiation striking on tilted collectors. Average hourly, daily, and monthly local insolation data are usually used; the most common insolation measurements are local horizontal global or beam. The global insolation is the most important input to estimate accurately insolation over tilted surfaces. Many mathematical models have been proposed to estimate hourly and daily global solar radiation on tilted surfaces from that measured on horizontal surfaces that include information such as level of cloudiness, pollution, temperature, and humidity, among other variables. Although these methods work well at local levels, there is not yet a general highly accurate method for predicting insolation.

At any time, the ratio of beam radiation on a tilted surface to that on horizontal surface is related by the geometric factor R_b. Figure 2.12 shows the geometric relationship between the solar angles for a horizontal and a tilted surface. The ratio R_b can be calculated by

$$R_b = \frac{I_b}{H_b} = \frac{I_{b,n}\cos\theta}{H_{b,n}\cos\theta_z} = \frac{\cos\theta}{\cos\theta_z} = \frac{\cos\theta}{\sin\alpha_s}$$

(2.48)

where $\cos\theta$, $\cos\theta_z$, and $\sin\alpha_s$ can be calculated from Equations 2.19, 2.8, and 2.9, respectively, and the subscripts b and n.

Figures 2.27A through 2.27E present the graphical representation of Equation 2.48 for surfaces tilted toward the equator. Each figure helps to calculate both the $\cos\theta$ and $\cos\theta_z$ as a function of $(\phi - \beta)$ and ϕ, respectively, in such a way that R_b is easily obtained for specific dates and latitudes. Each figure applies for two specific solar times symmetrical from solar noon, and they were calculated from the midpoint solar time for one particular hour. For example, to produce the R_b chart for 11 a.m. to 12 p.m., the calculation was made at 11:30 a.m. In particular cases where the surface is facing the equator, the following equations apply:

$$R_b\left(\gamma=0^\circ\right) = \frac{\cos(\phi-\beta)\cos\delta\sin\omega+\sin(\phi-\beta)\sin\delta}{\cos\phi\cos\delta\sin\omega+\sin\phi\sin\delta}$$

(2.49)

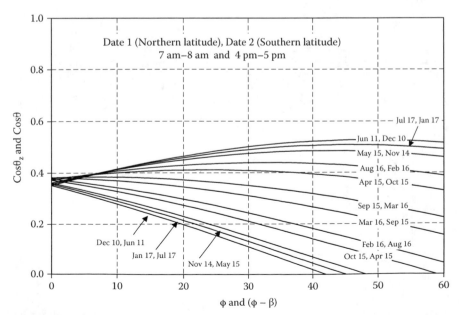

FIGURE 2.27A Graphical method to calculate R_b. $\cos\theta_z$ versus ϕ and $\cos\theta$ versus $(\phi - \beta)$ from 7 to 8 a.m. and from 2 to 3 p.m.

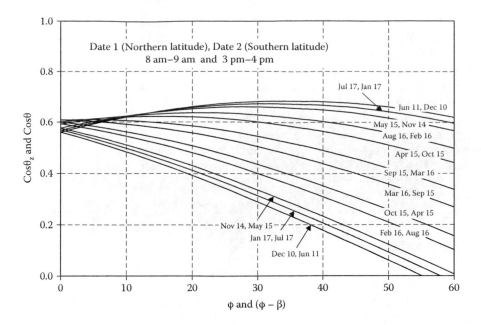

FIGURE 2.27B   Graphical method to calculate R_b. $\cos\theta_z$ versus ϕ and $\cos\theta$ versus $(\phi - \beta)$ from 8 to 9 a.m. and from 3 to 4 p.m.

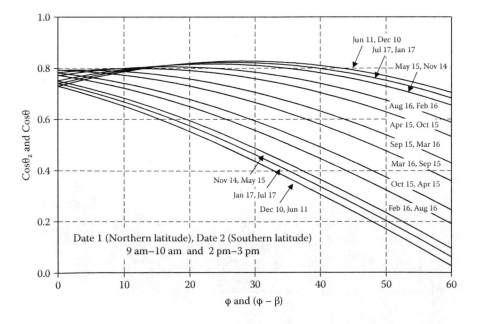

FIGURE 2.27C   Graphical method to calculate R_b. $\cos\theta_z$ versus ϕ and $\cos\theta$ versus $(\phi - \beta)$ from 9 to 10 a.m. and from 2 to 3 p.m.

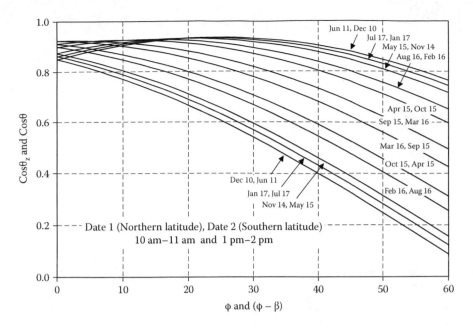

FIGURE 2.27D   Graphical method to calculate R_b. $\cos\theta_z$ versus ϕ and $\cos\theta$ versus $(\phi - \beta)$ from 10 to 11 a.m. and from 1 to 2 p.m.

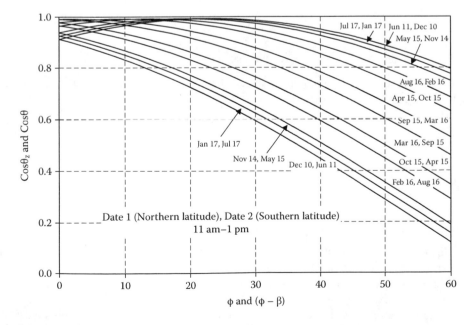

FIGURE 2.27E   Graphical method to calculate R_b. $\cos\theta_z$ versus ϕ and $\cos\theta$ versus $(\phi - \beta)$ from 11 a.m. to 1 p.m.

$$R_b\left(\gamma=180^\circ\right)=\frac{\cos(\phi+\beta)\cos\delta\sin\omega+\sin(\phi+\beta)\sin\delta}{\cos\phi\cos\delta\sin\omega+\sin\phi\sin\delta}$$

$$(2.50)$$

at noon,

$$R_{b,\text{noon}}\left(\gamma=0^\circ\right)=\frac{\cos\left|\phi-\delta-\beta\right|}{\cos\left|\phi-\delta\right|}$$

$$(2.51)$$

$$R_{b,\text{noon}}\left(\gamma=0^\circ\right)=\frac{\cos\left|-\phi+\delta-\beta\right|}{\cos\left|-\phi+\delta\right|}$$

$$(2.52)$$

Example 2.8

Calculate R_b for a surface facing south with $\beta = 30^\circ$, at $\phi = 31.8^\circ$ north for the hour 9 to 10 solar time on March 3.

Solution:

From Figure 2.27C,

$$R_b=\frac{\cos\theta}{\cos\theta_z}=\frac{0.78}{0.60}=1.3$$

or, calculated from Equation 2.49,

$$R_b\left(\gamma=0^\circ\right)=\frac{\cos(31.8-30)\cos(-7.5)\sin(-37.5)+\sin(31.8-30)\sin(-7.5)}{\cos(31.8)\cos(-7.5)\sin(-37.5)+\sin(31.8)\sin(-7.5)}=1.3$$

2.11.1 Instantaneous and Hourly Radiation

To calculate the average radiation in a tilted surface from total horizontal and beam radiation data, first the diffuse contribution on the horizontal must be calculated:

$$H_d = H - H_b$$

$$(2.53)$$

where H_b and H_d are the beam and diffuse contributions to the total radiation on a horizontal surface, respectively. Then,

$$H_d = H - I_b \sin\alpha_s = H - I_b \cos\theta_z$$

$$(2.54)$$

where the beam radiation I_b is known (measured with a pyrheliometer) and $\cos\theta_z$ or $\sin\alpha_s$ can be calculated from Equations 2.8 and 2.9, respectively.

Assuming the isotropic model proposed by Hottel and Woertz (1942), the sums of diffuse radiation and ground-reflected radiation on a tilted surface is the same regardless of its orientation. For which the solar radiation on a tilted surface is the sum of the beam contribution and the diffuse on the horizontal surface. In 1960, Liu and Jordan presented the isotropic diffused model, improving prior predictions. The total solar radiation incoming into a tilted surface is estimated as

$$I = H_b R_b + H_d R_d + H \rho_r R_r = HR$$

(2.55)

where

$$R_d = \cos^2 \frac{\beta}{2}$$

(2.56)

$$R_r = \sin^2 \frac{\beta}{2}$$

(2.57)

R_b, given by Equation 2.48 and ρ_r, is the effective diffuse ground reflectance of the total radiation. Table 2.5 presents values of reflectance integrated over the solar spectrum and incidence angle for different surfaces and landscapes. Substituting Equations 2.56 and 2.57 into Equation 2.55,

$$I = H_b R_b + H_d \cos^2 \frac{\beta}{2} + H \rho_r \sin^2 \frac{\beta}{2}$$

(2.58)

The ratio of total radiation on a tilted surface to that in the horizontal surface is determined by

$$R = \frac{I}{H} = \frac{H_b}{H} R_b + \frac{H_d}{H} \cos^2 \frac{\beta}{2} + \rho_r \sin^2 \frac{\beta}{2}$$

(2.59)

To estimate the same information when only the horizontal total radiation is known, H_b and H_d must still be calculated. This can be done by taking into account the clearness index K_T, which is related to sunshine duration for a particular location.

$$K_T = \frac{H}{H_o}$$

(2.60)

where H_o is the extraterrestrial radiation on a horizontal surface given by Equation 2.42. A simple correlation between the clearness index K_T and the total beam radiation I_b in W/m² was developed by Boes et al. (1976) from measurements taken in the United States:

$$I_b = \begin{cases} -520 + 1800K_T & 0.85 > K_T \geq 0.30 \\ 0 & 0.30 > K_T \end{cases}$$

(2.61)

Orgill and Hollands (1977); Erbs, Klein, and Duffie (1982); and Reindl, Beckman, and Duffie (1990) have reported similar correlations for K_T and H_d/H_o, although they were derived from different radiometric stations; data from Canada, United States, Europe, and Australia were processed. The Orgill and Hollands correlation is expressed by

$$\frac{H_d}{H} = \begin{cases} 1.0 - 0.249K_T & K_T < 0.35 \\ 1.557 - 1.84K_T & 0.35 \leq K_T \leq 0.75 \\ 0.177 & 0.75 < K_T \end{cases}$$

(2.62)

Example 2.9

Calculate the instantaneous irradiance for a surface facing south with $\beta = 30°$, at $\phi = 31.8°$ north at 10 a.m. solar time on March 3, when the global radiation measured was 750 W/m² and the beam radiation was 650 W/m². The surface is located within a government area. On March 3, $\delta = -7.5°$. From Table 2.5 $\delta_r = 0.38$.

TABLE 2.5
Effective Reflectance

Surfaces	δ_r
Snow	0.75
Fields with snow cover	0.6
Water	0.07
Open water	0.16
Soil	0.14
Earth roads	0.04
Dry grass	0.2
Green leaves	0.26
Dark building surfaces	0.27
Light building surfaces	0.06
Urban commercial	0.16
Urban institutional	0.38
Residential areas	0.2–0.4

Source: Hunn, B. D. and D. O. Calafell (1977). *Solar Energy*, 19(1) 87–89.

Solution:

From Equation 2.8,

$$\cos\theta_z = \sin(31.8)\sin(-7.5)+\cos(31.8)\cos(-7.5)\cos(-30)$$
$$= 0.66$$

The incidence angle for a surface facing south, Equation 2.24, applies:

$$\cos\theta = \sin(-7.5)\sin(31.8)\cos(30)$$
$$- \sin(-7.5)\cos(31.8)\sin(30)$$
$$+ \cos(-7.5)\cos(31.8)\cos(30)\cos(-30)$$
$$+ \cos(-7.5)\sin(31.8)\sin(30)\cos(-30)$$
$$= 0.854$$

From Equation 2.54,
$$H_d = 750-650\times0.66$$
$$= 321 \text{ W/m}^2$$

Substituting Equation 2.48 into Equation 2.58, the radiation on the tilted surface is given by

$$I_T = I_b\cos\theta+H_d\cos^2\frac{\beta}{2}+H\rho_r\sin^2\frac{\beta}{2}$$
$$= 650\times0.854+321\times\cos^2(15)+750\times0.38\times\sin^2(15)$$
$$873.7 \text{ W/m}^2$$

Example 2.10

Solve Problem 2.9 at the end of the chapter, assuming that only global solar radiation is available.

Solution:
Because $\sin \alpha_s = \cos \theta_z = 0.66$, then from Equation 2.43,

$$H_o = 1366.1 \left(1 + 0.033\cos\frac{360 \times 62}{365}\right) \times 0.66$$
$$= 915 \text{ W/m}^2$$

From Equation 2.60,

$$K_T = \frac{750}{915} = 0.82$$

From Equation 2.62,

$$H_d = 0.177 \, H = 0.177 \times 750$$
$$= 132 \text{ W/m}^2$$

Then, from Equation 2.54,

$$H_d = H - I_b \sin\alpha_s = H - I_b \cos\theta_z$$
$$I_b = \frac{H - H_d}{\sin\alpha_s} = \frac{750 - 132}{0.66}$$
$$= 935 \text{ W/m}^2$$

2.11.2 MONTHLY AVERAGE DAILY INSOLATION

Radiation data are most commonly available on a daily basis. Monthly average daily insolation is helpful for estimating the long-term performance evaluation of solar energy systems. This model uses an analogous clearness index to that given by Equation 2.60; this is the monthly average daily clearness index $\overline{K}_T$:

$$\overline{K}_T = \frac{\overline{H}}{\overline{H}_o} \tag{2.63}$$

where $\overline{H}$ is the monthly average daily global terrestrial insolation on a horizontal surface and the $\overline{H}_o$ is the monthly average daily extraterrestrial insolation given by Table 2.4, Figure 2.22, or Equation 2.43.

The monthly average daily insolation on a tilted surface is

$$\overline{I} = \overline{H}_b \overline{R}_b + \overline{H}_d \overline{R}_d + \overline{H}\rho_r R_r = \overline{H}R \tag{2.64}$$

where

$$\overline{R}_d = R_d, \overline{R}_r = R_r$$

then,

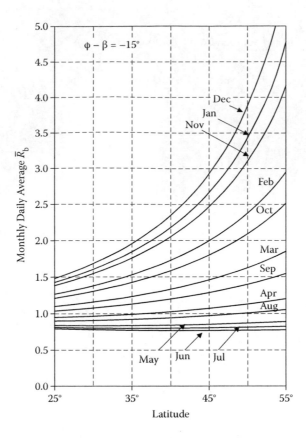

FIGURE 2.28 Estimated $\bar{R}_b$ for surfaces facing the equator as a function of latitude for $\phi - \beta = -15°$. For the Southern Hemisphere, assign the month according to the table included in Figure 2.11A.

$$\bar{I} = \overline{H}_b \overline{R}_b + \overline{H}_d \cos^2 \frac{\beta}{2} + \overline{H} \rho_r \sin^2 \frac{\beta}{2} = \overline{H}\,\overline{R}$$

(2.65)

and $\overline{R}_b$ is the monthly average of R_b:

$$\overline{R}_b \approx \frac{\overline{\cos\theta}}{\overline{\cos\theta_z}} = \frac{\overline{\cos\theta}}{\overline{\sin\alpha_s}}$$

(2.66)

This gives

$$\overline{R} = \frac{\bar{I}}{\overline{H}} = \frac{\overline{H}_b}{\overline{H}} \overline{R}_b + \frac{\overline{H}_d}{\overline{H}} \cos^2 \frac{\beta}{2} + \rho_r \sin^2 \frac{\beta}{2}$$

(2.67)

for surfaces in the Northern Hemisphere facing the equator with $\gamma = 0°$,

$$\overline{R}_b \left(\gamma = 0° \right) = \frac{\cos(\phi - \beta)\cos\delta\sin\omega'_{sunset} + (\pi/180)\,\omega'_{sunset}\,\sin(\phi - \beta)\sin\delta}{\cos\phi\cos\delta\sin\omega'_{sunset} + (\pi/180)\,\omega'_{sunset}\,\sin\phi\sin\delta}$$

(2.68)

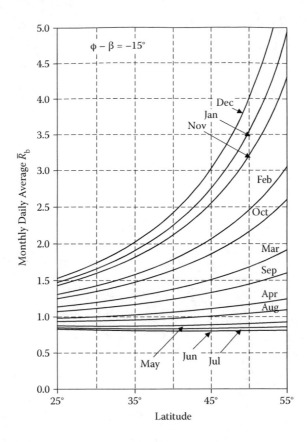

FIGURE 2.29 Estimated $\bar{R}_b$ for surfaces facing the equator as a function of latitude for $\phi - \beta = 0°$. For the Southern Hemisphere, assign the month according to the table included in Figure 2.11A.

where ω'_{sunset} is the effective surface sunset hour angle and corresponds to the smaller value from

$$\omega'_{sunset} = \cos^{-1}(-\tan\phi\tan\delta)$$
$$\omega'_{sunset} = \cos^{-1}(-\tan(\phi-\beta)\tan\delta) \tag{2.69}$$

and, for surfaces in the Southern Hemisphere facing the equator with $\gamma = 180°$,

$$\bar{R}_b\left(\gamma=180°\right) = \frac{\cos(\phi+\beta)\cos\delta\sin\omega'_{sunset} + (\pi/180)\omega'_{sunset}\sin(\phi+\beta)\sin\delta}{\cos\phi\cos\delta\sin\omega'_{sunset} + (\pi/180)\omega'_{sunset}\sin\phi\sin\delta} \tag{2.70}$$

where ω'_{sunset} now corresponds to the smaller value from

$$\omega'_{sunset} = \cos^{-1}(-\tan\phi\tan\delta)$$
$$\omega'_{sunset} = \cos^{-1}(-\tan(\phi+\beta)\tan\delta) \tag{2.71}$$

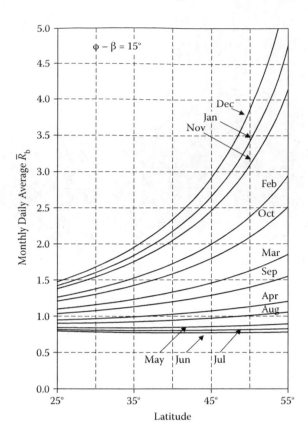

FIGURE 2.30   Estimated $\bar{R}_b$ for surfaces facing the equator as a function of latitude for $\phi - \beta = 0°$, $\phi - \beta = 15°$, and $\phi - \beta = -15°$. For the Southern Hemisphere, assign the month according to the table included in Figure 2.11A.

REFERENCES

Blanco-Muriel, M., D. C. Alarcon-Padilla, T. Lopez-Moratalla, and M. Lara-Coira. 2001. Computing the solar vector. *Journal of Solar Energy* 70 (5): 431–441.

Boes, E. C., I. J. Hall, R. R. Prairie, R. P. Stromberg, and H. E. Anderson. 1976. Distribution of direct and total solar radiation available for the USA. *Proceedings of the 1976 Annual Meeting of the American Section of ISES, Sharing the Sun,* vol. 1, Winnipeg, August 15–20, pp. 238–263.

Braun, J. E., and J. C. Mitchell. 1983. Solar geometry for fixed and tracking surfaces. *Solar Energy* 31 (5): 439–444.

Chapman, G. A., A. M. Cookson, and J. J. Dobias., 1996. Variations in total solar irradiance during solar cycle 22. *Journal of Geophysical Research* 101:13,541–13,548.

Cooper, P. I. 1969. The absorption of solar radiation in solar stills. *Solar Energy* 12 (3): 333–346.

Dewitte, S., D. Crommelynck, S. Mekaoui, and A. Joukoff. 2004. Measurement and uncertainty of the long-term total solar irradiance trend. *Solar Physics* 224:209–216.

Duffie, J. A., and W. A. Beckman. 1991. *Solar engineering of thermal processes,* 2nd ed., 919. New York: John Wiley & Sons.

Erbs D. G., S. A. Klein, and J. A. Duffie. 1982. Estimation of the diffuse radiation fraction for hourly, daily and monthly average global radiation. *Solar Energy* 28:293–902.

Ermolli, I., F. Berrilli, and A. Florio. 2003. A measure of the network radiative properties over the solar activity cycle. *Astronomy and Astrophysics* 412:857–864.

Fligge, M., and S. K. Solanki, 1998. Long-term behavior of emmission from solar faculae: Steps towards a robust index. *Astronomy and Astrophysics* 332: 1082–1086.

Fligge, M., S. K. Solanki, and Y. C. Unruh. 2000. Modeling irradiance variations from the surface distribution of the solar magnetic field. *Astronomy and Astrophysics* 353:380–388.

Foukal, P., and J. Lean. 1986. The influence of faculae on total solar irradiance and luminosity. *Astrophysical Journal* 302:826–835.

Foukal, P. and J. Lean. 1988. Magnetic modulation of solar luminosity by photospheric activity. *Astrophysical Journal* 328:347–357.

Fröhlich, C. 2006. Solar irradiance variability since 1978. *Space Science Reviews* 90:1–13.

Hickey, J. R., L. L. Stowe, H. Jacobowitz, P. Pellegrino, R. H. Maschhoff, F. House, and T. H. Vonder Haar. 1980. Initial solar irradiance determinations from *NIMBUS-7* cavity radiometer measurements. *Science* 208:281–283.

Hottel, H. C., and B. B. Woertz. 1942. Performance of flat plate solar heat collectors. *Transactions of the ASME* 64:91–104.

Hunn, B. D., and D. O. Calafell. 1977. Determination of average ground reflectivity for solar collectors. *Solar Energy*, 19 (1): 87–89.

Ibrahim, R., and A. Afshin. 2004. Solar position algorithm for solar radiation applications. *Solar Energy* 76 (5): 577–589.

Iqbal, M. 1983. *An introduction to solar radiation,* 390. New York: Academic Press.

Liu, B. H. Y., and R. C. Jordan. 1960. The interrelationship and characteristic distribution of direct, diffuse and total solar radiation. *Solar Energy* 4:1–19.

Meeus, J. 1998. *Astronomical algorithms,* 2nd ed. Richmond, VA: Willmann–Bell, Inc.

Michalsky, J. J. 1988. The astronomical almanac's algorithm for approximate solar position (1950–2050). *Journal of Solar Energy* 40 (3): 227–235.

NSRDB. 1994. NREL/TP-463-5784. National Renewable Energy Lab, Golden, CO.

Orgill J. F., and K. G. T. Hollands. 1977. Correlation equation for hourly diffuse radiation on a horizontal surface. *Solar Energy* 19:357–359.

Perrin de Brichambaut, C. 1975. *Estimation des ressources énergétiques solaires en France.* Edition Européennes Termiqué et Industrie, Paris, p. 63.

Reindl, D. T., W. A. Beckman, and J. A. Duffie. 1990. Diffuse fraction correlations. *Solar Energy* 45:1–7.

Spencer, J. W. 1971. Fourier series representation of the position of the Sun. *Search* 2 (5): 172.

Tayler, R. J. 1997. *The Sun as a star.* New York: Cambridge University Press.

Wehrli, C. 1985. Extraterrestrial solar spectrum. Publication no. 615, Physikalisch-Meteorologisches Observatorium + World Radiation Center (PMO/WRC) Davos Dorf, Switzerland, July.

Wenzler, T., S. K. Solanki, and N. A. Krivova. 2005. Can surface magnetic fields reproduce solar irradiance variations in cycles 22 and 23? *Astronomy and Astrophysics* 432:1057–1061.

Wenzler, T., S. K. Solanki, N. A. Krivova, and D. M. Fluri. 2004. Comparison between KPVT/SPM and SoHO/MDI magnetograms with an application to solar irradiance reconstructions. *Astronomy and Astrophysics* 427:1031–1043.

Wenzler, T., S. K. Solanki, N. A. Krivova, and C. Fröhlich. 2006. Reconstruction of solar irradiance variations in cycles 21–23 based on surface magnetic fields. *Astronomy and Astrophysics* 460:583–595.

Willson, R. C., and A. V. Mordvinov. 2003. Secular total solar irradiance trend during solar cycles 21–23. *Geophysics Research Letters* 30:3–1.

Zelenka, A., G. Czeplak, V. D'Agostino, W. Jossefsson, E. Maxwell, and R. Perez. 1992. Techniques for supplementing solar radiation network data. Final report of International Energy Agency Solar Heating and Cooling Program, Task 9, Subtask 9d, Paris.

PROBLEMS

2.1   Determine the location of the Sun at solar noon in El Paso, Texas, on March 21.

2.2   Determine the local standard time for the preceding problem.

2.3   Determine the sunrise azimuth angle for El Paso, Texas, on March 21.

2.4   For latitude of 31.8° south and longitude 106.4° west, determine the position of the Sun at 10:15 a.m. on March 21.

2.5   What is the optimal β slope for a fixed solar collector facing south on June 21 for the preceding problem?

2.6   Determine the local standard time for sunrise for Problem 2.3.

2.7   What is the monthly average daily extraterrestrial insolation on June 21 for latitudes of 25° north and 25° south?

2.8   Determine the solar incidence angle for a surface tilted at 32° for three azimuth angles: 15° west, 0°, and 15° east.

2.9   How long are the days in Problems 2.1 and 2.4?

2.10   What is the difference between zenith angle and incidence angle?

2.11   Calculate the hourly ratio R_b for a surface located at 35° north latitude, facing south and tilted at 20° from the horizontal.

2.12   Determine the March 21 location of the Sun at solar noon for where you live.

3 Fundamentals of Engineering
Thermodynamics and Heat Transfer

3.1 INTRODUCTION

This chapter provides an introduction to heat transfer and engineering thermodynamics. The science of thermodynamics deals with energy interaction between a system and its surroundings. These interactions are called heat transfer and work. Thermodynamics deals with the amount of heat transfer between two equilibrium states and makes no reference to how long the process will take. However, in heat transfer, we are often interested in rate of heat transfer. Heat transfer processes set limits to the performance of environmental components and systems. The content of this chapter is intended to extend the thermodynamics analysis by describing the different modes of heat transfer. It also provides basic tools to enable the readers to estimate the magnitude of heat transfer rates and rate of entropy destruction in realistic environmental applications, such as solar energy systems.

The transfer of heat is always from the higher temperature medium to the lower temperature medium. Therefore, a temperature difference is required for heat transfer to take place. Heat transfer processes are classified into three types: conduction, convection, and radiation.

Conduction heat transfer is the transfer of heat through matter (i.e., solids, liquids, or gases) without bulk motion of the matter. In other words, conduction is the transfer of energy from the more energetic to less energetic particles of a substance due to interaction between them. This type of heat conduction can occur, for example, through the wall of a boiler in a power plant. The inside surface, which is exposed to gases or water, is at a higher temperature than the outside surface, which has cooling air next to it. The level of the wall temperature is critical for a boiler.

Convection heat transfer is due to a moving fluid. The fluid can be a gas or a liquid; both have applications in an environmental process. In convection heat transfer, the heat is moved through the bulk transfer of a nonuniform temperature fluid. This type of heat transfer can occur in a flow of air over a lagoon or a waste-water treatment system.

Radiation heat transfer is energy emitted by matter in the form of photons or electromagnetic waves. Radiation can take place through space without the presence of matter. In fact, radiation heat transfer is highest in a vacuum environment. Radiation can be important even in situations in which there is an intervening medium; a familiar example is the heat transfer from a glowing piece of metal or from a fire.

3.2 CONDUCTION HEAT TRANSFER

To examine conduction heat transfer, it is necessary to relate the heat transfer to mechanical, thermal, or geometrical properties. Consider steady-state heat transfer through a wall of thickness Δx that is placed between two reservoirs: hot (T_H) and cold (T_C), respectively. Figure 3.1 shows the process pictorially.

As shown by the figure, heat transfer rate, $\dot{Q}$, is a function of the hot and cold temperatures, the slab geometry, and the following properties:

$$\dot{Q} = f(T_H, T_C, \text{geometry, properties}) \tag{3.1}$$

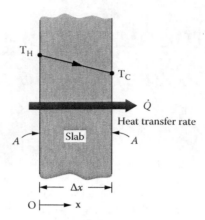

FIGURE 3.1 Conduction heat transfer through a slab of thickness Δx and area A.

It is also possible to express the heat transfer rate, $\dot{Q}$ (w), based on the hot and cold temperature difference, $T_H - T_C$, where the heat transfer rate is zero when there is no temperature difference. The temperature dependence can therefore be expressed as

$$\dot{Q} = f((T_H - T_C), T_H, \text{geometry, properties}) \tag{3.2}$$

Fourier has shown that heat transfer rate is proportional to the temperature difference across the slab and the heat transfer area and inversely proportional to the slab thickness. That is, (Bejan and Kraus)

$$\text{Heat Transfer Rate} \propto \frac{(\text{Area})(\text{Temperature difference})}{\text{Thickness}} \tag{3.3}$$

or

$$\dot{Q} \propto \frac{(A)(\Delta T)}{\Delta x} \tag{3.4}$$

In Equation 3.4, proportionality factor is a transport property (k) and is called thermal conductivity (W/mK). Therefore, Equation 3.4 becomes

$$\dot{Q} = kA \frac{(T_H - T_C)}{\Delta x} = -kA \frac{(T_C - T_H)}{\Delta x} = -kA \frac{\Delta T}{\Delta x} \tag{3.5}$$

Thermal conductivity is the measure of ability of a material to conduct heat. Parameter A is the cross-sectional area (m²) and Δx is the thickness of the slab (m). Thermal conductivity is a well tabulated property for a large number of materials. Some values for familiar materials are given in Table 3.1; others can be found in the references. (Bejan, Osizk, Ghassemi, et al.).

In the limit Equation 3.5, for any temperature difference ΔT across a length Δx as both approach zero becomes

$$\dot{Q} = -kA \frac{dT}{dx} \tag{3.6}$$

where

$$\frac{dT}{dx} \left(\frac{K}{m} \right)$$

is the temperature gradient.

TABLE 3.1
Thermal Conductivity at Room Temperature for Some Metals and Nonmetals

Metal	K (W/mK)	Nonmetal	K (W/mK)
Silver	420	Water	0.6
Copper	390	Air	0.026
Aluminum	200	Engine oil	0.15
Iron	70	Hydrogen	0.18
Steel	50	Brick	0.4–0.5

Source: Kaviany, K. 2001. *Principles of Heat Transfer*. New York: John Wiley & Sons.

Equation 3.6 is the one-dimensional form of Fourier's law of heat conduction. Temperature gradient shows the slope of the line and is negative based on the second law of thermodynamics (see Figure 3.2).

A more useful quantity to work with is the heat transfer per unit area,

$$q'' \left(\frac{W}{m^2} \right)$$

which is called the heat flux. That is,

$$q'' = \frac{\dot{Q}}{A}$$

(3.7)

3.3 ONE-DIMENSIONAL CONDUCTION HEAT TRANSFER IN A RECTANGULAR COORDINATE

Heat transfer through wall, window, and many other objects can easily be evaluated as being one dimensional, as shown by Figure 3.3. For steady-flow, one-dimensional heat conduction with no work, no mass flow, and no generation, based on Taylor expansion, the first law reduces to

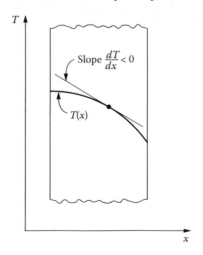

FIGURE 3.2 Temperature gradient, dT/dx.

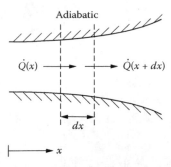

FIGURE 3.3   One-dimensional heat transfer.

$$\frac{d\dot{Q}(x)}{dx} = 0$$

(3.8)

Combining Equations 3.6 and 3.8 gives

$$\frac{d}{dx}(kA\frac{dT}{dx}) = 0$$

(3.9)

When properties are assumed constant and the chain rule is used, the energy equation is

$$\frac{d^2T}{dx^2} + (\frac{1}{A}\frac{dA}{dx})\frac{dT}{dx} = 0$$

(3.10)

Solving Equation 3.10 provides the temperature field in a plane wall. If the cross-sectional area stays constant, Equation 3.10 reduces to

$$\frac{d^2T}{dx^2} = 0$$

(3.11)

The solution to Equation 3.11 leads to a linear temperature variation

$$T(x) = T_H + \left(\frac{T_H - T_C}{L}\right)x$$

(3.12)

where l is the wall thickness.

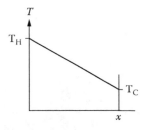

FIGURE 3.4   Temperature distribution through a slab.

Furthermore, the heat flux (q'') is constant and is as follows (Figure 3.1):

$$q'' = k\frac{(T_H - T_C)}{\Delta x}$$

(3.13)

3.4   THERMAL RESISTANCE CIRCUITS

For a slab with thickness L, Equation 3.5 can be rearranged as follows:

$$\dot{Q} = \frac{\Delta T}{L / kA}$$

(3.14)

Also, the electric current flow (I) equation is

$$I = \frac{\Delta V}{R_e}$$

(3.15)

where V is voltage and R_e is electrical resistance.

   Comparing Equations 3.14 and 3.15, the analog for heat transfer rate ($\dot{Q}$) is current and the analog for the temperature difference (ΔT) is the voltage difference. Based on this analogy, the electrical resistance is analogous to

$$R_e = \frac{L}{kA} = R_{cond}$$

(3.16)

R_{cond} is conduction thermal resistance and is the measure of wall resistance against heat flow. It is obvious that the thermal resistance R_{cond} increases as l increases, as A and k decrease. The concept of a thermal resistance circuit can be best used for problems such as a composite slab (see Figure 3.5).

   It can be easily shown that total heat transfer through the slab is

$$\dot{Q} = \frac{\Delta T}{R_{cond}}$$

(3.17)

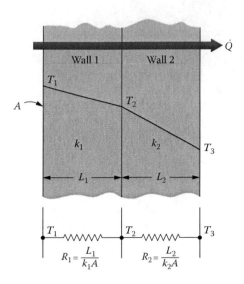

FIGURE 3.5   Heat transfer across a composite slab.

Heat transfer through a slab with several layers of material is analogous to the previous case and is as follows:

$$\dot{Q} = \frac{\Delta T}{\sum R} = \frac{\Delta T}{R_1 + R_2 + R_3 + ... + R_n}$$

(3.18)

For example, heat conduction through a brick slab with an interior and exterior layer, as shown by Figure 3.6, is given by

$$\dot{Q} = \frac{\Delta T}{\sum R_{cond}} = \frac{T_1 - T_4}{R_1 + R_2 + R_3}$$

(3.19)

Conduction thermal resistance for a slab with a dissimilar material can also be derived based on electrical concepts. Figure 3.6 shows the physical configuration. The conduction heat transfer resistances are in parallel and are

$$\frac{1}{R_{cond}} = \frac{1}{R_1} + \frac{1}{R_2}$$

(3.20)

3.5 ONE-DIMENSIONAL CONDUCTION HEAT TRANSFER IN A CYLINDRICAL COORDINATE

Most problems in heat transfer are in nonplanar geometry. Figure 3.7 depicts one of the nonplanar geometry problems, a long cylindrical shell.

For steady-flow, one-dimensional heat conduction with no work, no mass flow, and no generation, based on Taylor expansion, Equation 3.8 reduces to

$$\frac{d\dot{Q}(r)}{dr} = 0$$

(3.21)

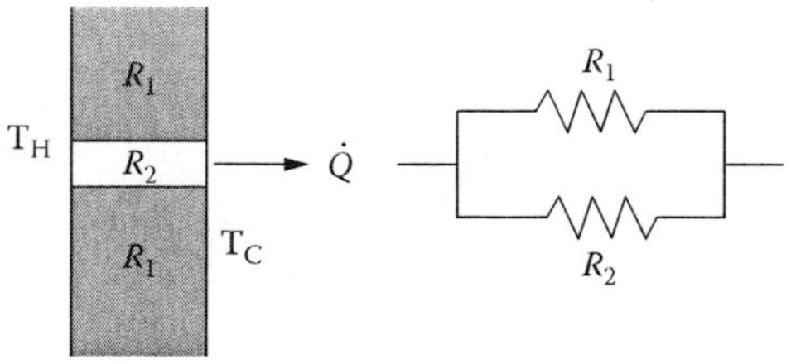

FIGURE 3.6 Heat transfer across a dissimilar material.

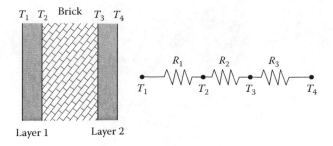

FIGURE 3.7 Heat transfer across a brick wall.

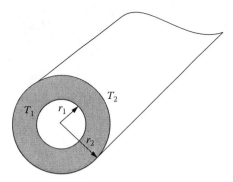

FIGURE 3.8   Conduction heat transfer through a cylinder.

where conduction heat transfer per unit length is

$$\dot{Q} = -k(2\pi r)\frac{dT}{dr}$$

(3.22)

Combining Equations 3.21 and 3.22 gives

$$\frac{1}{r}\frac{d}{dr}(kr\frac{dT}{dr}) = 0$$

(3.23)

For constant properties, Equation 3.23 becomes

$$\frac{1}{r}\frac{d}{dr}(r\frac{dT}{dr}) = 0$$

(3.24)

It can easily be shown that

$$\frac{dT}{dr} = \frac{T_2 - T_1}{r\ln(r_2 / r_1)}$$

(3.25)

Using Equation 3.25, Equation 3.22 becomes

$$\dot{Q} = -k(2\pi)\frac{T_2 - T_1}{\ln(r_2 / r_1)}$$

(3.26)

where conduction thermal resistance in the cylindrical coordinate is (Figure 3.8)

$$R = \frac{\ln(r_2 / r_1)}{2\pi k}$$

(3.27)

We can use the same steps to come up with the steady constant property and one-dimensional spherical heat conduction equation with no generation:

$$\dot{Q} = k(4\pi r_1 r_2)\frac{T_1 - T_2}{r_2 - r_1}$$

(3.28)

Figure 3.9 shows the spherical configuration.

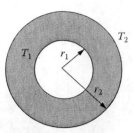

FIGURE 3.9   Conduction heat transfer through a sphere.

Again, conduction thermal resistance in spherical coordinate is (Figure 3.9)

$$R = \frac{\frac{1}{r_1} - \frac{1}{r_2}}{4\pi k}$$

(3.29)

The general steady one-dimensional conduction heat transfer equation with no generation for slab ($m = 0$), cylinder ($m = 1$), and sphere ($m = 2$) can be written as (Cengel):

$$\frac{1}{r^m}\frac{d}{dr}(r^m k \frac{dT}{dr}) = 0$$

(3.30)

The general one-dimensional conduction heat transfer equation for slab ($m = 0$), cylinder ($m = 1$), and sphere ($m = 2$) can be written as (Cengel):

$$\frac{1}{r^m}\frac{\partial}{\partial r}(r^m k \frac{\partial T}{\partial r}) + q''' = \rho C_p \frac{\partial T}{\partial t}$$

(3.31)

where

q''' (W/m³) is the heat generation
ρ (kg/m³) is density
C_p (kJ/kg.K) is heat capacity
t (s) is time

If properties are constant, then Equation 3.31 becomes

$$\frac{1}{r^m}\frac{\partial}{\partial r}(r^m \frac{\partial T}{\partial r}) + \frac{q'''}{k} = \frac{1}{\alpha}\frac{\partial T}{\partial t}$$

(3.32)

where $\alpha = k/\rho C_p$ (m²/s) is thermal diffusivity.

By doing an energy balance on a control volume and using the appropriate heat transfer rate equation, the three-dimensional constant properties heat conduction equations for rectangular (x, y, z), cylindrical (r, ϕ, z), and spherical (r, ϕ, θ) coordinating are, respectively,

$$\frac{\partial^2 T}{\partial x^2} + \frac{\partial^2 T}{\partial y^2} + \frac{\partial^2 T}{\partial z^2} + \frac{q'''}{k} = \frac{1}{\alpha}\frac{\partial T}{\partial t}$$

(3.33)

$$\frac{1}{r}\frac{\partial}{\partial r}\left(r\frac{\partial T}{\partial r}\right) + \frac{1}{r^2}\frac{\partial}{\partial \phi}\left(r\frac{\partial T}{\partial \phi}\right) + \frac{\partial^2 T}{\partial z^2} + \frac{q'''}{k} = \frac{1}{\alpha}\frac{\partial T}{\partial t}$$

(3.34)

$$\frac{1}{r^2}\frac{\partial}{\partial r}\left(r^2\frac{\partial T}{\partial r}\right)+\frac{1}{r^2\sin^2\theta}\frac{\partial^2 T}{\partial\phi^2}+\frac{1}{r^2\sin^2\theta}\frac{\partial}{\partial\theta}\left(\sin\theta\frac{\partial T}{\partial\theta}\right)+\frac{q'''}{k}=\frac{1}{\alpha}\frac{\partial T}{\partial t}$$

(3.35)

3.6 CONVECTION HEAT TRANSFER

The second type of heat transfer to be examined is convection, also used in many solar thermal processes. This describes energy transfer between a surface and fluid moving over the surface, as shown by Figure 3.10. The goal is to determine the flow and temperature behavior in convection heat transfer.

In convective heat transfer, it is necessary to examine some features of the fluid motion near a surface. When flow passes over a surface, a thin layer of slowly moving fluid, called the "boundary layer," exists close to the wall. In this region, fluid experiences velocity and temperature differences. The boundary layer thickness (δ) is not known. It is not a property and depends on flow velocity (Reynolds number), structure of the wall surface, pressure gradient, and Mach number. Outside this layer, temperature and velocity are roughly uniform. Diffusion contributes to convection heat transfer. However, the dominant contribution comes from the advection effect, which is the bulk motion of fluid properties.

It is customary to calculate the rate of heat transfer ($\dot{Q}$) from the surface to the fluid by Newton's law of cooling, which is

$$\dot{Q}=hA(T_s-T_\infty)$$

(3.36)

The quantity h (W/m²k) is known as the convective heat transfer coefficient, T_s is surface temperature, and T_∞ is the fluid temperature. For many situations of practical interest, the quantity h is known mainly through experiments. The average heat transfer rate ($\dot{Q}$) is obtained by integrating Equation 3.36 over the entire surface. This leads to an average convection coefficient ($\bar{h}$) for the entire surface as follows:

$$\bar{h}=\frac{1}{L}\int_{A_s}h\,dA_s$$

(3.37)

A thermal resistance may also be associated with heat transfer by convection at a surface using Equation 3.36:

$$\dot{Q}=\frac{\Delta T}{1/hA}=\frac{\Delta T}{R_{conv}}$$

(3.38)

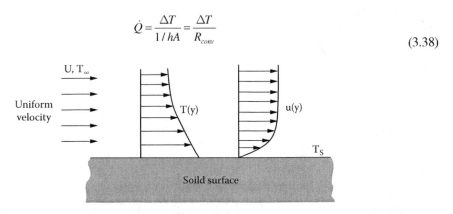

FIGURE 3.10   Convection heat transfer from a hot surface.

The convective thermal resistance (R_{conv}) is

$$R_{conv} = \frac{1}{hA}$$

(3.39)

The following sections discuss the methods of finding convective heat transfer for external and internal flow.

Another important factor in convective heat transfer is the friction coefficient (C_f). It is the characteristic of the fluid flow and is as follows (Incropera and DeWitt):

$$C_f = \frac{\tau_s}{\rho_\infty u_\infty^2 / 2}$$

(3.40)

where τ (N) is the shear stress.

Reynolds analogy. There is an approximate relation between flow and temperature field. It is called Reynolds analogy and is a relation between skin friction (momentum flux to the wall) and convective heat transfer coefficient. It provides a useful way to estimate heat transfer rates in situations in which the skin friction is known. The relation is expressed by

$$St = \frac{C_f}{2}$$

(3.41)

where

$$St = \frac{h}{\rho_\infty C_p u_\infty}$$

From dimensionless boundary layer conservation equations, the local and average convection coefficients for a surface in low-speed, forced convection with no phase change are correlated by equation of the form:

$$Nu_x = f(x^*, Re_x, Pr)$$

(3.42)

$$\overline{Nu} = f(Re_x, Pr)$$

(3.43)

For a prescribed geometry, such as a flat plate in a parallel flow, under a variety of test conditions (i.e., varying velocity, u_∞, plate length, L, and fluid nature such as air, water, or oil), there will be many different values of the Nusselt ($Nu_x = hx/k$) number corresponding to a wide range of Reynolds ($Re = u_\infty x/v$) and Prandtl numbers. Here, k is thermal conductivity of the fluid. The results on a log–log scale are presented by

$$\overline{Nu} = C\,Re^m\,Pr^n$$

(3.44)

C, m, and n are constants and vary with the nature of the surface geometry and type of flow. For instance, for fully developed laminar flow in a circular tube with constant surface heat flux, the value of $C = 4.36$, $m = 0$, and $n = 0$. For constant surface temperature, these values become $C = 3.66$, $m = 0$, and $n = 0$. Therefore, the Nusselt number for laminar flow in a circular tube with constant surface heat flux and constant surface heat transfer is calculated by, respectively,

$$Nu_D = \frac{hD}{k} = 4.36$$

(3.45)

$$Nu_D = \frac{hD}{k} = 3.66$$

(3.46)

For internal flow, Newton's law of cooling is expressed as

$$Q_s^{''} = h(T_s - T_m)$$

(3.47)

Where T_s is the pipe surface temperature, note that, as opposed to T_∞ that was constant, the mean temperature (T_m) varies with flow direction ($dT_m/dx \neq 0$). From the energy equation, the mean temperature at any location, $T_m(x)$, for *constant surface heat flux* is calculated by

$$T_m(x) = T_{m,i} + \frac{Q_s^{''} P}{\dot{m} C_p} x$$

(3.48)

P is the perimeter ($P = \pi D$ for a circular tube). The mean free temperature at any location, $T_m(x)$, for *constant surface temperature* is calculated by

$$\frac{T_s - T_m(x)}{T_s - T_{m,i}} = \exp\left(-\frac{Px}{\dot{m} C_p} \overline{h}\right)$$

(3.49)

The total heat transfer rate is expressed by

$$Q = \overline{U} A \Delta T_{lmtd}$$

(3.50)

The log mean temperature difference, ΔT_{lmtd}, is given by

$$\Delta T_{lmtd} = \frac{\Delta T_o - \Delta T_i}{\ln(\Delta T_o / \Delta T_i)}$$

(3.51)

ΔT_o and ΔT_i are the differences between the hot and cold fluid at the inlet and at the exit.

3.7 RADIATION HEAT TRANSFER

A solid body in a vacuum with a temperature (T_s) greater than the surrounding temperature (T_{sur}) becomes cool until it finally reaches thermal equilibrium with its surrounding. The cooling is due to thermal radiation exchange between the hot and cold surfaces. Figure 3.11 shows the actual process.

The origin of radiation is emission by matter and its transfer does not require the presence of any matter. Therefore, it is maximized in a vacuum. The nature of radiation heat transfer is by photons, according to some theories, or by electromagnetic emissions according to others. In both cases, wave standard properties like frequency (f) and wavelength (λ) are attributed to radiation. Thus, solar energy (light) has both wave and particle (photon) components.

Electromagnetic waves appear in nature for wavelengths over an unlimited range and radiation takes on different names (optics, thermal, radio, x- and γ-rays, etc.) depending on the wavelength. The type of radiation pertinent to heat transfer is thermal radiation—the portion that extends from 0.1 to 100 μm. All bodies at a temperature above absolute zero emit radiation in all directions over a wide range of wavelengths. The amount of emitted energy from a surface at a given wavelength depends

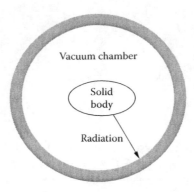

FIGURE 3.11   Cooling a body by radiation.

on the material, condition, and temperature of the body. A surface is said to be diffuse if its surface properties are independent of direction and gray if its properties are independent of wavelength.

3.7.1 Surface Property

When radiated photons or electromagnetic waves reach another surface, they may be absorbed, reflected, or transmitted, as shown by Figure 3.12. From an energy stand point, the sum of the absorbed, reflected, and transmitted fraction of radiation energy must be equal to unity:

$$\alpha + \rho + \tau = 1 \qquad (3.52)$$

where

α is absorptivity (fraction of incident radiation that is absorbed)
ρ is reflectivity (fraction of incident radiation that is reflected)
τ is transmissivity (fraction of incident radiation that is transmitted)

Reflective energy may be either diffuse or specular (mirror-like). Diffuse reflections are independent of the incident radiation angle. For specular reflections, the reflection angle equals the angle of incidence.

3.7.2 Blackbody Radiation

Another key concept for solar energy is blackbody radiation. A blackbody is an ideal thermal radiator. It absorbs all incident radiation, regardless of wavelength and direction (absorptivity, $\alpha = 1$). It also emits maximum radiation energy in all directions (diffuse emitter). The emitted energy by

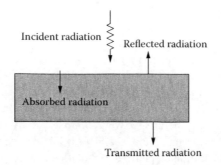

FIGURE 3.12   Radiation surface properties.

a blackbody (blackbody emissive power, W/m^2) is given by the Stefan–Boltzmann law (Incropera and DeWitt):

$$E_b = \sigma T^4 \tag{3.53}$$

where σ is the Stefan–Boltzmann constant ($\sigma = 5.67 \times 10^{-8}$ W/m^{2}K^4) and T is the absolute temperature.

3.7.3 REAL BODY RADIATION

The emitted energy of a real surface is given by

$$E_b = \varepsilon \sigma T^4 \tag{3.54}$$

where ε is the emissivity and is given by

$$\varepsilon = \frac{E}{E_b} \tag{3.55}$$

E is radiation from a real body and E_b is radiation from a blackbody at absolute temperature T. Based on Kirchhoff's law, for a small non-blackbody in the cavity, $\varepsilon = \alpha$.

Radiative exchange between two or more surfaces depends on the surface geometry, orientation, radiative properties, and temperatures. In general, energy interchange between any two surfaces in space at different temperatures is given by the radiation shape factor or view factor, F_{ij}. F_{ij} is the fraction of energy that leaves surface i and reaches surface j and it is given by

$$F_{ij} = \frac{q_{i \to j}}{A_i J_i} \tag{3.56}$$

J is radiosity and is the total energy that leaves a surface. $q_{i \to j}$ is the rate at which radiation leaves surface i and is intercepted by surface j. Net radiation exchange between two black surfaces is given by

$$A_i F_{ij} E_{bi} - A_j F_{ji} E_{bj} = \dot{Q}_{ij} \tag{3.57}$$

If both surfaces are at the same temperature, Equation 3.57 simplifies to the shape factor reciprocity relation:

$$A_i F_{ij} = A_j F_{ji} \tag{3.58}$$

Using Equation 3.58, the net heat exchange between the two black surfaces is then

$$\dot{Q}_{ij} = A_i F_{ij} \left(E_{bi} - E_{bj} \right) = A_j F_{ji} (E_{bi} - E_{bj}) = A_i F_{ij} \sigma (T_i^4 - T_j^4) \tag{3.59}$$

Radiation exchange between diffuse and gray surfaces is given by

$$\dot{Q}_{ij} = \frac{\sigma (T_i^4 - T_j^4)}{\dfrac{1 - \varepsilon_i}{\varepsilon_i A_i} + \dfrac{1}{A_i F_{ij}} + \dfrac{1 - \varepsilon_j}{\varepsilon_j A_j}} \tag{3.60}$$

For two infinite large parallel plates with thick surfaces ($\tau = 0$), Equation 3.60 reduces to

$$\dot{Q}_{ij} = \frac{\sigma(T_i^4 - T_j^4)}{\dfrac{1}{\varepsilon_i} + \dfrac{1}{\varepsilon_j} - 1}$$

(3.61)

where $\alpha = \varepsilon$.

3.8 INTRODUCTION TO THERMODYNAMICS

The word *thermodynamics* consists of two words: thermo (heat) and dynamics (power). Thermodynamics is a branch of science concerned with the nature of heat and its conversion to work. Historically, it grew out of the fact that a hot body can produce work and the efforts to construct more efficient heat engines—devices for extracting useful work from expanding hot gases. Today, thermodynamics deals with energy and the relationship between properties of substances.

Thermodynamics starts with the definition of several basic concepts that leads to fundamental laws. These laws govern the conversion of energy from one form to another, the direction in which heat will flow, and the availability of energy to do work. Therefore, for the laws of thermodynamics to be expressed, certain properties and concepts must be defined.

A thermodynamic system is defined by its boundary. Everything that is not within the boundary is part of the surroundings or environment. The environment often contains one or more idealized heat reservoirs—heat sources with infinite heat capacity enabling them to give up or absorb heat without changing their temperature. There are basically two types of systems: closed and open. A closed system (control mass) has a fixed quantity of matter. Thus, no mass crosses the boundary of the system. In an open system (control volume), the quantity of mass is not constant and mass can cross the boundary.

Each system is characterized by its properties. Three important independent properties that usually are used to describe a system are temperature, pressure, and specific volume. It is not easy to give an exact definition for temperature, which is the measure of the relative warmth or coolness of an object. Based on our physiological sensations, we know the level of temperature qualitatively with words like cold, warm, hot, etc. Because our senses may be misleading, we cannot assign numerical values to temperatures based on our sensations alone.

The zero*th* law of thermodynamics serves as a basis for the validity of temperature measurement. This law indicates that if two bodies are in thermal equilibrium with a third body, they are also in thermal equilibrium with each other. Replacing the third body with a thermometer or other instrument having a scale calibrated in units called degrees helps in measuring the temperature of a system. The size of a degree depends on the particular temperature scale used.

3.8.1 THE FIRST LAW OF THERMODYNAMICS

Based on experimental observation, energy can be neither created nor destroyed; it can only change form. The first law of thermodynamics (or the conservation of energy principle) states that during an interaction between a system and its surroundings, the amount of energy gained by the system must be exactly equal to the amount of energy lost by the surroundings.

For a closed system (control mass), the first law of thermodynamics may be expressed as (Moran and Shapiro):

$$\begin{pmatrix} \text{Net amount of energy transfer as heat} \\ \text{and work to/or from the system} \end{pmatrix} = \begin{pmatrix} \text{Net change in amount of energy} \\ \text{(increase or decrease) within the system} \end{pmatrix}$$

or

$$Q - W = \Delta E \tag{3.62}$$

where

Q is the net energy transfer by heat across the system boundary (J)
W is the net energy transfer by work across the boundary (J)
ΔE is the net change of total energy within the system (J)

Total energy, E, of a system consists of internal energy (U), kinetic energy (KE), and potential energy (PE). Therefore, from Equation 3.62, the change in total energy of the system may be expressed as

$$\Delta E = \Delta U + \Delta KE + \Delta PE \tag{3.63}$$

where each term is

$$\Delta U = (U_2 - U_1), \Delta KE = \frac{1}{2}m(V_2^2 - V_1^2), \Delta PE = mg(z_2 - z_1)$$

Using Equation 3.63, the first law of thermodynamics becomes

$$Q - W = \Delta U + \frac{1}{2}m(V_2^2 - V_1^2) + mg(z_2 - z_1) \tag{3.64}$$

or the instantaneous time rate form of the energy balance is as follows:

$$\frac{dE}{dt} = \dot{Q} - \dot{W} \tag{3.65}$$

where $\dot{Q}$ is the rate of heat transfer across the boundary and $\dot{W}$ is the rate of work across the boundary.

The first law of thermodynamics for an open system (control volume) may be expressed as

$$\left(\begin{array}{c} \text{Net rate of energy transfer} \\ \text{as heat and work to/from} \\ \text{control volume at time } t \end{array} \right) + \left(\begin{array}{c} \text{Net rate of energy transfer} \\ \text{by mass entering the} \\ \text{control volume} \end{array} \right) = \left(\begin{array}{c} \text{Time rate of change of the energy} \\ \text{within the control volume} \end{array} \right)$$

or

$$\frac{dE_{cv}}{dt} = \dot{Q}_{cv} - \dot{W}_{cv} + \sum_i \dot{m}_i \left(h_i + \frac{V_i^2}{2} + gz_i \right) - \sum_e \dot{m}_e \left(h_e + \frac{V_e^2}{2} + gz_e \right) \tag{3.66}$$

where h is expressed as an intensive property of the fluid that is called enthalpy and $\dot{m}$ is mass flow rate of the fluid.

3.8.2 The Second Law of Thermodynamics

The first law of thermodynamics states the fact that energy (a useful concept) is conserved. It says nothing about the way this occurs or even whether one form of energy can be converted to another. However, the second law of thermodynamics is concerned with the usefulness of energy or, more specifically, with the direction in which energy transfers may occur.

There are many statements of the second law. The second law states that it is impossible to construct a device operating in a cycle and producing as its sole effect net positive work, while exchanging heat with only one reservoir (Kelvin–Planck statement of the second law). It also states that a transformation whose final result is to transfer heat from a body at a given temperature to a body at a higher temperature is impossible. In general, the second law states that the net work is always less than the heat supplied.

For a closed system, the second law of thermodynamics may be expressed as

$$\dot{S}_2 - \dot{S}_1 = \int_1^2 \left(\frac{\delta\dot{Q}}{T}\right)_b + \dot{S}_{gen} \tag{3.67}$$

where

b is boundary
T is absolute temperature
$\dot{Q}$ is rate of heat transfer
$\dot{S}_{gen}$ is the amount of entropy generated by system irreversibility

By combining Equations 3.67 and 3.64, the irreversibility associated with a process (I) may be expressed by

$$I = W_{rev} - W_u = T_0 S_{gen} \tag{3.68}$$

where

W_{rev} is reversible work
W_u is the useful work and is expressed as the difference between the actual work and work of surroundings
T_0 is temperature of the surroundings

The second law of thermodynamics for an open system may be expressed as

$$\frac{dS_{cv}}{dt} = \sum_j \frac{\dot{Q}_j}{T_j} + \sum_i \dot{m}_i S_i - \sum_e \dot{m}_e S_e + \dot{S}_{gen} \tag{3.69}$$

3.8.3 THE THIRD LAW OF THERMODYNAMICS

A postulate related to but independent of the second law is that it is impossible to cool a body to absolute zero by any finite process. Although one can approach absolute zero as closely as one desires, one cannot actually reach this limit. The third law of thermodynamics, formulated by Walter Nernst and also known as the Nernst heat theorem, states that if one could reach absolute zero, all bodies would have the same entropy. In other words, a body at absolute zero could exist in only one possible state, which would possess a definite energy, called the zero-point energy. This state is defined as having zero entropy.

REFERENCES

Bejan, A. and A. D. Kraus. 2003. *Heat transfer handbook*, 1st ed. New York: John Wiley & Sons.
Cengel, Y. S. 2007. *Heat transfer a practical approach*, 3rd ed. New York: McGraw-Hill.
Ghassemi, M. and Y. Mollayi Barsi. 2005. *Effect of liquid film (indium) on thermal and electromagnetic distribution of an electromagnetic launcher with new armature*, IEEE Transactions on Magnetics, 41(1) 1 – 6.

Incropera, F. P. and D. P. DeWitt. 2008. *Fundamentals of heat and mass transfer*, 7th ed. New York: John Wiley & Sons.

Kaviany, M. 2001. *Principles of heat transfer,* 1st ed. New York: John Wiley & Sons.

Moran, M. J. and H. N. Shapiro. 2007. *Fundamentals of engineering thermodynamics*, 6th ed. New York: John Wiley & Sons.

Necati- Osicik, M. 1993. *Heat conduction*, 2nd ed. New York: John Wiley & Sons.

PROBLEMS

3.1   Define and explain different modes of heat transfer.

3.2   Drive the energy equation for a two dimensional plane wall with a uniform energy source.

3.3   Drive the unsteady energy equation for a circular pipe with no heat generation.

3.4   Draw the thermal resistance network and the electrical analogy for heat transfer through a circular pipe subjected to convection heat transfer on both sides, see Figure 3.8.

3.5   Draw the thermal resistance network and the electrical analogy for heat transfer through a spherical pipe subjected to convection heat transfer from interior and convection and radiation effect from outside, see Figure 3.9.

3.6   Find the Nusselt number for a 1 inch diameter circular tube made from stainless steel with constant surface heat flux.

3.7   Draw the radiation network and the net rate of radiation heat transfer for a two surface enclosure.

3.8   Define and explain 1st and 2nd laws of thermodynamics.

3.9   Write the 1st and 2nd law of thermodynamics for a piston-cylinder when the system is in thermal communication with the surrounding.

3.10   Write the 1st and 2nd law of thermodynamics for an adiabatic turbine working at steady state.

3.11   Write the 1st law of thermodynamics for a tank that is being filled by water.

4 Solar Thermal Systems and Applications

4.1 INTRODUCTION

Solar thermal energy has been used for centuries by ancient people's harnessing solar energy for heating and drying. More recently, in a wide variety of thermal processes solar energy has been developed for power generation, water heating, mechanical crop drying, and water purification, among others. Given the range of working temperatures of solar thermal processes, the most important applications are

- for less than 100°C: water heating for domestic use and swimming pools, heating of buildings, and evaporative systems such as distillation and dryers;
- for less than 150°C: air conditioning, cooling, and heating of water, oil, or air for industrial use;
- for temperatures between 200 and 2000°C: generation of electrical and mechanical power; and
- for less than 5000°C: solar furnaces for the treatment of materials.

For processes where more than 100°C are required, the solar energy flux is not enough to elevate the working fluid temperature to such a high level; instead, some type of concentration of the energy flux using mirrors or lenses must be used. Then the ratio of the energy flux received for the energy absorber to that captured by the collector must be greater than one, and designs often easily achieve a concentration of hundreds of suns.

4.2 SOLAR COLLECTORS

Solar collectors are distinguished as low-, medium-, or high-temperature heat exchangers. There are basically three types of thermal solar collectors: flat plate, evacuated tube, and concentrating. Although there are great geometric differences, their purpose remains the same: to convert the solar radiation into heat to satisfy some energy needs. The heat produced by solar collectors can supply energy demand directly or be stored. To match demand and production of energy, the thermal performance of the collector must be evaluated. The instantaneous useful energy collected ($\dot{Q}_u$) is the result of an energy balance on the solar collector.

To evaluate the amount of energy produced in a solar collector properly, it is necessary to consider the physical properties of the materials. Solar radiation, mostly short wavelength, passes through a translucent cover and strikes the energy receiver. Low-iron glass is commonly used as a glazing cover due to its high transmissivity; the cover also greatly reduces heat losses. The optical characteristics of the energy receiver must be as similar as possible to those of a blackbody, especially high absorbtivity.

Properties of high thermal conductivity can be improved by adding selective coatings. Together with the radiation absorption, an increase of the receiver's temperature is experienced; the short-wave radiation is transformed then into long-wave radiation. The glazing material essentially becomes opaque at the new wavelength condition favoring the greenhouse effect. A combination of high transmissivity toward the solar radiation of the cover and high absorbtivity of the receiver brings great performance for a well-designed solar collector.

4.2.1 FLAT-PLATE COLLECTORS

A flat-plate solar collector consists of a waterproof, metal or fiberglass insulated box containing a dark-colored absorber plate, the energy receiver, with one or more translucent glazings. Absorber plates are typically made out of metal due to its high thermal conductivity and painted with special selective surface coatings in order to absorb and transfer heat better than regular black paint can. The glazing covers reduce the convection and radiation heat losses to the environment. Figure 4.1 shows the typical components of a classic flat-plate collector. These systems are always mounted in a fixed position optimizing the energy gain for the specific application and particular location. Flat collectors can be mounted on a roof, in the roof itself, or be freestanding.

The collector gains energy when the solar radiation travels through the cover; both beam and diffuse solar radiation are used during the production of heat. The greater the transmittance (τ) of the glazing is, the more radiation reaches the absorber plate. Such energy is absorbed in a fraction equal to the absorbtivity (α) of the blackened-metal receiver. If this were a perfect absorber, such as a blackbody discussed in Chapter 3, absorbtivity would be one. The instantaneous energy gained by the receiver ($\dot{Q}_r$ or $\dot{q}_r$) is given by

$$\dot{Q}_r = \dot{q}_r A_c = (\tau\alpha)_{\text{eff}} I_T A_c \qquad (4.1)$$

where $(\tau\alpha)_{\text{eff}}$ is the effective optical fraction of the energy absorbed, I_T is the solar radiation incident on the tilted collector, and A_c is the collector aperture area. The aperture is the frontal opening of the collector that captures the Sun's rays. Once such radiation is absorbed, it is converted into thermal energy heating up the absorber plate. In general terms, solar collectors present great heat losses. Although the glazing does not allow infrared-thermal energy (long wavelength) to escape, the temperature difference between the absorber plate and the ambient causes heat losses by convection ($\dot{Q}_{\text{conv}}$ or $\dot{q}_{\text{conv}}$) to the surroundings according to the following equation:

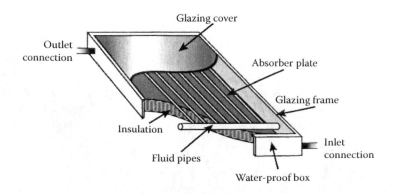

FIGURE 4.1 Main components of a flat-plate collector.

$$\dot{Q}_{\mathrm{conv}} = \dot{q}_{\mathrm{conv}} A_{\mathrm{r}} = UA_{\mathrm{r}}(T_{\mathrm{r}} - T_{\mathrm{a}}) \tag{4.2}$$

where A_{r} is the area of the receiver, U is an overall heat loss coefficient, T_{r} is the receiver's temperature and T_{a} is the ambient temperature. Also, some heat is lost by radiation ($\dot{Q}_{\mathrm{rad}}$ or $\dot{q}_{\mathrm{rad}}$) due to the difference of temperature between the collector and the sky dome. For simplicity, the last is assumed to be the same as the ambient temperature:

$$\dot{Q}_{\mathrm{rad}} = \dot{q}_{\mathrm{rad}} A_{\mathrm{r}} = \varepsilon_{\mathrm{eff}} \sigma A_{\mathrm{r}}(T_{\mathrm{r}}^4 - T_{\mathrm{a}}^4) \tag{4.3}$$

where $\varepsilon_{\mathrm{eff}}$ is the effective emissivity of the collector and σ is the Stefan–Boltzmann constant. The heat losses from the bottom and from the edges of the collector always exist. Their contribution, however, is not as significant as the convective and radiative losses from the top. The energy balance in the collector results from combining the energy gain stated by Equation 4.1 and the heat losses represented in Equations 4.2 and 4.3 as

$$\dot{Q}_{\mathrm{u}}(t) = \dot{q}_{\mathrm{u}} A_{\mathrm{c}} = (\tau\alpha)_{\mathrm{eff}} I_T A_{\mathrm{c}} - UA_{\mathrm{r}}(T_{\mathrm{r}} - T_{\mathrm{a}}) - \varepsilon_{\mathrm{eff}} \sigma A_{\mathrm{r}}(T_{\mathrm{r}}^4 - T_{\mathrm{a}}^4) \tag{4.4}$$

where $\dot{Q}_{\mathrm{u}}$ is the usable energy collected.

A heat-conducting fluid, usually water, glycol, or air, passes through pipes attached to the absorber plate. As the fluid flows through the pipes, its temperature increases. This is the energy to be utilized for productive activities (e.g., power generation). The amount of the energy taken by the working fluid corresponds to a fraction of the useful energy collected after the heat losses.

The instantaneous thermal efficiency corresponds to the fraction from the incoming solar radiation that is actually recovered to be used:

$$\eta = \frac{\dot{Q}_{\mathrm{u}}}{I_T A_{\mathrm{c}}} = (\tau\alpha)_{\mathrm{eff}} - \frac{UA_{\mathrm{r}}}{I_T A_{\mathrm{c}}}(T_{\mathrm{r}} - T_{\mathrm{a}}) - \frac{\varepsilon_{\mathrm{eff}} \sigma A_{\mathrm{r}}}{I_T A_{\mathrm{c}}}(T_{\mathrm{r}}^4 - T_{\mathrm{a}}^4) \tag{4.5}$$

The overall efficiency in a specific period of time is

$$\eta = \frac{\displaystyle\int_{t_1}^{t_2} \dot{Q}_{\mathrm{u}} dt}{A_{\mathrm{c}} \displaystyle\int_{t_1}^{t_2} I_T dt} \tag{4.6}$$

For low-temperature collectors such as flat plate, heat losses by radiation are very small compared to convection losses, for which the efficiency equation is reduced to

$$\eta = \frac{\dot{Q}_{\mathrm{u}}}{I_T A_{\mathrm{c}}} = (\tau\alpha)_{\mathrm{eff}} - \frac{UA_{\mathrm{r}}}{I_T A_{\mathrm{c}}}(T_{\mathrm{r}} - T_{\mathrm{a}}) \tag{4.7}$$

The device is a hermetic and isolated box, so it is difficult to measure the temperature of the receiver; thus, Equation 4.7 must be in terms of the inlet and outlet temperatures of the circulating fluid. Being congruent with techniques used in heat exchanger's design, the effectiveness removal factor F_R is introduced. This relates the collector's actual performance directly to a reference performance. Then, the efficiency and the useful heat gain ($\dot{Q}_u$) equations become

$$\eta = F_{R}\left[(\tau\alpha)_{\text{eff}} - \frac{UA_{r}}{I_{T}A_{c}}(T_{\text{in}} - T_{a})\right]$$ (4.8)

$$\dot{Q}_{u} = \eta I_{T}A_{c} = I_{T}A_{c}F_{R}\left[(\tau\alpha)_{\text{eff}} - \frac{UA_{r}(T_{\text{in}} - T_{a})}{I_{T}A_{c}}\right]$$ (4.9)

as a function of the fluid inlet temperature T_{in}. On the other side, the usable heat gained by the working fluid is given by

$$\dot{Q}_{u} = \dot{m}C_{P}\left(T_{\text{out}} - T_{\text{in}}\right)$$ (4.10)

where T_{out}, C_{P}, and $\dot{m}$ are the fluid outlet temperature, heat capacity at constant pressure, and mass flow rate of the working fluid, respectively.

4.2.1.1   Flat-Plate Collector Thermal Testing

To determine the performance of flat-plate solar collectors, the thermal F_{R}, U, and $(\tau\alpha)_{\text{eff}}$ parameters for each collector must be calculated by applying a standard testing method. The most widely used methods are those documented in ASHRAE 93 (2003), ISO 9806-1 (1994), and EN12975-2 (2001). In the United States and the European Union, only certified collectors following exclusive standards are required for solar installations. The Solar Ratings Certification Corporation (SRCC) of Florida is the key certifying body in the United States for solar thermal collectors following the ASHRAE 93 standard.

In the three methods, the parameters are obtained from a collector time constant (τ) test, an instantaneous thermal efficiency (η) test, and an incident angle modifier ($K\theta_{b}(\theta)$) test. Rojas et al. (2008) compared the ASHRAE 93 and EN12975-2 standard testing methodologies and the thermal results obtained for a single-glazed flat-plate collector and found good agreement, although the methodologies are very different. The ASHRAE 93 standard test is a steady-state thermal method that must be conducted outdoors under suitable weather conditions. This is quite complicated because a combination of the values of irradiance, temperature, and wind speed conditions must fall into rather narrow ranges. The test requires a minimum total solar irradiance of 790 W/m², maximum diffuse fraction of 20%, wind speed between 2.2 and 4.5 m/s, and an incidence angle modifier between 98 and 102% (normal incidence value). However, such environmentally prescribed conditions do not often occur in some locations. In contrast, the EN12975-2 test provides an alternative transient test method that can be conducted over a larger range of environmental conditions.

For the τ-test, first, the environmental conditions must comply with standard requirements, and also variation in irradiance and ambient temperature must fall within ± 32 W/m² and ± 1.5 K, respectively. The collector is exposed to the Sun while water is circulating. The inlet water temperature must be controlled to be the same as the outdoor-air ambient dry bulb temperature with an allowed variation of the greater of 1 K or $\pm 2\%$. For the volumetric flow rate, the variation could be the greater of ± 0.0005 gal/min or $\pm 2\%$. Under such steady-state conditions, the collector is abruptly covered with an opaque surface for restricting any irradiance absorption. Immediately, the inlet-controlled and outlet-uncontrolled temperatures are continuously recorded. Because there is no energy gain, the uncontrolled-outlet temperature starts decreasing. The time in which the temperature difference between outlet and inlet decreases up to 0.368 ($1/e$) of its initial value is the so-called time constant.

The instantaneous thermal efficiency of a collector η is estimated according to Equation 4.7 as the ratio between the useful energy gain (Equation 4.10) and the actual solar irradiance, I, captured by the collector area A_{c}. The thermal efficiency test is performed at near normal incidence condition (i.e., almost null variation of the incidence angle), for which $(\tau\alpha)_{\text{eff}}$ remains constant throughout the test. Also, both F_R and U are constants for the tested temperature because all other variables—radiation, flow rate, ambient temperature, and inlet temperature—are restricted to little variation.

By plotting η against $(T_{\text{in}} - T_a)/I_T$ for a given collector, the efficiency is a linear function as stated in Equation 4.8 and represented in Figure 4.2. ASHRAE 93 efficiency tests are conducted for four different collector inlet temperatures. In addition, steady-state test standards require a minimum of 16 data points for the four different inlet temperatures to obtain the efficiency curve. The lowest inlet temperature corresponds to the ambient temperature and the highest is established upon the maximum operating temperature recommended by the manufacturer. Only data taken during steady state are used to calculate the efficiency.

The slope of the efficiency plot $(F_R U)$ represents the rate of heat loss from the collector; collectors with glazing covers present less of a slope than those without them. When the inlet temperature is the same as the ambient temperature, the maximum collection efficiency, known as the optical efficiency $(\tau\alpha)_{\text{eff}}$, is found. For this condition, the $(T_{\text{in}} - T_a)/I_T$ value is zero and the intercept corresponds to $F_R(\tau\alpha)_{\text{eff}}$. Another point of interest is the intercept of the curve with the $(T_{\text{in}} - T_a)_T/I_T$ axis. This point of operation is reached when useful energy is no longer removed from the collector due to stagnation of the working fluid. In this case, the incoming optical energy equals the heat losses, requiring that the temperature of the absorber increase until this balance occurs. This temperature is called the stagnation temperature. For well-insulated collectors, the stagnation temperature can reach very high levels and cause fluid to boil.

The optical efficiency of a parabolic trough collector decreases with incidence angle for several reasons: the decreased transmission of the glazing and the absorption of the absorber, the increased width of the solar image on the receiver, and the spillover of the radiation from troughs of finite length.

Instantaneous thermal efficiency of a solar collector decreases with the incidence angle of the irradiance. At low incidence angles, light transmission decreases through the glazing, and the width of the solar image on the receiver increases. For the incidence angle modifier $(K\theta_b(\theta))$ test, one inlet temperature at steady-state conditions is fixed throughout the whole test to determine the collector efficiency at the incidence angles of 0, 30, 45, and 60°. Changing the azimuth angle of the collector

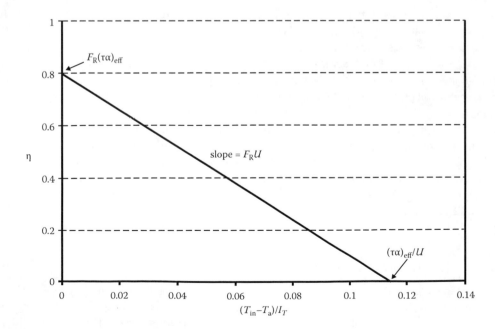

FIGURE 4.2 Instantaneous efficiency of a flat-plate collector.

modifies the incidence angles. The angular dependence of the incidence angle modifier upon the incidence angle θ is approximately given by

$$K\theta_b(\theta) = 1 - b_o\left(\frac{1}{\cos\theta} - 1\right)$$

(4.11)

The parameter b_o is called the incidence angle modifier coefficient. The effect of the modifier angle on the efficiency is then given by

$$\eta = \eta_n\left[1 - b_o\left(\frac{1}{\cos\theta} - 1\right)\right]$$

(4.12)

where η_n is the efficiency value for normal incidence when there are no optical losses through the gap between the receiver and the reflector.

4.2.1.2 Collector Efficiency Curve

For other than flat-plate geometries, curves for collector efficiency and incidence modifier angle could be generalized by means of the second-order Equations 4.8 and 4.11 as follows:

$$\eta = a_o K\theta_b(\theta) - a_1\frac{T_{in} - T_a}{I_T} - a_2\frac{\left(T_{in} - T_a\right)^2}{I_T}$$

(4.13)

$$K\theta_b(\theta) = 1 - b_o\left(\frac{1}{\cos\theta} - 1\right) - b_1\left(\frac{1}{\cos\theta} - 1\right)^2$$

(4.14)

where a_o is the intercept of the performance curve, a_1 and b_o are the first-order coefficients for the respective equations, and a_2 and b_1 are the second-order coefficients. For fixed collectors, the second-order parts do not represent a significant contribution; these terms are dropped. For fixed collectors, representation of η against $\Delta T/I_T$ results in straight lines as shown in Figure 4.2 for flat-plate collectors. The intercept corresponds to the optical properties of the collector and the slope is a heat-loss coefficient. A high-performance collector has high optical properties and low a_1 value.

Table 4.1 presents the most general classification of solar-thermal collectors, including the operating temperature ranges. High temperatures are obtained by concentrating solar irradiance via reflecting surfaces. Efficiency curves for typical solar collectors are presented in Figure 4.3. It should be noted that the efficiency for the flat-plate collector design drops very quickly in comparison with other designs. Flate-plate collectors are widely used due to their simplicity, low cost, minimal maintenance, and suitability for a broad number of applications regarding their temperature ranges. They are typically the most economical choice for regions with high direct sunlight (e.g., deserts).

4.2.2 EVACUATED-TUBE SOLAR COLLECTORS

Evacuated-tube solar collectors have better performance than flat plate for high-temperature operation in the range of 77–170°C. They are well suited to commercial and industrial heating applications and also for cooling applications by regenerating refrigeration cycles. They can also be an effective alternative to flat-plate collectors for domestic space heating, especially in regions where it is often cloudy (e.g., New England, Germany, etc.).

TABLE 4.1
Solar Thermal Collectors

Collector type	Temperature range (°C)	Concentration ratio
Flat-plate collector	30–80	1
Evacuated-tube collector	50–200	1
Compound parabolic collector	60–240	1–5
Fresnel lens collector	60–300	10–40
Parabolic trough collector	60–250	15–45
Cylindrical trough collector	60–300	10–50
Parabolic dish reflector	100–500	100–1,000
Heliostat field collector	150–2,000	100–1,500

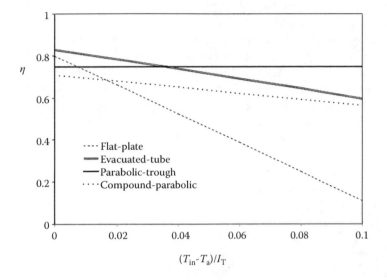

FIGURE 4.3 Efficiency curves for the most common types of solar collectors.

An evacuated-tube solar collector consists of rows of parallel glass tubes connected to a header pipe as shown in Figure 4.4(a, b). The air within each tube is removed reaching vacuum pressures around 10^{-3} mbar. This creates high insulation conditions to eliminate heat loss through convection and radiation, for which higher temperatures than those for flat-plate collectors can be attained. A variant to the vacuum is that the tube can use a low thermal conductivity gas such as xenon. Each evacuated tube has an absorber surface inside. Depending on the mechanism for extracting heat from the absorber, evacuated-tube solar collectors fall into either a direct-flow or heat-pipe classifications.

In direct-flow tubes, the working fluid flows through the absorber. (Figure 4.4c–e) These collectors are classified according to their connecting-material joints as glass–metal or glass–glass and, further, by the arrangement of the tubes (such as concentric or U-pipe). Inside each evacuated tube, a flat or curved metallic fin is attached to a copper or glass absorber pipe. The fin is coated with a selective thin film whose optical properties allow high absorbance of solar radiation and impede radiative heat loss. The glass–metal collector type is very efficient, although it can experience loss of vacuum due to the junction of materials with very different heat expansion coefficients. Within this type, the fluid can follow either a concentric or a U-shape path; for both, the working fluid flows in and out at the same end (the header pipe). The concentric configuration could incorporate a mechanism to rotate each single-pair fin pipe up to the optimum tilt incidence angle, even if the collector is mounted horizontally. However, the U-pipe configuration is the most typical direct-flow/evacuated-tube solar collector.

For the glass–glass type, tubes consist of two concentric glass tubes fused together at one end. The space between the tubes is evacuated. The inner tube is also covered with a selective surface coating to absorb solar energy while inhibiting heat losses by radiation. These collectors perform well in cloudy and low-temperature conditions. Glass–glass solar tubes may be used in heat pipe or U-pipe configurations. They are not generally as efficient as glass–metal tubes but are cheaper and tend to be more reliable. For high-temperature applications, glass–glass tubes can be more efficient than glass–metal tubes.

In a heat-pipe-evacuated tube collector, each vacuum-sealed glass tube allocates one metal pipe, usually copper, attached to an absorber plate. The heat pipe is also at vacuum pressure. Inside the heat pipe is a small quantity of water. Because water boils at a lower temperature when pressure is decreased, the purpose of the vacuum is to easily change from the liquid phase to a vapor. Vaporization is achieved around 25–30°C, so when the heat pipe is heated above this, vapor rapidly rises to the top of the heat pipe, transferring heat. As the heat is lost, the vapor condenses and returns to the bottom for the process to be repeated. Even though the boiling point has been reduced due to the vacuum, the freezing point remains the same (some additives will prevent freezing at overnight low temperatures). A schematic of a heat-pipe-evacuated tube is shown in Figure 4.4(f). The copper used for heat pipes must have a low content of oxygen; otherwise, it will leach out into the vacuum, forming pockets in the top of the heat pipe and causing detrimental performance.

In comparing heat-pipe and U-pipe configurations, the two have close efficiency ratings; however, the U-pipe has some advantages, such as being more economical and compact than heat-pipe collectors. Additionally, U-pipe collectors can be installed perfectly vertical or horizontal, allowing for a wider variety of installation options, which permits these solar collectors to be used where other collectors cannot be used. Heat-pipe collectors must be mounted with a minimum tilt angle of around 25° so that the internal fluid of the heat pipe can return to the hot absorber. Installation and maintenance of heat pipes are simpler than for direct-flow collectors; individual tubes can be exchanged without emptying the entire system.

4.2.3 Concentrating Collectors

There are two ways of classifying solar thermal collectors according to their concentration ratio (C). In the most general terms, solar collectors are classified as flat-plate collectors with a concentration ratio $C = 1$ and as concentrating collectors with $C > 1$. The existing types of concentrating collectors are parabolic-compound, parabolic-trough, parabolic-dish, Fresnel, and central tower concentrators, among others. Two definitions of concentration ratio for these systems are used. In the first, the concentration ratio depends on geometric characteristics, and it is given by

$$C = \frac{A_a}{A_r} \tag{4.15}$$

where A_a is the area of the collector aperture, and A_r is the energy absorber or receiver area. The geometric concentration ratio is a measure of the average concentration for the case where energy flux in the receiver is homogeneous, although this is not what actually happens. In contrast, very

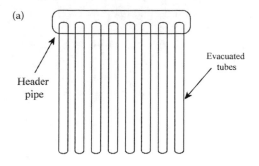

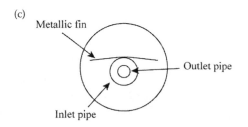

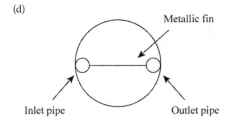

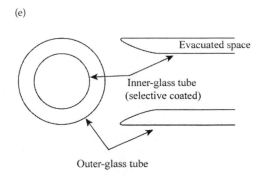

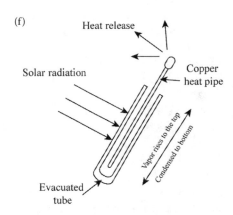

FIGURE 4.4   (a) Arrangement of evacuated tubes on a solar collector. (b) Evacuated-tube solar collectors. (c) Glass-metal evacuated tubes with direct concentric flow. (d) Glass-metal evacuated tubes with U-shape pipe. (e) Glass-glass evacuated tubes and (f) Heat-pipe evacuated tubes solar collector.

complex flux distribution reaches the receiver; in general, high-intensity concentration occurs in the center and decreases to the ends of the receiver.

The second definition corresponds to the ratio of the average energy flux received on the energy absorber to that captured by the collector aperture; this is called the flux concentration ratio. The local flux concentration ratio at any point of the receiver is

$$C = \frac{I_r}{I_a}$$

(4.16)

where I_r and I_a are the energy flux at any point of the receiver and the energy flux on the aperture, respectively.

The concentration ratio depends upon the concentrator's geometry. When the concentrator is a channel trough, the receiver geometrically represents a line; when it is a dish, the radiation is redirected to one point. The heat transfer analysis in the focal line for the channel trough must be undertaken as a two-dimensional object. For the dish, where radiation is coming from all directions, the analysis corresponds to a three-dimensional object.

According to Equation 4.16, there is no restriction to the maxim concentration ratio. If the receiver area tends to zero, then the concentration ratio will tend to infinity. According to the second law of thermodynamics, there exists a maximum work limit for any process; in contrast to the first law, this accounts for the inefficiencies inherent to specific processes. The thermodynamic limit for the concentration ratio is ideally found by considering the two interchanging bodies as blackbodies in such a way that the collector's surface will capture all the energy emitted by the Sun. Of course, for this to happen, it also must be assumed that the space between the two blackbodies' surfaces is in a vacuum at the zero-absolute temperature (0 K). Then the energy irradiated by both surfaces is expressed by the Stefan–Boltzmann law in terms of view factors as follows:

$$Q_{s \to r} = A_s F_{s \to r} \sigma T_s^4 \tag{4.17}$$

$$Q_{r \to s} = A_r F_{r \to s} \sigma T_r^4 \tag{4.18}$$

where

$Q_{s \to r}$ and $Q_{r \to s}$ are the interchanging irradiated energy from the Sun to the receiver and from the receiver to the Sun, respectively
A_s and A_r are the heat-interchanging areas of the Sun with the receiver
$F_{s \to r}$ and $F_{r \to s}$ are the fractions of the energy that actually reach each other's surfaces
T_r and T_s are the receiver and Sun temperatures, respectively
σ is the Stefan–Boltzmann constant $\sigma = 5.67 \times 10^{-8}$ W/(m^2 K^4)

The maximum energy interchange between two surfaces is achieved when both surfaces' temperatures are at thermal equilibrium as stated by the zero law of thermodynamics. At thermal equilibrium,

$$Q_{s \leftrightarrow r} = Q_{s \to r} - Q_{r \to s} = 0 \tag{4.19}$$

Then, the relationship between the geometrical factors can be found as

$$A_s F_{s \to r} = A_r F_{r \to s} \tag{4.20}$$

By applying a reciprocity relationship,

$$A_s F_{s \to a} = A_a F_{a \to s} \tag{4.21}$$

Then the geometric concentration ratio can be related as

$$C = \frac{A_a}{A_r} = \frac{F_{r \to s} F_{s \to a}}{F_{a \to s} F_{s \to r}} \tag{4.22}$$

For the ideal case, all the energy emitted by the Sun that is intercepted by the collector aperture is captured by the receiver,

$$F_{s \to r} = F_{s \to a} \tag{4.23}$$

Now, the correlation ratio is in terms of the fraction of energy interchanged by the two surfaces:

$$C = \frac{F_{r \to s}}{F_{a \to s}} \tag{4.24}$$

Because the fraction of the energy is always $F_{r \to s} \leq 1$, the maximum concentration is

$$C \leq C_{max} = \frac{1}{F_{a \to s}} \tag{4.25}$$

The concentration ratio cannot exceed the reciprocal of the geometric factor between the collector aperture and the Sun. Figure 4.5 schematizes the geometric relationship between the concentrator surface and the Sun.

$$F_{a \to s} = \sin\left(\frac{\alpha}{2}\right) \tag{4.26}$$

The angle subtended by the Sun (α) viewed by an observer on Earth is 32′. Then, the maximum concentration ratios for linear and circular concentrators are

$$C_{linear,max} = \frac{1}{\sin\left(\dfrac{\alpha}{2}\right)} = 215 \tag{4.27}$$

$$C_{circular,max} = \frac{1}{\sin^2\left(\dfrac{\alpha}{2}\right)} = 46000 \tag{4.28}$$

When the maximum concentration ratios are known, it is possible to calculate the maximum temperature that can be obtained by solar concentrators. An energy balance between the Sun and the receiver, including all thermal and optical terms, gives

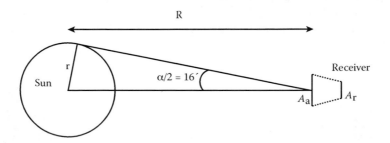

FIGURE 4.5 Geometric relationships between the concentrator surface and the Sun.

$$T_r = T_s \left[\frac{(1-\eta)\eta_o C}{\varepsilon_r C_{ideal}} \right] \tag{4.29}$$

where

η is the fraction of the energy that enters the collector that is extracted for usable heat as well as the heat losses per conduction and convection

η_o is the optical efficiency resulting from the product of transmittance, reflectance, and absorbance

ε_r is the receiver emissivity

C is the geometric concentration ratio

For a linear receiver, the maximum temperature that can be reached when no extraction of heat and no losses occur (i.e., $\eta = 0$) and the absorber is nonselective ($\eta_o \approx \varepsilon_r$) is

$$C_{linear,max} = 215 \quad T_r = T_s \left[\frac{C}{C_{ideal}} \right]^{\frac{1}{4}} = 5800 \left[\frac{215}{46000} \right]^{\frac{1}{4}} = 1600 \; ^\circ C$$

For a three dimensional concentrator it is

$$C_{max,3D} = \frac{1}{\sin^2 \left(\dfrac{\alpha_a}{2} \right)} = 46000$$

$$C = C_{max,3D} = 46000 = C_{ideal} \quad T_r = T_s \left[\frac{C}{C_{ideal}} \right]^{\frac{1}{4}} = 5800 \; ^\circ C$$

4.2.3.1　Optic Fundamentals for Solar Concentration

In the ray approximation, transmission of light energy is supposed to travel in straight lines except when it encounters an obstruction; then, reflection and refraction occur. The assumption works well when the sizes of obstructions are large compared to the wavelength of the traveling light. When a light ray strikes against a transparent surface (as seen in Figure 4.6), part of the incident ray may be reflected from the surface with an angle equal to the angle formed between the incident ray and the normal to the surface. Some other part crosses the surface boundary and the difference of the refractive index (n) of the two materials, resulting in a change in the direction and the speed of the light (v). In Figure 4.6, the splitting of the incoming light ray into reflected and refracted light is presented when $n_2 > n_1$. Because the velocity of light is lower in the second medium ($v_2 < v_1$), the angle of refraction, θ_2, is less than the angle of incidence θ_1. The refraction process is described by Snell's law:

$$n_1 \sin\theta_1 = n_2 \sin\theta_2 \tag{4.30}$$

Simple optical instruments such as mirrors and lenses are used to focus energy in a receiver to absorb as much energy as possible to convert it into usable energy. The geometrics of the optical focusing surfaces can be plane, parabolic, or spherical. For solar energy applications, these instruments are used only to converge energy onto the receiver.

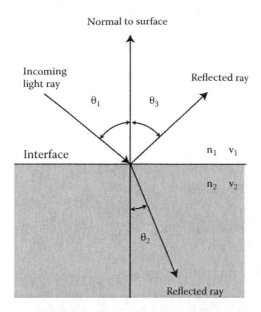

FIGURE 4.6 Optical processes experienced by a light ray when intercepting an obstacle.

Mirrors are made out of a conducting material for the reflection to be close to 100%, and are used to redirect light. Figure 4.7 presents concave and convex-spherical mirrors where optical angles are defined. Concave mirrors are called converging or positive, and the convex are called diverging or negative. The symmetry axis for both mirrors is the line along their diameters; the point C represents the center of the spherical truncated surface and R is the radius. In a concave mirror, the reflection of two incident rays—parallel to the symmetry axis and close to it so that the angles of incidence and reflection are small and cross each other at a point on the axis—is called the focal point of the mirror. The distance f from the mirror is the focal length. The two right triangles with the opposite side d give

$$\alpha \approx \tan\alpha = \frac{d}{R} \qquad (4.31)$$

and

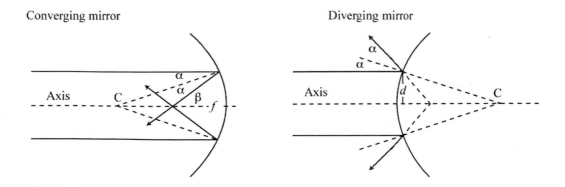

FIGURE 4.7 Reflection of light and focal point in concave and convex mirrors for small incidence angles.

$$\beta \approx \tan\beta = {d}/{f} \qquad (4.32)$$

The angle of reflection is equal to the angle of incidence α, so $\beta = 2\alpha$. Then,

$$f = {R}/{2} \qquad (4.33)$$

In a convex mirror, the center of the sphere is on the side opposite from where light rays go and to where light is reflected. Keeping the assumption that the angles are small with respect to the surface normal and close to it, the reflected rays diverge as if they came from a point behind the mirror. Such a fake point corresponds to the focal point of the mirror, but because it is not a real sinking point, this is known as a virtual focal point. The main interest for solar energy is to concentrate energy by forming images, so reflecting surfaces with virtual focal points are not of interest within this text.

When concave spherical mirrors reflect all incoming parallel rays to the axis rather than only the ones close to it, the rays cross the symmetry axis and form an image line from the focal point up to the interception of the axis with the mirror, as seen in Figure 4.8. When the light rays are not parallel to the axis, the focal line rotates symmetrically with respect to its center, maintaining the pattern of the reflected rays. For these characteristics, the receiver design for solar energy applications is strongly affected by the reflecting surface dimensions.

Lenses are made of a transparent material and the purpose of using them is to manipulate light by refraction to create images. Figure 4.9 presents three typical glass lenses. Lenses that are thicker in the middle have a positive focal length; after passing through the lens, incident rays parallel to the axis converge to a point. The thickness of lenses is small compared to the radii of curvature of the surfaces. For paraxial rays, using the law of refraction and small angle approximations, it can be shown that the focal length is given by the following formula:

$$\frac{1}{f} = \left(\frac{n}{n_o} - 1\right)\left(\frac{1}{R_1} - \frac{1}{R_2}\right) \qquad (4.34)$$

where

n is the index of refraction of the substance from which the lens is made, usually glass or plastic
n_o is the index of refraction of the transparent medium on either side of the lens, usually air, for which $n_o = 1$

Solar rays parallel to vertical axis   Solar rays 20° to vertical axis

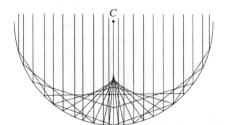 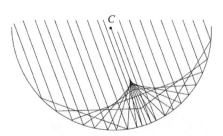

FIGURE 4.8   Reflection of light for a spherical reflecting mirror for two different incidence angles.

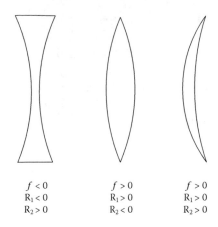

$$\begin{array}{ccc} f < 0 & f > 0 & f > 0 \\ R_1 < 0 & R_1 > 0 & R_1 > 0 \\ R_2 > 0 & R_2 < 0 & R_2 > 0 \end{array}$$

FIGURE 4.9 Lenses.

R_1 and R_2 are the radii of the two lens surfaces—positive for convex surfaces and negative for concave; R_1 is the radius of the first surface encountered by the traveling light and R_2 is the radius of the other surface

The procedure for locating images with lenses is similar to that for mirrors. In Figure 4.10, a lens with two parallel surfaces receives two rays from a faraway object. The ray that points to the center of the lens passes through essentially without deflection. A parallel ray to the axis is refracted and passes through the focal point for a positive lens. A ray passing through or toward a focal point emerges parallel to the axis. For a negative lens, the ray is reflected away.

The analysis of paraxial ray approximation gives the same formulas for location of the images as for mirrors. When the object distance is greater than $2f$, the image distance is less than $2f$. The image is real, inverted, and reduced. To form a real and enlarged image, the object distance must be between f and $2f$. As with the positive mirror, an object placed closer to the lens than f will form a virtual image. The image is upright and enlarged.

4.2.3.2 Parabolic Concentrators

The parabola is found in numerous situations in the physical world. In three dimensions, a parabola traces out a shape known as a paraboloid of revolution when it is rotated about its axis and as a parabolic cylinder, when it moves along the axis normal to its plane. Solar collectors whose reflecting surfaces follow such geometrics are called parabolic dish concentrators and parabolic troughs, respectively. If a receiver is mounted at the focus of a parabolic reflector, the reflected light will be absorbed and converted into a useful form of energy. The reflection to a point or a line and subsequent absorption by a receiver constitute the basic functions of a parabolic concentrating collector.

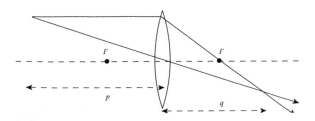

FIGURE 4.10 Positive lens.

Figure 4.11 shows a representation of a parabola. It has a single axis of reflective symmetry, which passes through its focus (F) and is perpendicular to its directrix. The point of intersection of this axis and the parabola is called the vertex (V); it is exactly at the middle between the focus and the directrix. In parabolic geometry, the length $\overline{FR}$ is always equal to the length $\overline{RD}$. In parabolic surfaces, the angle of reflection equals the angle of incidence, according to Snell's law, for which all radiation parallel to the axis of the parabola is reflected to the focal point.

Taking the origin at the vertex, V, the equation for a parabola symmetrical about the x-axis is

$$y^2 = 4fx \tag{4.35}$$

where f is the focal length. In polar coordinates, the equation becomes

$$\frac{4f}{r} = \frac{\sin^2 \theta}{\cos \theta} \tag{4.36}$$

r is the distance from the origin to any point of the parabola $\overline{VR}$, and θ is the angle between the parabola axis and the line VR. In solar applications, it is useful to shift the parabola's origin to the focal point F; in the Cartesian coordinate system, this parabola is represented by

$$y^2 = 4f(x+f) \tag{4.37}$$

In polar coordinates, a functional equation is

$$p = \frac{2f}{1+\cos \psi} \tag{4.38}$$

where p is the distance from the origin F to any point of the curve R ($\overline{FR}$) and the angle ψ is measured between the lines VF and FR.

The extent of a solar concentrator is usually defined in terms of the rim angle, ψ_{rim}, or the ratio of the focal length to aperture diameter, f/d (Figure 4.12). Flat parabolas are characterized by a small rim angle because the focal length is large compared to the aperture diameter. The height (h)

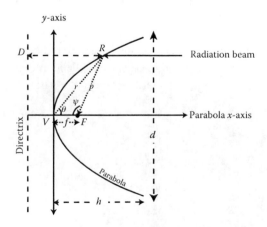

FIGURE 4.11 Angular and distance description of a parabola.

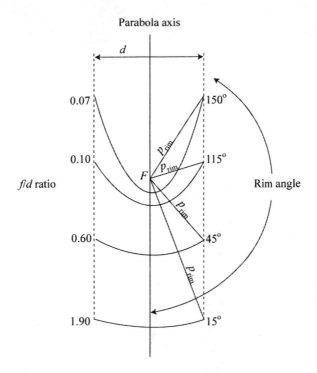

FIGURE 4.12 Rim angle and *f/d* ratio for parabola segments with common focal point.

of the parabolic concentrator corresponds to the vertical distance from the vertex to the aperture of the parabola. Mathematical expressions correlating focal length, aperture diameter, height, and rim angle in a parabola are as follow:

$$h = \frac{d^2}{16f} \tag{4.39}$$

$$\tan \psi_{rim} = \frac{1}{\left(\frac{d}{8h}\right) - \left(\frac{2h}{d}\right)} \tag{4.40}$$

$$\frac{f}{d} = \frac{1}{4 \tan \left(\frac{\psi_{rim}}{2}\right)} \tag{4.41}$$

Another useful property of the parabola is the arc length (*s*), which is given by

$$s = \left[\frac{d}{2}\sqrt{\left(\frac{4h}{d}\right)^2 + 1}\right] + 2f \ln\left[\frac{4h}{d} + \sqrt{\left(\frac{4h}{d}\right)^2 + 1}\right] \tag{4.42}$$

A parabolic trough collector corresponds to a linear translation of a two-dimensional parabolic reflector; as a result, the focal point becomes a focal line (Figure 4.13). When the parabolic reflector is aligned parallel to the solar rays, all the incoming rays are redirected toward the focal line. The

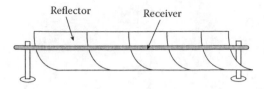

FIGURE 4.13 Parabolic trough collector.

parabolic trough must accurately track the motion of the Sun to maintain the parabola axis parallel to the incident rays of the Sun. Otherwise, if the incident beam is slightly off to the normal to the concentrator aperture, beam dispersion occurs, resulting in spreading of the image at the focal point. For a parabolic trough collector of length l and an aperture distance d, the collector aperture area is given by

$$A_a = ld \tag{4.43}$$

Its reflective surface area is

$$A_s = ls \tag{4.44}$$

where s is the arc length of the parabola and is given by Equation 4.39.

In contrast with the parabolic trough, the aperture of a low-rim cylindrical trough need not track at all to maintain focus. As presented in Figure 4.12, a high-rim-angle cylindrical trough would have a focal plane rather than a focal line. This effect of rim angle on the focus of a cylindrical trough can be seen by observing the path of an individual ray as it enters the collector aperture. For practical applications, if the rim angle of a cylindrical trough is kept lower than 30°, spherical aberration is small and a virtual line focus trough is achieved. The advantage of a cylindrical reflector is that it does not need to track the Sun as long as some means are provided to intercept the moving focus.

4.2.4 Compound Parabolic Concentrators (CPCs)

Unlike the trough and dish concentrators that clearly present a focal line or point, the compound parabolic is a nonimaging concentrator. This design does not require the light rays to be parallel to the concentrator's axis. A CPC collector is composed of two truncated parabolic reflectors; neither one keeps its vertex point but both rims must be tilted toward the Sun. Figure 4.14 shows the geometric relationship between the two parabola segments for the construction of a CPC. The two parabolas are symmetrical with respect to reflection through the axis of the CPC and the angle in between them is defined as the acceptance angle (θ_{accep}). In a parabola, light rays must always be parallel to the parabola's axis; otherwise, it is out of focus and the image is distorted. When the rim of a parabola is tilted toward the Sun, the light rays are redirected on the reflecting surface somewhere below the focus; in contrast, when it is pointing away, the rays are reflected somewhere above the focus.

In CPC designs, the half parabola tilted away from the Sun is replaced with a similarly shaped parabola whose rim points toward the Sun. All incoming rays fall into a region below the focal point of the parabola segments. Figure 4.15 shows the ray tracing for a CPC collector. Light with an incidence angle less than one-half the acceptance angle is reflected through the receiver opening; for greater angles, light rays are not directed to the receiver opening but rather to some

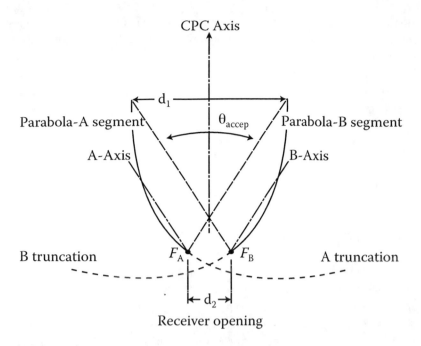

FIGURE 4.14   Cross-section of a CPC collector.

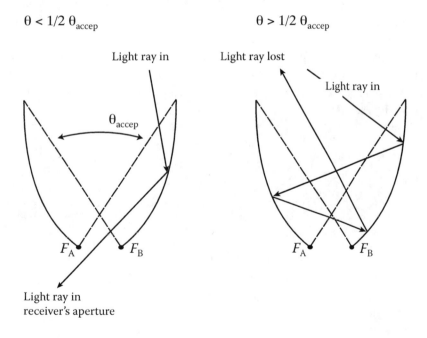

FIGURE 4.15   Ray tracing for single light rays in a CPC collector.

other point of the reflecting surface. The light ray is eventually reflected back out through the CPC aperture.

By translating the cross section shown in Figure 4.14 over a line, the CPC structure is obtained. The receiver is positioned in the region below the focus of the two parabolic surfaces to capture the incoming solar rays. Receivers also might take different geometries, such as flat plates at the base of the intersection of the two surfaces or cylindrical or U-tubes passing through the region below the focus. Moreover, evacuated tubes can be integrated with CPC collectors. Figure 4.16 presents the arrangement of several CPC collectors.

CPC collectors provide a geometric concentration ratio (CR_g) within the range of 1.5 up to 10 times the solar radiation with no tracking during the day. The geometric concentration ratio of a CPC is related to the acceptance angle, θ_{accep}, by

$$CR_g = \frac{1}{\sin\left(\frac{1}{2}\theta_{accep}\right)} \tag{4.45}$$

The CR_g must be increased as an attempt to increase performance at elevated temperatures; then, according to Equation 4.45, the acceptance angle of the CPC must be reduced. Typically, CPC receivers are aligned in the east–west direction and their apertures are tilted toward the south. They need no hourly tracking but must be adjusted periodically throughout the year. The narrowing of the acceptance angle results in a requirement for increasing the number of tilt adjustments throughout the year as presented in Table 4.2. A $\theta_{accep} = 180°$ corresponds to the geometry of a flat-plate collector and for 0° is equivalent to a parabolic concentrator. Temperatures in the range of 100–160°C have been reached with CR_g greater than six, showing efficiencies of around 50% (Rabl, O'Gallagher, and Winston 1980). At lower CR_g, the collector performance is better than that for a double-glazed flat-plate collector at about 70°C; yet, its output remain competitive for lower temperatures.

Only a few studies have been conducted to investigate instantaneous efficiency for CPCs. Carvalho et al. (1995) tested the performance of a CPC to determine efficiency curves for both north–south and east–west orientations. As expected, results are different for each orientation because the convection regime is different in both cases. The linear and second-order least-squares fits obtained in both cases are

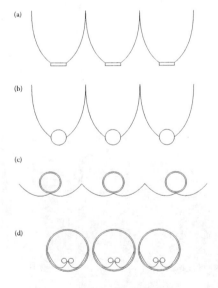

FIGURE 4.16 Cross sections of nontracking collectors with CPC reflectors: (a) external reflector with flat absorber; (b) external reflector with large absorber tube; (c) external reflector with large absorber tube surrounded by an evacuated glass tube; (d) CPC with U-tube absorber inside evacuated tubes.

TABLE 4.2
Tilt Requirements of CPCs during the Year at Different Acceptance Angles

Acceptance half–angle (°)	Collection time average over year (h/day)	Number of adjustments per year	Average collection time if tilt is adjusted every day (h/day)
19.5	9.22	2	10.72
14	8.76	4	10.04
11	8.60	6	9.52
9	8.38	10	9.08
8	8.22	14	8.82
7	8.04	20	8.54
6.5	7.96	26	8.36
6	7.78	80	8.18
5.5	7.60	84	8.00

Source: Rabl, A. et al. 1980. *Solar Energy* 25 (4): 335–351.

$$\eta_{N-S} = \left(0.74 \pm 0.01\right) - \left(4.3 \pm 0.2\right)\frac{\Delta T}{I_{col}} + \left(1 \pm 4\right) \times 10^{-3} I_{col} \left(\frac{\Delta T}{I_{col}}\right)^2 \tag{4.46}$$

$$\eta_{N-S} = \left(0.74 \pm 0.01\right) - \left(4.3 \pm 0.2\right)\left(\frac{\Delta T}{I_{col}}\right) \tag{4.47}$$

$$\eta_{E-W} = \left(0.72 \pm 0.01\right) - \left(1.5 \pm 0.2\right)\frac{\Delta T}{I_{col}} + \left(4.9 \pm 0.4\right) \times 10^{-2} I_{col} \left(\frac{\Delta T}{I_{col}}\right)^2 \tag{4.48}$$

$$\eta_{E-W} = \left(0.74 \pm 0.01\right) - \left(4 \pm 0.2\right)\left(\frac{\Delta T}{I_{col}}\right) \tag{4.49}$$

where $I_{col} = (I_b + I_d)/C$ in Watts per square meter, where C is the concentration ratio after truncation, and $\Delta T = T_{avg,f} - T_a$ in Kelvin for $T_{avg,f}$ is the arithmetic mean between inlet and outlet fluid temperatures. On the other hand, the U.S. National Renewable Energy Laboratory (NREL) has proposed the following linear equation to determine the instantaneous efficiency of an east–west-orientated CPC:

$$\eta_{CPC} = 0.73 - 0.64 \left[\frac{T_r - T_a}{I_a}\right] \tag{4.50}$$

where T_r is the temperature of the average temperature of a receiver, T_a is the ambient temperature, and I_a is the global solar irradiance entering the collector aperture in Watts per square meter. This equation is for a CPC with a concentration ratio of five and an acceptance angle of about 19°.

4.2.5 Fresnel Lens Concentrators

Fresnel lenses have been also incorporated into solar thermal energy systems. These solar collectors reduce the amount of material required compared to a conventional spherical lens by breaking the lens into a set of concentric annular sections, as shown in Figure 4.17. Although such canted facets are brought to the plane, discontinuities exist between them. The volume is greatly reduced while keeping close optical properties to a corresponding normal lens. The more facets created, the better the optical approximation is. A high-quality linear Fresnel lens should have more than 1,000 sections per centimeter. The flatness results in great savings in material, thus reducing production costs.

The effectiveness of Fresnel lenses can be reduced by the sharpness of the facets. Any ray striking the back side of a facet or the tip or valley of a facet is not directed to the receiver. To maintain the refracted image focused on a receiver that is fixed with respect to the lens, the Fresnel collector or any other lens system requires at least one single-axis tracking system to keep the incident light rays normal to the lens aperture.

4.2.6 Heliostats

The energy collection in a large-scale solar-thermal power plant is based on the concentration of the Sun's rays onto a common focal point to produce high-temperature heat to run a steam turbine generator. The radiation concentration is achieved by using hundreds of large sun-tracking mirrors called heliostats. Each heliostat directs the solar radiation toward the highest point in a tower where the receiver is located to absorb the heat. Central receivers are distinguished by large power levels (1–500 MW) and high temperatures (540–840°C). High-quality heat transfer fluids are used to transport the energy to a boiler on the ground to produce the steam to be used in a traditional power plant.

The tracking angles for each heliostat, along with the corresponding incidence angle can be derived using vector techniques where the zenith, east, and north (z, e, n) directions are the appropriate coordinates whose origin, **O,** is located at the base of the receiving tower. Figure 4.18 shows the proposed Cartesian coordinate system; point **A** ($z_0, 0, 0$) corresponds to the location in space where the receiver is placed and point **B** (z_1, e_1, n_1) is the location of a heliostat close to the ground. Each heliostat presents a unique value pair for the altitude (α_H) and azimuth (γ_H) angles depending on its location with regard to the energy receiver. To determine such angles, three vectors must be defined: a vector representing the direction of the Sun's ray hitting the heliostat ($\vec{s}$), one corresponding to

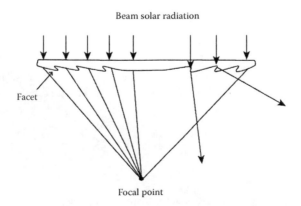

Beam solar radiation

Facet

Focal point

FIGURE 4.17   Ray trace on a Fresnel lens.

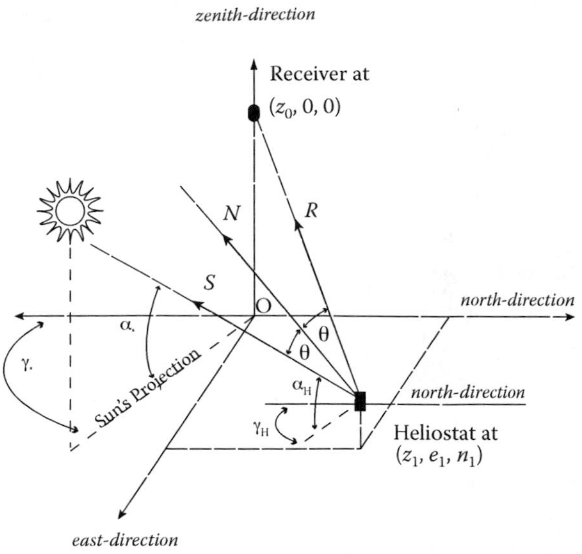

FIGURE 4.18 Geometric relationships between heliostat and receiver in a zenith–east–north Cartesian coordinate system.

the heliostat normal ($\vec{N}$), and the third physically representing the redirection of the Sun's ray toward the point **A,** receiver ($\vec{R}$). These three vectors are represented respectively by the following equations:

$$\vec{S} = S_z \hat{i} + S_e \hat{j} + S_n \hat{k} \tag{4.51}$$

$$\vec{N} = N_z \hat{i} + N_e \hat{j} + N_n \hat{k} \tag{4.52}$$

$$\vec{R} = R_z \hat{i} + R_e \hat{j} + R_n \hat{k} \tag{4.53}$$

where $\hat{i}$, $\hat{j}$, and $\hat{k}$ are the unit vectors along the z, e, and n axes, respectively. The $\vec{S}$-components can be written in terms of solar altitude (α_s) and azimuth (γ_s) as

$$
\begin{aligned}
S_z &= \sin\alpha_s \\
S_e &= \cos\alpha_s \sin\gamma_s \\
S_n &= \cos\alpha_s \cos\gamma_s
\end{aligned}
\tag{4.54}
$$

and the $\vec{R}$ vector is defined as

$$\vec{R} = \frac{(z_0 - z_1)\hat{i} - e_1\hat{j} - n_1\hat{k}}{\sqrt{(z_0 - z_1)^2 + e_1^2 + n_1^2}} \tag{4.55}$$

To redirect the Sun's rays, the law of specular reflection must be applied: The angle of incidence is equal to the angle of reflection. The scalar point between the vectors of $\vec{S}$ and $\vec{R}$ results in a practical expression that involves the incidence angle as follows:

$$\cos 2\theta = \vec{S} \cdot \vec{R} \tag{4.56}$$

Substituting Equations 4.51 and 4.53 into Equation 4.56, the angle of incidence or reflection can be calculated when the position of the Sun and position of the receiver relative to the heliostat are known:

$$\cos 2\theta = R_z \sin\alpha_s + R_e \cos\alpha_s \sin\gamma_s + R_n \cos\alpha_s \cos\gamma_s \tag{4.57}$$

The mirror normal can be found by adding the incidence and reflection vectors and dividing by the appropriate scalar quantity. This gives

$$\overline{N} = \frac{(R_z + S_z)\hat{i} + (R_e + S_e)\hat{j} + (R_n + S_n)\hat{k}}{2\cos\theta} \tag{4.58}$$

Substituting Equation 4.57, the altitude and azimuth of the reflecting surface (α_H and δ_H, respectively) in terms of the orthogonal coordinates are given by,

$$\sin\alpha_H = \frac{R_z + \sin\alpha_s}{2\cos\theta} \tag{4.59}$$

and

$$\sin\gamma_H = \frac{R_z + \cos\alpha_s \sin\gamma_s}{2\cos\theta\cos\alpha_H} \tag{4.60}$$

or

$$\cos\gamma_H = \frac{R_n + \cos\alpha_s \cos\gamma_s}{2\cos\theta\cos\alpha_H} \tag{4.61}$$

Central receiver technology for generating (Figure 4.19) electricity has been demonstrated at the Solar One pilot power plant in Barstow, California. This system consists of 1,818 heliostats, each with a reflective area of 39.9 m² covering 291,000 m² of land. The receiver is located at the top of a 90.8 m high tower and produces steam at 516°C (960°F) at a maximum rate of 42 MW (142 MBtu/h).

4.3 TRACKING SYSTEMS

As explained before, the purpose of using reflecting surfaces or lenses is to redirect the incoming solar light to the surface focal point in order to collect as much energy as possible. The angle between the surface axis and the solar rays must be kept at zero; to achieve this, a sun-tracking system must be implemented to keep the collector's aperture always perpendicular to the light rays during the day. For the particular geometrics of the spherical surface with symmetrical rotation about its axis, the collector might not move during the day, but the receiver can. For nonconcentrator collectors such as PV modules to produce electricity directly, sun trackers are used to maximize the solar energy gain throughout the day.

The tracking systems are divided into two types according to their motions. The following of the Sun can be done either with one single rotation axis (east–west or north–south) or by two rotation axes where the array points directly at the Sun at all times and is capable of rotating independently

FIGURE 4.19 Primary and secondary focal points seen in the air at the Barstow power tower in California (Courtesy DOE).

about two axes. Two-axis tracking arrays capture the maximum possible daily energy, although they are more expensive and require extensive maintenance that may not be worth the cost, especially for smaller scale solar energy systems.

4.4 SOLAR THERMAL SYSTEMS

The purpose of using any type of solar thermal collector is to convert the solar radiation into heat to be used in a specific application, whether domestic or industrial. The main components of the most general solar thermal system are the solar collection system, a storage tank, pumps, and the load, as shown in Figure 4.20. A real system includes all the necessary controlling systems and relief valves. The load can be used in any particular application and will vary with production of heat, cold, drying, or mechanical work. The useful energy extracted from the collectors is given by Equation 4.4, which accounts for the energy gathered by the collector minus the heat losses by convection and radiation. In terms of the inlet temperature T_{in}, this equation becomes

$$\dot{Q}_u = \eta I A_c = I_T A_c F_R \left[(\tau \alpha)_{eff} - \frac{U A_r (T_{in} - T_a)}{I_T A_c} - \frac{\varepsilon_{eff} \sigma A_r (T_{in}^4 - T_a^4)}{I_T A_c} \right] \qquad (4.62)$$

When heat loss by radiation is unimportant, Equation 4.62 is reduced to Equation 4.9. The energy obtained from the solar collector field depends on the inlet temperature, and this depends on the load pattern and the losses from the storage tank, pipes, and relief valves. Using a strict estimation of T_{in} when simulating a solar thermal process, energy losses from pipes could be estimated by solving the following differential equation for any pipe segment j:

$$\dot{Q}_j = m_j C_p \frac{dT_j}{dt} = -(UA)_j (T_j - T_a) \qquad (4.63)$$

where

m_j is the mass of the fluid in the pipe segment j
C_p is the heat capacity at constant pressure of the fluid
T_j is the average temperature of the fluid in the same segment
t is time

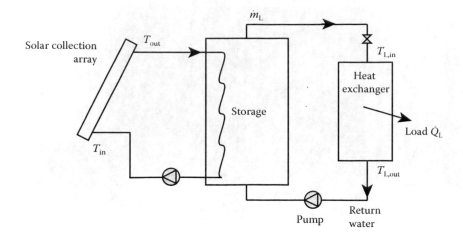

FIGURE 4.20   General diagram for solar thermal systems.

The total energy loss rate from the pipes to the environment is the summation of the individual losses from each element of pipe, given as

$$\dot{Q}_{pipe} = \sum_{j=1}^{n} \dot{Q}_{j} \tag{4.64}$$

By assuming that change of phase does not occur in the storage tank and that temperature is perfectly homogeneous, the rate of change in the amount of energy stored is

$$\dot{Q}_{st} = m_{st} C_{p} \frac{dT_{st}}{dt} \tag{4.65}$$

where m_{st} is the mass of the storage medium and T_{st} is its temperature. Typically, only solids and liquids are used in thermal storage because gases require large volumes. More sophisticated equations are needed for stratification within the storage tank. Another form of calculating $\dot{Q}_{st}$ is

$$\dot{Q}_{st} = \dot{Q}_{u} - \dot{Q}_{L} - \dot{Q}_{pipe} - \dot{Q}_{st,loss} \tag{4.66}$$

where $\dot{Q}_{L}$ is the rate at which heat is taken for the useful application and $\dot{Q}_{st,loss}$ is the rate of heat loss in the storage tank. The rate at which energy is taken from storage and provided to load is $\dot{Q}_{L}$:

$$\dot{Q}_{L} = \delta_{L} \dot{m}_{L} C \left(T_{L,in} - T_{L,out} \right) \tag{4.67}$$

where

δ_{L} is a variable of control, which takes the values zero or one corresponding to the supply of the load
$\dot{m}_{L}$ is the mass flow rate at which the fluid is pumped back to the storage tank
$T_{L,in}$ is the storage temperature
$T_{L,out}$ is the temperature of the fluid leaving the heat exchanger

4.4.1 Passive and Active Solar Thermal Systems

Passive solar technologies are means of using sunlight for useful energy without use of active mechanical systems. In such technologies, thermal energy flow occurs by radiation, conduction, or natural convection. To be used directly, distributed, or stored with little use of other energy sources, the heat obtained from sunlight is managed through some type of thermal mass medium such as water, air, rock, or oil. Some passive systems use a small amount of conventional energy to control dampers, shutters, night insulation, and other devices that enhance solar energy collection, storage, and use, and reduce undesirable heat transfer. Passive systems have the advantage that electricity outage and electric pump breakdown are not issues. This makes such systems generally more reliable, easier to maintain, and possibly longer lasting than active systems.

Passive solar technologies include direct and indirect solar gain. Both systems use the same materials and design principles. However, an indirect gain system positions the solar collectors separated from the space where energy is needed, for which the thermal mass medium is circulating between the two places.

Active systems use electric pumps, valves, and controllers to circulate water or other heat-transfer fluids through the collectors. Although they are usually more expensive than passive systems, they are generally more efficient. Active systems are often easier to retrofit than passive systems because their storage tanks do not need to be installed above or close to the collectors. If installed using a PV panel to operate the pump, an active system can operate even during a power outage.

4.4.1.1 Solar Thermal Application: Water Heating for Domestic Use

The main components of a solar water heater are the solar collector, storage, and heat distribution. Several configurations differ on the heat transport between the solar collector and the storage tank, as well as on the type of freeze protection. The most successful solar heaters are the integrated collector and storage (ICS), thermosiphon, drain-back, and drain-down systems (Table 4.3). These are habitually assisted in backup by a conventional system. In some countries, the installation of solar equipment must comply with local, state, and national building codes, roofing codes, plumbing codes, and national electrical codes.

The ICS and *thermosiphon* are passive solar water heaters where fluid circulation occurs by natural convection, as shown in the diagram of Figure 4.21. The absorber's energy gained by solar radiation is transferred to the copper pipes. The inlet fluid is located at the bottom of the collector; as heat is captured, the water inside the pipes warms up. The hotter the water is, the less dense and better it is for circulation. When hot water travels toward the top, the cooler and denser water within the storage tank falls to replace the water in the collector. Under no or low insolation, circulation stops; the warm and less dense fluid stagnates within the tank. The ICS is a self-contained integration of a solar collector and solar heated water storage, usually holding 30–40 gallons in a tank.

Both the ICS and the thermosiphon heaters are a low-cost alternative to an active-open-loop solar water system for milder climates. These systems have 40- to 120-gal storage tanks installed vertically or horizontally above the collector.

In open-loop systems, the water that is pumped through the collectors is the same hot water to be used. These systems are not recommended for sites where freezing occurs. These *active open-loop systems* are called drain-down systems and they can operate in either manual or automatic mode (Figure 4.22). The drain-down system relies on two solenoid valves to drain water. It requires two temperature sensors, a timer and a standard controller. The controller is wired to the freezing sensor in the back of the collector and to another placed at the exit of the collector, as well as to the solenoid valves and the pump. When the pump starts, the system fills, the valves remain open, and, when the pump stops, the system drains.

This design is efficient and lowers operating costs; however, it is not appropriate for hard or acidic water because corrosion and scale formation eventually disable the valves. During hard freezes, it is

TABLE 4.3
Summary of Solar Water Heaters for Domestic Use

System type	Characteristics and use	Advantages	Disadvantages
ICS batch (30–40 gal)	Integrated collector and storage Limited to regions that have more than 20 freezes per year	No moving parts Little to no maintenance May arrange two in series	Hot water availability from 12 p.m. to 8 p.m.
Thermosiphon (40–120 gal)	Higher performance than ICS but more difficult to protect from freezing	Lasts for years in locations with few freezes No mechanical and electrical parts	Tanks must be located above the collectors Collector may need descaling in hard water
Drain-down (80–120 gal)	Open loop Designed to drain water in freezing climates	Can drain all the water out of the collector Useful in areas with light freezes	Freeze protection is vulnerable to numerous problems Collectors and piping must have appropriate slope to drain Collector may need descaling in hard water
Glycol antifreeze (80–120 gal)	Active closed loop Cold climates Most freeze proof Can be used when drain-back systems are not possible Higher maintenance and shorter collector absorber plate life than drain-back systems	Very good freeze protection Can be powered by PV modules or by AC power	Most complex, with many parts Antifreeze reduces efficiency Heat exchanger may need descaling in hard water Anti-freeze turns acidic after 3–5 years of use and must be replaced or will corrode pipes.
Drain-back Highly recommended	Active closed loop Cold climates If pump fails, does not damage any part of the system More efficient than pressurized glycol antifreeze systems	Good freeze protection The simplest of reliable freeze protection systems Fluid not subject to stagnation No maintenance on the heat transfer fluid	Piping must have adequate slope to drain Requires a high-pressure AC pump Heat exchanger may need descaling in hard water

Source: Adapted from Lane 2004.

not unusual for utility companies to shut down some sections for hours; this causes a serious problem because the system uses electrical valves. Also, if a spool valve has not been operated for quite some time in an area with hard water, it may be cemented stuck in a closed position from mineral deposits and may not open when needed. Manual freeze protection depends on the occupants to pay attention and to stop circulation and drain the system. The drain-down systems usually force air into the storage tank when temperatures are high; an air vent must be placed at the highest point in the collector loop. The timer is used to power down the system when there is no solar radiation.

Drain-down refers to draining the collector fluid out of the system; drain-back refers to draining the collector fluid back into the storage tank. Although either method can be used for unpressurized systems, drain-back cannot be used in a pressurized applications such as a solar domestic water heater because storage invariably pressurizes.

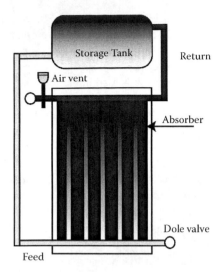

FIGURE 4.21 Thermosiphon water heater.

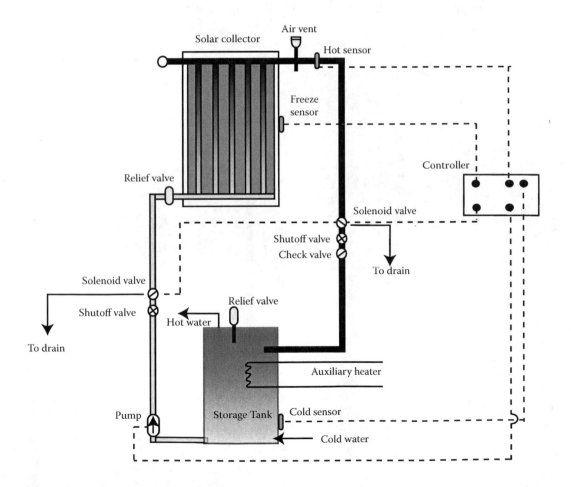

FIGURE 4.22 Automatic drain-down open-loop water heater for domestic use.

Closed-loop or active indirect systems pump a heat-transfer fluid, usually water or a glycol–water antifreeze mixture, through the solar water heater. These systems are popular in locations subject to extended subzero temperatures because they offer good freeze protection. However, glycol anti-freeze systems are more expensive to purchase and to install. Propylene glycol is normally used in domestic situations because it is not toxic. This is an unpressurized system, so the glycol does not need to be changed—unlike in pressurized systems.

The main components of a drain-back system are the solar collector, the storage tank, and the closed loop, where the water–glycol mixture is pumped through the collectors and a heat exchanger is located inside the storage tank (Figure 4.23). The closed loop is unpressurized but not open to the atmosphere. The heat transfer fluid transfers part of the collected solar heat to the water stored in the tank. The water in the storage tank is allowed to pressurize due to the high temperatures experienced. For this system, only a one-function controller is used to turn on the pump. When the hot sensor registers lower temperature than the cold sensor, the pump is turned off. Then, all the water in the collector and pipes above the storage tank is drained back, ensuring freeze protection. Drain-back systems must use a high-head AC pump to start up at full speed and full head. The pump must be located below the fluid level in the tank and have sufficient head capability to lift the fluid to the collector exit at a low flow rate.

Another type of closed-loop solar water heater is the *pressurized glycol antifreeze system* (Figure 4.24). Within the closed loop, a water–glycol mixture circulates as protection from freezing. Glycol percentage in the mixture varies from 30 to 50% depending on the typical high temperatures for the

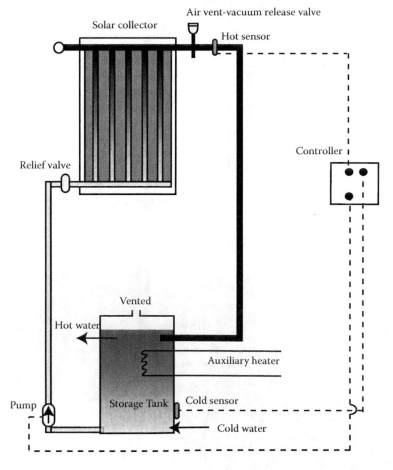

FIGURE 4.23 Drain-back system for domestic water heating.

region. Basically, this system comprises the same main components as the drain-back systems: solar collectors, circulation system, storage tank, and heat exchanger. The heat exchanger can be integrated in the wall of the storage tank or immersed as a coil, or it can be an external exchanger. The glycol will need to be changed every 3–5 years because it eventually turns acidic from heating.

The pressurized system is much more complicated than the drain-back system because it requires the implementation of auxiliary components to protect the main equipment. The antifreeze circulation system consists of a differential controller, temperature sensors, and AC pumps. If a blackout occurs, a major problem arises. One-hour stagnation of the antifreeze under high solar intensity makes glycol acidic. Then, it must be replaced sooner than it normally would be. To avoid stagnation, the pump must be working properly during the day. To ensure this, a DC photovoltaic pump should be integrated. AC and DC pumps can be connected in parallel in the same system. Other essential parts for this system include a pressure gauge to measure the amount of antifreeze within the circulating system, an expansion tank, a check valve above the pump to prevent reverse-flow thermosiphoning at night, a pressure relief valve, and an air vent at the highest point in the system.

4.4.1.2 Solar Thermal Application: Water Heating for Industrial Use

Temperature requirements for heat production in industrial processes range from 60 to 260°C. In this temperature range, solar thermal systems have great applicability. However, the challenge lies in the integration of a periodic, dilute, and variable solar input into a wide variety of industrial processes. Issues in the integration are selection of collectors, working fluid, and sizing of components. Application-specific configurations are required to be adopted and designed. The specific configuration consists of concentrating collectors, pressurized hot-water storage, and a load heat exchanger.

Table 4.4 summarizes the potential industrial processes with favorable conditions for application of solar technologies in congruence with their heat-quality production. An important measure to fit adequately within the current energy transition is to meet such great energy demands by incorporating solar technologies in both developed and developing countries. Moreover, replacing of technologies must occur to some extent, along with improvement of process efficiencies.

Despite the great success of solar energy for domestic applications—particularly water heating, almost no implementation has occurred for industrial processes, mainly due to the high initial capital costs involved, and lack of understanding of the expected benefits. The heat supply in industry usually consists of hot water or low-pressure steam. Hot water or steam at medium temperatures less than 150°C is used for preheating fluids or for steam generation of a fluid with smaller working temperatures. High thermal efficiencies are always experienced when the working temperatures are low due to the elimination of heat losses by radiation and great reduction in the convective and conductive areas. When temperatures higher than 100°C are required, the solar collection system is pressurized.

Figure 4.25 shows a diagram where solar collectors and a conventional system for producing heat in industrial uses are combined. The industrial system includes the solar collection array, circulating pumps, a storage tank, and the necessary controls and thermal relief valve. When the temperature of the water in the storage tank is greater than that required in the process, the water is mixed with the cooler source water; when it is less, an auxiliary heater is used.

4.4.2 Case of Active Solar Drying: Sludge Drying

The handling and disposition of the hundreds of tons of sludge generated per day in wastewater treatment plants all over the world represent not only an enormous problem for human health and the environment but also economical and technological challenges. In addition to high water content, sludge is compressed of high concentrations of bacteria, viruses, and parasites (U.S. EPA 1989, 1999; Carrington 2001; Sahlströma et al. 2004); organic compounds (Abad et al. 2005; Mantis et

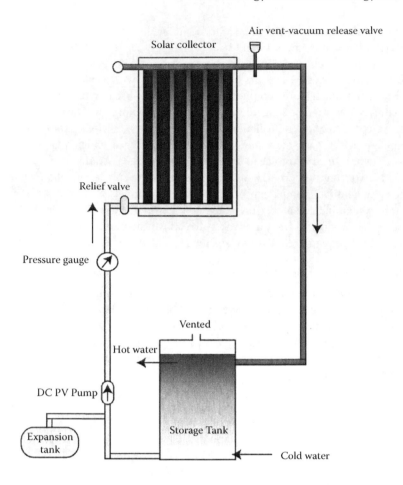

FIGURE 4.24   Pressurized antifreeze system for domestic water heating.

al. 2005); and heavy metals (Díaz Aguilar et al. 2001; Mantis et al. 2005; Bose and Bhattacharyya 2008). Several studies have proven the potential for use of sludge to improve soil fertility due to its high content of macronutrients for flora—particularly nitrogen and phosphorous and some organic substances that improve physicochemical characteristics of soil (Cooker 1983; Abad et al. 2005). However, its use can cause problems for human health and the environment.

In order to lower the costs of handling and disposition of the great sludge volumes, first, mechanical methods are applied to reduce 20–40% of the water; beyond this, water removal can only be achieved by thermal methods (Metcalf and Eddy 2003). This implies tremendous fuel consumption and greenhouse gas emissions. Luboschik (1999) reported a solar sludge dryer design able to evaporate 800 kg of water per square meter per year, with low cost of operation and maintenance and energy consumption as well. Bux et al. (2002) developed a solar dryer with continuous mixing that reduced from 3 to 93% of total solids within 64 days. The energy consumption was 78% less than that for a conventional system. Salihoglu et al. (2007) calculated a recovery time of 4 years for a system located in Bursa, Turkey.

The operation of a solar sludge dryer begins when solar radiation enters the drying chamber through a transparent cover. A great part of such energy is absorbed by the sludge. Due to the greenhouse effect, caused by selection of the construction materials and the hermeticity of the system,

TABLE 4.4
Temperature Ranges for Different Industrial Processes

Industry	Process	Temperature (°C)
Dairy	Pressurization	60–80
	Sterilization	100–120
	Drying	120–180
	Concentrates	60–80
	Boiler feed water	60–90
Tinned food	Sterilization	110–120
	Pasteurization	60–80
	Cooking	60–90
	Bleaching	60–90
Textile	Bleaching, dyeing	60–90
	Drying, degreasing	100–130
	Dyeing	70–90
	Fixing	160–180
	Pressing	80–100
Paper	Cooking, drying	60–80
	Boiler feed water	60–90
	Bleaching	130–150
Chemical	Soaps	200–260
	Synthetic rubber	150–200
	Processing heat	120–180
	Preheating water	60–90
Meat	Washing, sterilization	60–90
	Cooking	90–100
Beverages	Washing, sterilization	60–80
	Pasteurization	60–70
Flours and by-products	Sterilization	60–80
Timber by-products	Thermodiffusion beams	80–100
	Drying	60–100
	Preheating water	60–90
	Preparation pulp	120–170
Bricks and blocks	Curing	60–140
Plastics	Preparation	120–140
	Distillation	140–150
	Separation	200–220
	Extension	140–160
	Drying	180–200
	Blending	120–140

Source: Kalogirou, S. 2003. *Applied Energy* 76:337–361.

sludge and air temperatures tend to increase. Such an increment generates diffusion transport of water from the sludge surface to the air content within the chamber. The driving force for this process consists of the difference of vapor pressure between the sludge surface and the chamber. Vapor pressure in the air rises when water content in the air also increases. To accelerate water removal, vapor pressure equilibrium must be avoided and moisturized air must be removed. The farther the

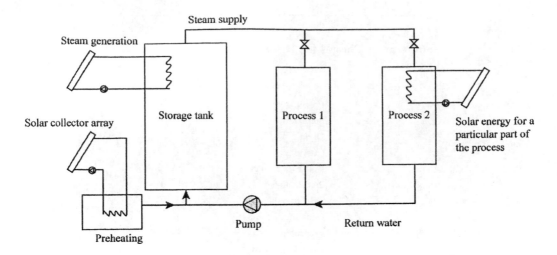

FIGURE 4.25   Hybrid solar/conventional system for industrial use.

saturation condition of water in air is, the greater is the potential for mass transport from sludge surface to air in the chamber. On the other hand, the hotter the system is, the greater is the vapor transport. To avoid stratification in temperature and humidity, the dryer should have a ventilation system. The moisturized air is removed via an extractor. When the air in the chamber has reached low water content, the system returns to a closed system with respect to mass. Because of the harmful characteristics of the material to be dried, the system must be controlled automatically. The automatic operation is controlled by temperature and humidity differences between internal and external conditions.

According to Cota and Ponce (2008), solar sludge drying represents an alternative and inexpensive method for disinfection of sludge with a high content of pathogenic microorganisms. In their studies, the overall effectiveness of the solar dryer was determined by assessing thermal and microbiological performance. Water content in sludge during the process was used as an indicator of thermal effectiveness; the results showed an exponential decay of water content that achieved up to a 99% reduction. Regarding microbiological removal effectiveness, there was a strong dependence between the number of bacteria present and the water content in the sludge. As a consequence, with the removal of 96% of water, it was verified that the elimination of fecal coliforms fell from 3.8×10^6 to 1.6 MPN (most probable number) per gram of dried sludge; for *Salmonella* spp., the reduction was from 1.5×10^{13} to 1.9×10^3 MPN per gram of dried sludge (see Table 4.5).

4.4.2.1   Solar Thermal Application: Solar Distillation

Distillation is a process that allows purifying some components of a solution based on differences of volatilities. In general terms, when solutes have much smaller volatilities than the solvent, distillation is carried out by evaporating the solvent in a particular region of the device and then condensing the vapor in a different region to obtain as pure a solvent as possible. When conventional energy supply is replaced by solar radiation, the process is called solar distillation. For the conventional process, the production rate remains constant under stable conditions of pressure, temperature, energy consumption, composition, and flow rate of the inlet stream. For the solar process, although predictable, it varies during the course of a day, showing a maximum during the hours with the highest irradiance. The variation is not only hourly but also daily over the whole year.

The most widely used application for solar water distillation has been for water purification. The advantage of solar over conventional systems in the purification of simple substances, such as brine or well waters, is that operation and maintenance are minimal because no moving parts are

TABLE 4.5
Experimental Findings during Active Solar Drying of Wastewater Sludge

Day	Residence time (h)	Accumulated global solar radiation (kWh/m²)	Water content in sludge inside dryer (%)	Water content in sludge outside dryer (%)	Fecal coliforms (NMP/g)	Eliminated fecal coliforms (%)	Salmonella (NMP/g)	Eliminated Salmonella (%)
06/302007	0	0.0	86.22	86.22	3.87E+06	0.0000	1.57E+13	0.0000
06/30/2007	7	4.5	82.00		1.34E+06	65.3747	6.03E+11	96.1651
07/01/2007	24	6.0	80.21		2.77E+06	28.4238	6.36E+08	99.9959
07/02/2007	50	14.7	77.10	77.10	1.34E+06	65.3747	4.29E+08	99.9972
07/03/2007	74	20.4	77.20	76.00	1.08E+06	72.0930	2.03E+08	99.9987
07/04/2007	98	25.9	64.10	76.40	5.78E+04	98.5078	8.08E+07	99.9994
07/09/2007	218	50.0	43.00	66.00	3.23E+04	99.1646	1.22E+05	99.9999
07/11/2007	269	59.0	6.67	55.00	1.60E+00	99.9999	1.92E+03	99.9999

involved. Also, there is no consumption of fossil fuels in solar distillation, leading to zero greenhouse-gas emissions. Most importantly, these types of systems can be installed in remote sites to satisfy freshwater needs of small communities that do not have conventional electric service.

Solar distillation represents one of the simplest yet most effective solar thermal technologies. Currently, several solar still prototypes exist; differences lie in their geometries and construction materials. All designs are distinguished by the same operation principles and three particular elements: solar collector, evaporator, and condenser. These elements can be identified in Figure 4.26.

The natural process of producing fresh water is copied by solar distillation. A solar still is an isolated container where the bottom is a blackened surfaced with high thermal absorbtivity and the cover is a transparent material, generally tempered glass. Purification is carried out when solar radiation crosses the glazing cover and reaches the solar collector, the black surface, and the majority of this energy is absorbed. During this process, the electromagnetic radiation is converted into heat, causing an increment in the temperature of the collector, which is then available to be transferred into the water. The heat is trapped within the system due to the greenhouse effect. The convective heat losses to the environment should be minimized by adequate insulation.

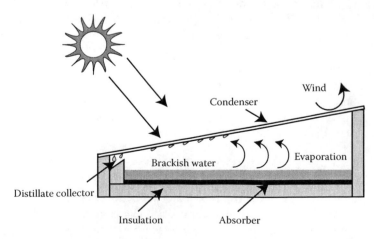

FIGURE 4.26 Basic operation of a solar still.

Because radiation is continuously entering the system, the temperature rises. As the water temperature rises, diffusion of water into the air starts to take place. Evaporation occurs; no boiling is involved because the maximum temperatures experienced are always below 80°C. These conditions favor the water not transporting components of higher solubilities or suspended solids. The glazing works as the condenser as well; because it is in direct contact with the environment, its temperature is lower than that of the collector and the water. The colder the surface is, the more easily condensation occurs. The glazing cover must be tilted for the distilled water to migrate toward a collection system. This process removes impurities such as salts and heavy metals, as well as destroys microbiological organisms. The most common solar still is a passive single basin solar distiller that needs only sunshine to operate.

The intensity of solar energy falling on the still is the single most important parameter affecting production. The daily distilled-water output (M_e [=] kg/m^2/day) is the amount of energy utilized in vaporizing water in the still (Q_e [=] J/m^2/day) over the latent heat of vaporization of water (L [=] J/ kg). Solar still efficiency (η) is amount of energy utilized in vaporizing water in the still over the amount of incident solar energy on the still (Q_t [=] J/m^2/day). These can be expressed as

$$M_e = \frac{Q_e}{L} \tag{4.68}$$

$$\eta = \frac{Q_e}{Q_t} \tag{4.69}$$

Typical efficiencies for single-basin solar stills approach 60%. Solar still production is a function of solar energy and ambient temperature. For instance, production rates for a square meter in sunny areas like the southwestern United States, Australia, or the Middle East can average about 6 l per day in the winter to over 15 l per day during the summer. Measured daily solar still performance for a year in liters per square meter of still per day is shown in Figure 4.27.

Distillation is the only stand-alone point-of-use (POU) technology with U.S. National Sanitation Foundation (NSF) international certification for arsenic removal, under Standard 62. Solar distillation removes all salts, as well as microbiological contaminants such as bacteria, parasites, and viruses. Table 4.6 shows results of tests conducted on single-basin solar stills by New Mexico State University and Sandia National Laboratories (SNL) (Zachritz, 2000; Zirzow, 1992). The results demonstrate that solar stills are highly effective in eliminating microbial contamination and salts. After the introduction of more than 10,000 viable bacteria per liter in the feed water, 4 and 25 viable cells per liter were found in the distillate. Introduction of a billion or more *Escherichia coli* viable cells each day over a period of 5 days did not change the number of viable cell numbers found in the distillate, nor was *E. coli* recovered in the distillate.

Table 4.7 presents the results obtained by SNL for single-basin solar stills. The SNL tests were conducted with supply water concentrations of 13 and 16% (standard saltwater). The stills effectively removed all salts. The total dissolved salts (TDS) concentration of the water fell from 36,000 and 48,000 TDS to less than 1 TDS.

4.4.3 Case of Passive Direct and Indirect Solar Distillation: Water Desalination

Passive solar distillation is a more attractive process for saline water desalination than other desalination methods. The process can be self-operating, of simple construction and relatively maintenance free, and avoid recurrent fuel expenditures. These advantages of simple passive solar stills, however, are offset by the low amounts of freshwater produced—approximately 2 L/m^2 for the

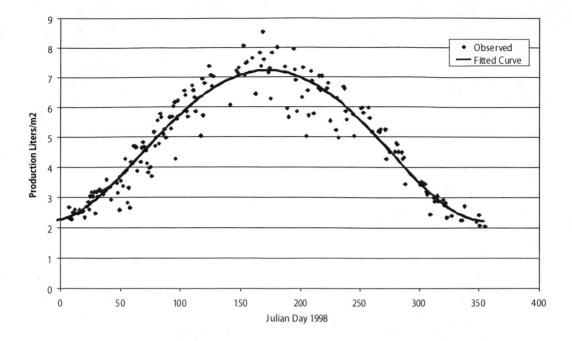

FIGURE 4.27 Measured basin solar still annual performance in Las Cruces, New Mexico, on a square-meter basis (Zachritz, 2000).

TABLE 4.6
Microbial Test Results for Solar Stills

Sample	Volume tested ml	Total organisms per liter
Supply	50	16,000
Distillate	1,000	4
E. coli seed	—	2,900,000,000
Distillate	750	11 (No *E. coli*)
E. coli seed	—	7,500,000,000
Distillate	1,000	18 (No *E. coli*)
Supply	10	24,000
Distillate	1,000	13
Supply	1	12,000
Distillate	1,000	6

Source: New Mexico State University, 1992.

simple basin type of solar still (Zaki, Radhwan, and Balbeid 1993)—and the need for regular flushing of accumulated salts (Malik et al. 1982). The performance of this type of solar still can be improved by integrating the unit with a solar collector. Studies by Zaki, Al-Turki, and Fattani (1992) show that yields can be increased by using a concentrating collector and report that, due to a smaller absorber surface area, thermal losses from the concentrating collector were significantly reduced and resulted in increased thermal efficiency and higher productivity.

TABLE 4.7
Sandia National Laboratories Still-Water Quality Test Results (Zirzow, SAND92-0100)

Sample type	13% Salinity feedwater	Distilled water (13% case)	16% Salinity feedwater	Distilled water (16% case)
Calcium (total)	340	1.5	371	<0.10
Iron (total)	0.27	<0.05	0.48	<0.06
Magnesium (total)	2.1	2.1	<0.005	<0.005
Manganese (total)	0.04	<0.02	0.07	<0.02
Ammonia as N	<0.1	0.1	<0.1	<0.1
Chloride	19,000	<1.0	25,000	2.6
Fixed solids	32,000	<1.0	41,000	31
Nitrate as NO_3	34	0.1	26	<0.1
Nitrate as NO_2	0.013	<0.01	0.02	<0.01
TDS	36,000	<1.0	48,000	<1.0
Volatiles and organics	4,200	<1.0	6,000	13

Passive collector systems remove the need for two separate components and adjoining pipe work by integrating the collector/concentrator with the solar still, leading to lower system costs and reduced thermal distribution losses. In a study investigating possible rural applications for the CPC, Norton et al. (1997) suggested the incorporation of a basin type of still with an inverted absorber line-axis asymmetric CPC. The inverted absorber configuration can achieve higher temperatures by minimizing thermal losses by convection suppression. In the study of an inverted absorber solar distillation unit conducted by Suneja, Tiwari, and Rai (1997), a double effect still was also used to improve output. Latent heat of vaporization in the lower vessel is reused to heat the water mass in the upper vessel. This also enhances the condensation process in the lower vessel through lower surface temperatures. Sol Aqua has built many stills for household and village level applications around the world (Figure 4.28; Sol Aqua, 2009)

4.4.4   CASE OF PASSIVE SOLAR INDIRECT DRYING: FOOD DRYING

Drying is the oldest method of food preservation and solar food dryers are an appropriate food preservation technology for a sustainable world. By reducing the moisture content of food to between 10 and 20%, bacteria, yeast, mold, and enzymes are all prevented from spoiling it. The flavor and most of the nutritional value is preserved and concentrated. Vegetables, fruits, meat, fish, and herbs can all be dried and preserved for several years in many cases. Solar dryers have the same basic components as do all low-temperature solar thermal energy conversion systems.

Three major factors affect food drying: temperature, humidity, and air flow. Increasing the vent area by opening vent covers will decrease the temperature and increase the air flow without having a great effect on the relative humidity of the entering air. In general, more air flow is desired in the early stages of drying to remove free water or water around the cells and on the surface. Reducing the vent area by partially closing the vent covers will increase the temperature and decrease the relative humidity of the entering air and the air flow. This would be the preferred setup during the later stages of drying, when the bound water needs to be driven out of the cells and to the surface.

4.4.5   CASE OF AN ACTIVE SOLAR CHEMICAL PROCESS: WATER DETOXIFICATION

The presence of corrosive substances, solvents, organic compounds, metals, etc. in surface waters is a serious problem worldwide (Halmann et al. 1992; Oeberg et al. 1994). Current methods for removal of polluting agents in water are in many cases expensive, inefficient, or attack just one side

FIGURE 4.28 SolAqua solar still village array under test at Sandia National Laboratories.

of the problem. For example, disinfection methods eliminate pathogenic organisms, but they do not degrade organic polluting agents (Larson 1990; Magrini and Webb 1990), and chlorination processes produce carcinogenic by-products (Glaze 1986; Shukairy and Summers 1992). An alternative method to destroy traces of priority organic pollutants in wastewater or underground waters is photocatalysis. Unlike other technologies, photocatalysis allows total mineralization of the organic compounds without creating intermediate toxic compounds.

The operation principle of the photocatalytic system is based on the collection of ultraviolet sunlight radiation by a parabolic trough concentrator, which focuses the Sun's radiation on a receiver transparent tube located along the focal line of the trough, acting as an axial chemical reactor. Along the photoreactor, the polluted water flows together with a catalyst—usually titanium dioxide (TiO_2). Thus, solar degradation takes place when the high flux of ultraviolet energy acts on the active sites in the surface of the catalyst. This produces very strong oxidant free radicals, which turn the organic molecules into water, carbon dioxide, and diluted acids through consecutive oxide-reduction reactions.

Research conducted by Jimenez et al. (2000) to study the solar photocatalytic phenomena showed how to optimize complete degradation of organic compounds. For this research, sodium dodecylbenzene sulfonate (DBSNa) was selected as the polluting agent because of its widespread use in the manufacturing of products such as toothpaste, bath soaps, shampoos, etc. The advantage of this synthetic agent compared to biodegradable soap is that it produces foam even with hard waters. This feature has caused extensive use of DBSNa, and therefore high concentrations of this surfactant in industrial and city effluents are common.

The experimental setup consisted of a parabolic trough reflector, the photoreactor, and the fluid circulation system. Table 4.8 lists the characteristics of the solar concentrator and photoreactor design.

To optimize the TiO_2 concentration during DBSNa photocatalytic degradation, catalyst concentrations were varied from 0.05 to 1.5 percentage weight (wt%) for a particular concentration of DBSNa. Also, the role of an oxidizing agent in the catalytic reaction was determined; H_2O_2 concentration was varied from 0 to 15,000 ppm at 0.2 weight percentage of TiO_2. In this study, a total of 37 tests were conducted and, in all of them, the initial concentration of DBSNa was 37 mg/L. The final concentrations of the anionic surfactant at different resident times were determined by the methylene blue active substances method. During these tests, the registered solar radiation was always between 850 and 945 W/m^2, and with the 41 suns concentrated, guaranteeing a minimum UV photonic density of 1.253×10^{22} photons/m^2-s.

TABLE 4.8

Components of the Solar Collector and Photoreactor for the Study of Photocatalysis

Component	Characteristics
Parabolic trough	Mechanical system to follow the Sun's movement
	Geometric concentration: 41 suns
	Aperture: 106 cm
	Focal length: 26.6 cm
	Aperture angle: 903
	Frontal length: 172 cm
	Reflective surface: aluminum
	Average reflectance in the UV region: 75%
Photoreactor	Pyrex reactor tube
	Diameter: 2.54 cm
	Free length: 183 cm
	Average transmittance in the UV region: 85%

Figure 4.29 shows results of the DBSNa degradation when the device was operated in (a) continuous recirculation and (b) stagnant operating modes. As shown in curve (a), the DBSNa concentration diminishes rapidly from 33 mg/L, the initial concentration, to 5 mg/L in the first 30-min exposure. But above an operation time of 30 min, DBSNa decreases slowly, and after 120 min of exposure, a 1.0 mg/L concentration of DBSNa is still observed, indicating an incomplete degradation. In the continuous operation mode, DBSNa concentration rapidly decreases in the starting minutes, and after 30 min a concentration of only 3.2 mg/L was measured. The degradation of DBSNa after a 30-min exposure continues, although not so rapidly as in the initial phase; after 70 min, a zero concentration of DBSNa was registered, indicating total degradation.

Comparing curves (a) and (b), it is found that DBSNa degradation is thoroughly achieved in the continuous recirculation mode rather than in the stagnant operating mode. This can be explained by the facts that better particle distribution, rather than particle precipitation, and greater OH production are possible under the continuous recirculation mode. Also, it must be remembered that the continuous recirculation mode involves stirring of the aqueous solution and its exposure to the open air.

Tests were run to determine the role of the catalyst without oxidant agent during the degradation process. Figure 4.30 plots percentage degradation curves of the DBSNa as function of the TiO_2 concentration after (a) 10-, (b) 20-, (c) 30-, (d) 40-, (e) 50-, and (f) 60-min exposure time. In curve (a), we observe a rapid degradation increase of 26% using only 0.05 wt% of TiO_2. Degradation reaches a sharp maximum of 69% at 0.3 wt%. At 0.5 wt%, degradation accounts for only 27%, resembling the value obtained at 0.05 wt%. Finally, above 0.5 wt%, the curve slope declines very slowly, reaching 13% at 1.5 wt%. For 0.3 wt% TiO_2 concentrations, the higher values were obtained regardless of exposure time: 71% (20 min), 79% (30 min), 93% (40 min), 93% (50 min), and 94% (60 min). Above 0.4 wt% of TiO_2, the degradation rate abruptly drops near the value achieved at 0.05 wt%. Curves (c)–(f) seem to be nearly symmetric and centered at 0.3 wt%. Above 0.5 wt%, all degradation curves show a very slow descent, arriving at the final values of 17, 19, 22, 32, 54, and 59% at 1.5 wt% (curves a–f, respectively).

It is important to note that curve (f) has a flatter behavior at its maximum than the other curves. This fact indicates that, for larger times, there is a range of TiO_2 concentrations that maximize the DBSNa degradation rather than a single value (0.3 wt%), as shown at the beginning of the process (curves a and b). Figure 4.30 shows clearly that a TiO_2 concentration between 0.2 and 0.4 wt% optimizes the photocatalytic reaction. Above 0.4 wt%, the catalyst itself could interfere with the

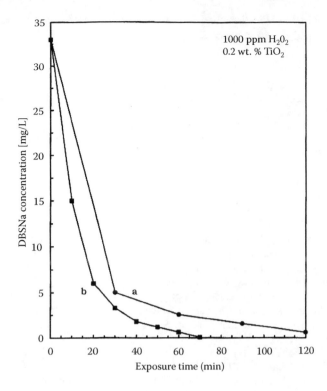

FIGURE 4.29 Photocatalytic degradation of DBSNa as function of the exposure time in (a) stagnant-operating and (b) continuous-recirculation modes.

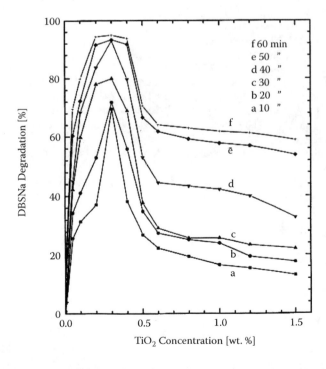

FIGURE 4.30 Photocatalytic degradation of DBSNa as function of the TiO_2 concentration without oxidant agent.

diffusion of the reactants and products and retard the reaction; below that value, there may not be enough catalyst for the degradation process.

REFERENCES

ANSI/ASHRAE Standard 93-2003. 2003. Methods of testing to determine thermal performance of solar collectors.

Cota-Espericueta, A. D., and C. Ponce-Corral. 2008. Removal of pathogenic bacteria in wastewater sludge during solar drying. *Review of International Contaminant Ambient* 24 (4): 161–170. ISSN-0188 4999.

Egbo, G., I. S. Sintali, and H. Dandakouta. 2008. Analysis of rim angle effect on the geometric dimensions of solar parabolic-trough collector in Bauchi, Nigeria. *International Journal of Pure and Applied Sciences* 2 (3): 11–20.

European Standard EN12975-2:2001. 2001. Thermal solar systems and components—Solar collectors—Part 2: Test methods.

Harris J. A., and T. G. Lenz. 1985. Thermal performance of solar concentrator/cavity receiver systems. *Solar Energy* 34 (2): 135–142.

ISO Standard 9806-1:1994(E). 1994. Test methods for solar collectors—Part 1: Thermal performance of glazed liquid heating collectors including pressure drop.

Jimenez, A. E., C. A. Estrada, A. D. Cota, and A. Roman. 2000. Photocatalytic degradation of DBSNa using solar energy. *Solar Energy Materials & Solar Cells* 60:85–95.

Kalogirou, S. 2003. The potential of solar industrial process heat applications. *Applied Energy* 76:337–361.

Larson, R. A. 1990. In *Biohazards of drinking water treatment,* 2nd ed., 3. Boca Raton, FL: Lewis Publishers Inc.

Malik, M. A. S., G. N. Tiwari, A. Kumar, and M. S. Sohda. 1982. *Solar distillation.* Oxford, England: Pergamon Press.

McCarthy, C. M. 1992. Solar still analysis letter. Department of Biology, New Mexico State University, Las Cruces, New Mexico, August 25, 1992.

Norton, B., P. C. Eames, Y. P. Yadav, and P. W. Griffith. 1997. Inverted absorber solar concentrators for rural applications. *Ambient Energy* 18 (3): 115–120.

Rabl, A., J. O'Gallagher, and R. Winston. 1980. Design and test of nonevacuated solar collectors with compound parabolic concentrators. *Solar Energy* 25 (4): 335–351.

Rojas, D., J. Beermann, S. A. Klein, and D. T. Reindl. 2008. Thermal performance testing of flat-plate collectors. *Solar Energy* 82:746–757.

Sol Aqua, Solar Still Basics 2009. www.solaqua.com

Suneja, S., G. N. Tiwari, and S. N. Rai. 1997. Parametric study of an inverted absorber double-effect solar distillation system. *Desalination* 109:177–186.

Zachritz, W. H., L. Mimbela, R. Polka, K. Stevens, L. Cisneros, H. Floyd, and A. Hanson. 2000. Application of solar stills for drinking water in border Colonias. Southwest Technology Development Institute, New Mexico State University, EDA project no. 08-39-03086, Austin, Texas, April 2000.

Zaki, G. M., A. Al-Turki, and M. Fattani. 1992. Experimental investigation on concentrator assisted solar stills. *Solar Energy* 11:193–199.

Zaki, G. M., A. M. Radhwan, and A. O. Balbeid., 1993. Analysis of assisted coupled solar stills. *Solar Energy* 51 (4): 277–288.

Zirzow, J. A. 1992. Solar stills complement photovoltaic systems. Sandia National Laboratories, SAND92-0100, Albuquerque, New Mexico, February 1992.

5 Photovoltaic Cells

Contributing Author Jeannette M. Moore

5.1 INTRODUCTION

Modern society's reliance on electrical power is so great that it is considered a basic need. It is usually supplied by the electrical grid; however, in some places that have no access to the electric grid, such as outer space, remote rural areas, or developing nations, power supplies are problematic and solar offers a least-cost power option. Photovoltaics provide practical solutions to many power supply problems in both space and remote terrestrial applications. In addition to larger power applications, portable electronic devices may charge their batteries using solar cells or get their power directly from solar cells.

Electricity can be produced from sunlight through a process called the PV effect, where "photo" refers to light and "voltaic" to voltage. The term describes a process that produces direct electrical current from the radiant energy of the Sun. The PV effect can take place in solid, liquid, or gaseous material; however, it is in solids, especially semiconductor materials, that acceptable conversion efficiencies have been found. Solar cells are made from a variety of semiconductor materials and coated with special additives. The most widely used material for the various types of fabrication is crystalline silicon, representing over 90% of global commercial PV module production in its various forms.

A typical silicon cell, with a diameter of 4 in., can produce more than 1 W of direct current (DC) electrical power in full sun. Individual solar cells can be connected in series and parallel to obtain desired voltages and currents. These groups of cells are packaged into standard modules that protect the cells from the environment while providing useful voltages and currents. PV modules are extremely reliable because they are solid state and have no moving parts. Silicon PV cells manufactured today can provide over 40 years of useful service life.

PV devices—or solar cells—are made from semiconductor materials. Semiconductor materials are those elements or compounds that have conductivity intermediate to that of metals or insulators.

5.2 CRYSTAL STRUCTURE

Silicon and many semiconductor materials on a microscopic scale are crystalline structures arranged in an orderly fashion from the atoms from which they are composed. The smallest subsection of this orderly arrangement, whereby the entire structure may be reproduced without voids or overlaps, is called the primitive cell. Often the primitive cells have awkward shapes, so we use a simpler unit cell, typically defined by three orthogonal axes such as x, y, and z, with unit vectors located along each axis. The length of the edge of the unit cell is called the lattice constant. The orientation of planes within the crystal can be expressed by using Miller indices, as shown in Figure 5.1.

Example 5.1

Find the Miller index of the illustrated crystal plane in Figure 5.1.

Solution:

Intercepts are 1, 3, and 2. Take inverses 1, 1/3, and 1/2. Smallest integrals with the same ratios are 6, 2, and 3 (6/6, 2/6, 3/6). Therefore, this plane is expressed as a Miller index of (6 2 3). Negative intercepts are indicated by a bar over the corresponding index.

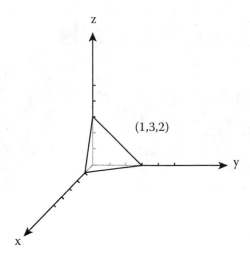

FIGURE 5.1   Illustration of Miller indices example.

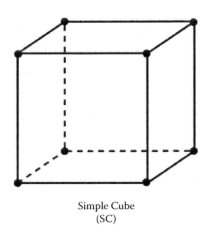

Simple Cube
(SC)

FIGURE 5.2   Simple cube crystal lattice.

To further complicate matters there are different types of crystal lattices, the most common of which are simple cubic, face-centered cubic, and body-centered cubic. These are shown in Figures 5.2, 5.3, and 5.4, respectively. The atomic arrangement of many of the semiconductors used in solar cell manufacturing is called a diamond lattice or zinc blend lattice structure. This structure consists of two interpenetrating face-centered cubic structures.

The three types of material structures commonly used to produce Photovoltaics are (1) amorphous, with no order or periodicity within the compound; (2) polycrystalline, with local order, visible grain boundaries, and bluish color; and (3) single (or mono) crystalline, which is typified by long-range order and periodicity and is almost black with no visible grain boundaries. As it turns out, the crystalline structure of a semiconductor is important because when small concentrations of impurities (other elements such as phosphorus or boron) are introduced into the structure, the conductive properties of the material can radically change. A discussion of how this conduction occurs includes the concept of energy bands within the crystalline structure.

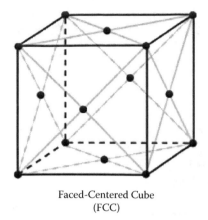

Faced-Centered Cube
(FCC)

FIGURE 5.3   Face-centered cube crystal lattice.

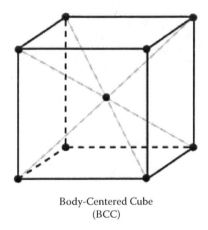

Body-Centered Cube
(BCC)

FIGURE 5.4   Body-centered cube crystal lattice.

5.3   CELL PHYSICS

Particles come in two types, fermions and bosons, which have very different properties in terms of number in an energy state. No two fermions can be in the same energy state when they are close together in a material, while bosons can be in the same energy state. Electrons are fermions and photons are bosons. Most properties of materials (mechanical, electrical, thermal, chemical, biological) can be explained by their electron structure and, in general, by outer electron structure. Electric fields, E, are created by charged particles, and if a charged particle is placed in an external electric field there is a force on it, which will make it move. Then, energy or work is available. The electric potential, V, is the energy/charge. Electrical terms include:

- charge, Q or q, Coulombs (C) 1 C is a very large number of electrons, 1 e = $1.6 * 10^{-19}$ C, positive or negative
- electric potential (V) = energy/charge; V = E/Q, Volt (V) = Joule (J)/Coulomb; energy = V * Q
- current (I) = dq/dt, number of charges moving past a point in 1 s, Ampere (A) = Coulomb/second
- resistance (R) = V/I Ohm = Volt/Ampere

- power (P) = V * I Watt = Volt * Ampere
- electron Volt (eV) = the energy one electron would acquire from moving through a potential of 1 V; a unit of energy defined as 1 eV = $1.6 * 10^{-19}$ J

5.4 ENERGY BANDS

To introduce the idea of energy bands (Figures 5.5 and 5.6), imagine that there are a large number of identical atoms located far enough apart from each other that little or no interaction takes place between them. As they are pushed closer in a uniform fashion, electrical interactions occur between them. Because of these electrical interactions and because of the Pauli exclusion principle (no two electrons can occupy the same quantum state), the quantum mechanical wave functions begin to distort—especially those of the outer (or valence) electrons. The valence electron wave functions extend over more and more atoms, thus changing the energy level of the substance from one that was sharp and distinctive to a collection of energy levels or bands. A wave function is a description

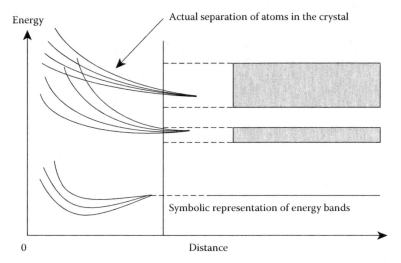

FIGURE 5.5   Energy bands in a solid.

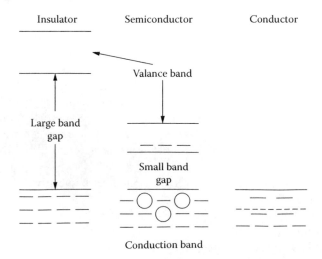

FIGURE 5.6   Electron energy states form bands in materials.

of an electron using kinematic rather than spatial point descriptors. It is similar to that used to describe sound and electromagnetic waves. Whereas some material medium is needed in order to propagate, the wave function describes the particle, but the function itself cannot be defined in terms of anything material. It can only describe how it is related to physically observable effects.

The nature of the energy bands determines whether the material is a conductor, an insulator, or a semiconductor. An insulator at absolute zero has a completely full valence band and a completely empty conduction band (the next higher band); a semiconductor at absolute zero also has a full valence band and empty conduction band, but the difference is that the gap between the conduction and the valence bands is much smaller in the semiconductor. A band gap is the gap between a valence energy band and a conduction energy band. The smaller gap between bands means that application of thermal energy can cause an electron to "jump" from the valence band to the conduction band. As temperature increases, the conduction band rapidly fills and the conductivity of the material also increases.

In a conductor, electrons are in the conduction band even at absolute zero. The conduction band is thus named because it is usually partially occupied with electrons; this means that it is conducive to electron mobility and therefore conduction of electricity. The valence band can be defined as the band that contains electrons at lower energy levels. Between the valence and the conduction bands is a band called the forbidden energy gap or, simply, forbidden gap. It is a range of energy levels that an electron is not allowed to occupy based on quantum mechanics.

Energy bands can explain the three main types of material: Conductors have free electrons that can move, semiconductors have a few electrons that can move, and insulators have no free electrons. In a conductor, the conduction band is partially filled with electrons, so energy states are available for free electrons. Metals are good conductors. In a semiconductor at room temperature, some electrons have enough energy to get into the valance band, which leaves a hole in the conduction band. The band gap is small. Another way the electron can obtain that energy is by the absorption of light: a photon. In an insulator such as glass, there are not any free electrons because all electron states are filled and the band-gap energy is large.

5.5 MORE ABOUT ELECTRONS AND THEIR ENERGY

If we want to know the number of quantum states that have energy in a given range, we use the density of states equation:

$$g(E) = \frac{4\pi(2m)^{\frac{3}{2}}}{h^3}\sqrt{E} \qquad (5.1)$$

where

m = free electron rest mass = 9.11×10^{-31} kg
h = Planck's constant = 6.625×10^{-34} J-s = 4.135×10^{-15} eV-s
E = electronic charge = 1.6×10^{-19} C or J/eV

Note that this equation is for use in the crystal space volume of a^3 and is used to obtain the density of quantum states per unit volume of crystal.

To determine how the electrons are distributed among the quantum states at any given temperature, use the Fermi–Dirac distribution function:

$$f(E) = \frac{1}{1 + \exp\left(\dfrac{E - E_f}{kT}\right)} \tag{5.2}$$

where

k = Boltzmann's constant = 1.38×10^{-23} J/K = 8.62×10^{-5} eV/K
T = temperature in Kelvin scale
E_F = Fermi energy
E = energy level of a particular state

Note that the Fermi energy determines the statistical distribution of electrons. At energies above E_F, the probability of a state being occupied by an electron can be significantly less than unity. Note also that the preceding distribution function has a strong correlation to temperature.

Example 5.2

Calculate the probability that an energy level $5kT$ above the Fermi energy is occupied by an electron.

Solution:

$$f(E) = \frac{1}{1 + \exp\left(\dfrac{5kT}{kT}\right)} = 0.006693 \approx 0.7\%$$

5.6   ELECTRONS AND HOLES

Conduction in a semiconductor is due not only to electrons in the conduction band, but also to the movement of holes in the valence band. When an electron gains the energy needed to overcome its covalent bond and "jumps" from the valence to the conduction band, it leaves behind a hole in the valence band. Often the analogy of a two-level parking garage is used to illustrate the movement of electrons and holes in a semiconductor. If the bottom floor of the parking structure is full of cars, there is no room for movement until one or more cars move to the next level. Instead of thinking of the cars moving forward on the bottom level, think of the space as moving. This is similar to the lack of movement in a filled valence band until an electron moves to the conduction band.

Another analogy for hole movement is that of a soap bubble in liquid. Holes are vacancies that behave like positively charged particles in the valence band even though the charged particles are the electrons. When a hole in the valence band is created by the electron movement from the valence to the conduction band, the result is a combination of electron and hole called an electron–hole pair (EHP). Electrons are generally thought of as moving in one direction in the conduction band and holes as moving in the opposite direction in the valence band.

5.7   DIRECT AND INDIRECT BAND-GAP MATERIALS

Semiconductors can be either direct band gap or indirect band gap. In order to explain this phenomenon an E versus k (or k-space) diagram is sometimes used. A k-space diagram is the plot of electron energy in a crystal versus k, where k is a constant that takes into account the momentum of the crystal motion. An E versus k diagram of gallium arsenide (GaAs), which is considered a direct band-gap semiconductor material, is shown in Figure 5.7. Note that the minimum conduction

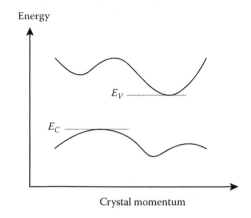

FIGURE 5.7   *k*-Space diagram of a direct band gap.

band energy and maximum valence band energy occur at the same *k* value, thus promoting a more efficient absorption of the photon. A semiconductor whose maximum valence band energy and minimum conduction band energy do not occur at the same *k* value is called an indirect band-gap material (Figure 5.8). Silicon is one such indirect band-gap material.

5.8   DOPING

Doping is the purposeful introduction of impurities into a semiconductor material in order to change its electronic properties by controlling the number of electrons in the conduction band. Impurity atoms can be introduced into a material in two ways. They may be squeezed into the interstitial spaces between the atoms of the host crystal (called interstitial impurities) or they may substitute for an atom of the host crystal while maintaining the regular crystalline atomic structure (substitutional impurities). The diagram in Figure 5.9 uses a bond model with a group V atom replacing a silicon (Si) atom to illustrate this concept. Note that a much smaller amount of energy is required to release this electron as compared to the energy needed to release one in a covalent bond.

The energy level of this fifth electron corresponds to an isolated energy level lying in the forbidden gap region. This level can be called a donor level and the impurity atom responsible is called a donor. A concentration of donors can increase the conductivity so drastically that conduction due to

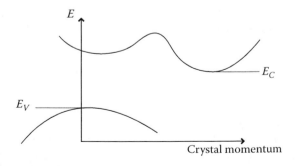

FIGURE 5.8   Indirect bandgap illustration.

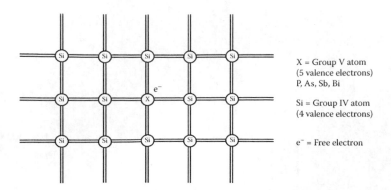

FIGURE 5.9   Silicon lattice with substitutional doping of group V, such as a larger phosphorous atom.

impurities becomes the dominant conductance mechanism. In this case, the conductivity is due almost entirely to negative charge (electron) motion and the material is called an n-type semiconductor.

Similarly, when a group III impurity (boron) is introduced, there are only three valence electrons and the material has an affinity to attract electrons from the material, thus leaving a hole. Hole movements collectively create an energy level in the forbidden gap close to the valence band. This level can be called an acceptor level and the impurity atom responsible is called an acceptor. The material is called a p-type positive semiconductor with p-type impurities.

5.9   TRANSPORT

A charge carrier is the electron and/or hole that moves inside the semiconductor and gives rise to electrical currents. The net flow of the carriers in a semiconductor will generate currents. The process by which these charged particles move is called transport. There are two basic transport mechanisms: drift and diffusion. Drift is the movement of charge due to electric fields and the total drift current density is the sum of the individual electron and hole drift current densities. Diffusion is the flow of charge due to density gradients and is the process where particles flow from a region of high concentration toward a region of low concentration.

5.10   GENERATION AND RECOMBINATION

The interaction between the Sun or source of light and a PV device can be complex. The transmitted energy of the photons in the light source must be greater than the energy needed by the electron to overcome the forbidden gap between the valence and the conduction bands and to generate EHPs. When the EHPs are created, the concentration of carriers in illuminated material is in excess of the values in the dark. If the light is suddenly removed, the concentrations decay back to their equilibrium values. The process by which this decay occurs is called recombination.

5.11   THE P–N JUNCTION

Just as there are diverse applications for solar cells, there are diverse methods for manufacturing them. The technology used in manufacturing and testing space solar cells is more advanced than that used for terrestrial applications, but the same basic theory of operation applies for all types.

The most common solar cells are basically large p–n (thought of as positive–negative) junction diodes that use light energy (photons) to produce DC electricity. No voltage is applied across the junction; rather, a current is produced in the connected load when the cells are illuminated. A diode is an electronic device that permits unidirectional current. The solar cell is fabricated by having n- and p-layers, which make up a junction (Figures 5.10 and 5.11). The p–n junction is formed by combining doped semiconductor materials such as Si or GaAs.

The energy conversion in a solar cell consists of two essential steps. First, absorption of light of an appropriate wavelength generates an electron–hole pair. Light absorption refers to the annihilation or absorption of photons by the excitation of an electron from the valence band up to the conduction band. Electrons flow readily through the n-type material and holes flow readily through p-type material. The light-generated electron and hole are separated by the electronic structure of the device: electrons to the negative terminal and holes to the positive terminal. The electrical power is collected by metal (ohmic) contacts on the front and back of the cell. Typically, the back contact is solid metal and the front is a metal grid (Figure 5.12). The presence of electrons and holes creates net negative and positive charges, which in turn induce an electric field in the region near the metallurgical junction. The electric field "sweeps out" the electrons and holes to create what is called the depletion region. These terms and methodologies are true for most p–n junction diodes.

To improve efficiency, the materials should be modified to have band-gap energies for photons in the visible range. The spectrum from infrared to ultraviolet covers a range from 0.5 to about 2.9 eV. For example, red light has an energy of about 1.7 eV, and blue light has an energy of about 2.7 eV. Effective PV semiconductors have band-gap energies ranging from 1.0 to 1.6 eV. Band gaps in semiconductors are in the 1 to 3 eV range. For example, crystalline silicon's band-gap energy is 1.1 eV.

Today terrestrial PV devices convert 7–22% of light energy into electric energy. About 55% of the energy of sunlight cannot be used by most PV cells because this energy is either below the band gap or carries excess energy. Another way to improve efficiency is using multiple p–n junctions (tandem cells), which have efficiencies as high as ≥35%. Cell efficiencies for silicon decrease as temperatures

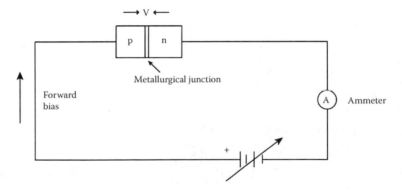

FIGURE 5.10 p–n junction in a simple circuit.

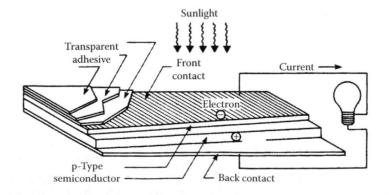

FIGURE 5.11 PV cell formed by n- and p-layers.

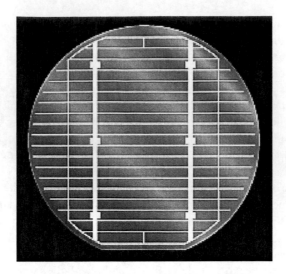

FIGURE 5.12   Example of a PV silicon cell. The horizontal and vertical lines are thin metal conductors.

increase (because voltage drops), and higher temperatures also threaten long-term stability and life. Therefore, PV cells are best operated when they are cool, especially for concentrating collectors.

5.12 SOLAR CELL EQUATIONS

In order to derive the ideal current-voltage characteristics of a p–n junction diode when illuminated by light, mathematical results from the ideal diode equation are combined with the illuminated characteristics of the solar cell. The ideal diode law is expressed as the following equation:

$$I = I_o(\exp(\frac{qV}{k_B T}) - 1) \tag{5.3}$$

where I_o, the saturation current density, is given by

$$I_o = A(\frac{qD_e n_i^2}{L_e N_A} + \frac{qD_h n_i^2}{L_h N_D}) \tag{5.4}$$

where

 N_A = the acceptor concentration in the p-region
 N_D = the donor concentration in the n-region
 n_i = intrinsic carrier concentration
 D_h = diffusion coefficient of the holes
 D_e = diffusion coefficient of the electrons
 L_e = diffusion length (how far into the material before electrons diffuse) of the electrons
 L_h = diffusion length of the holes
 A = the cross-sectional area of the diode
 q = the magnitude of the electronic charge (same as in Equation 5.2)
 k_B = Boltzmann's constant (k_B) (can also simply use "k")

The assumption is made for solar cells that the generation rate of electron–hole pairs by the light (g_{op}) is constant throughout the device. The typical diode equations for determining the excess minority carrier concentrations include an additional term relating g_{op} as follows:

$$\frac{d^2(\Delta p)}{dx^2} = \frac{\Delta p}{L_h^2} - \frac{g_{op}}{D_h}$$

(5.5)

Because g_{op} is constant, $\dfrac{g_{op}}{D_h}$ is also constant. The general solution to this differential equation is

$$\Delta p = g_{op}\tau_h + C\exp(\frac{x}{L_h}) + D\exp(\frac{-x}{L_h})$$

(5.6)

where τ_h is the minority carrier lifetime. The particular solution to the equation is

$$p_n(x) = p_{no} + g_{op}\tau_h + [p_{no}\exp(\frac{qV}{k_BT}) - 1 - g_{op}\tau_h]\exp\frac{-x}{L_h}$$

(5.7)

where q is the magnitude of the electronic charge, k_B is Boltzmann's constant, and x is distance. A similar expression defines $n_p(x)$. The corresponding current density inclusive of g_{op} is given by the expression

$$J_h(x) = \frac{qD_hp_{no}}{L_h}(\exp(\frac{qV}{k_BT}) - 1)\exp(\frac{-x}{L_h}) - qg_{op}L_h\exp(\frac{-x}{L_h})$$

(5.8)

Again, a similar expression defines J_e. Adding J_h and J_e gives the following result:

$$I = I_o(\exp(\frac{qV}{k_BT}) - 1) - I_L$$

(5.9)

where I_o is the same as in Equation 5.4 and I_L is defined by the expression

$$I_L = qAg_{op}(L_e + W + L_h)$$

(5.10)

W is the width of the depletion region and is found by standard diode equations. I_L is related to the photon flux incident on the cell and is dependent on the wavelength of the incident light. The illuminated characteristics of the current are the same as the regular diode (or dark) characteristics, but shifted down by the current I_L.

A *majority carrier* is the charge carrier that determines current. Majority carriers in a p-type material are holes and therefore its minority carriers are electrons. Majority carriers in an n-type material are electrons and its minority carriers are holes. An *intrinsic carrier* is a semiconductor with valence band holes and conduction band electrons present in equal numbers

5.13   CHARACTERIZATION

Typically, three parameters are used to characterize solar cell output: short circuit current (I_{sc}), open circuit voltage (V_{oc}), and fill factor (FF). In order to find open circuit voltage, I in Equation 5.9 is set to 0 (meaning short circuit current) to give the ideal value as follows:

$$V_{oc} = \frac{k_BT}{q}\ln(\frac{I_L}{I_0} + 1)$$

(5.11)

This equation illustrates the interdependence of I_{sc} and V_{oc}. The fill factor uses I_{sc} and V_{oc} as well as maximum power points of both current and voltage (I_{mp} and V_{mp}) in the following expression:

$$FF = \frac{V_{mp}I_{mp}}{V_{oc}I_{sc}}$$

(5.12)

An empirical expression often used is the following:

$$FF = \frac{\upsilon_{oc} - \ln(\upsilon_{oc} + 0.72)}{\upsilon_{oc} + 1}$$

(5.13)

where υ_{oc} is a normalized voltage defined as

$$\upsilon_{oc} = \frac{V_{oc}}{(\frac{K_B T}{q})}$$

(5.14)

The energy conversion efficiency for solar cells is calculated using the following equation:

$$\eta = \frac{V_{mp}I_{mp}FF}{P_{in}}$$

(5.15)

where P_{in} is the total power in the light incident on the cell. This equation demonstrates the dependency of efficiency calculations on the fill factor. See Figure 5.13 for a graphical illustration of I_{sc}, V_{oc}, I_{mp}, and V_{mp}. Illustrated also is the difference between light and dark current.

Fill factor is defined by dividing the square area on the graph in Figure 5.13 by a larger outer square area formed by the intersection of I_{sc} and V_{oc}. Note from the figure that results from illumination are in a region in the fourth quadrant, indicating where electrical power can be extracted from the device. As Figure 5.13 illustrates, when no light is incident on the cell, a solar cell is equivalent to a diode or semiconductor current rectifier. Note that the concentration of carriers in

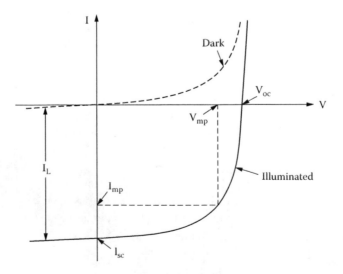

FIGURE 5.13   IV curve of a solar cells.

an illuminated cell exceeds that of the values with no illumination. When no light is incident on the cell, the carriers fall back to their equilibrium values via recombination. Standard methods of depicting the I-V curve for a solar cell are to include only the fourth quadrant and "flip" into the first quadrant as depicted in Figure 5.14. Note that the shaded rectangle illustrates the maximum power rectangle. The fill factor is calculated as a ratio of the maximum power rectangle over the rectangle formed by V_{oc} and I_{sc}.

5.14 EFFICIENCY

Values that affect the efficiency of solar cells are the band gap of the semiconductor, operating temperature, incident light, type and purity of the material, and parasitic resistances. See Figure 5.15 for some recent efficiency claims.

Construction of the current-voltage (I-V) curve is crucial to estimating efficiencies in the solar cell. As shown in the previous equations, I_o needs to be as small as possible for maximum V_{oc}. The generally accepted estimate of the minimum value of the saturation current density is

$$I_o = 1.5 \times 10^5 \exp(-\frac{E_g}{k_B T}) \tag{5.16}$$

As E_g increases, I_o decreases and V_{oc} gets larger. Put another way, the maximum value of V_{oc} increases with increasing band gap. Subsequently, V_{oc} has an effect on V_{mp}, which in turn affects the efficiency. This is one of the reasons why GaAs cells with a band gap of 1.4 are more efficient than Si cells with a band gap of 1.12. Another reason that GaAs cells are more efficient is due to their characteristics related to light absorption. In direct band-gap materials such as GaAs, the light is absorbed quickly and the risk of its passing straight out the back is reduced. Most terrestrial solar cells are constructed of silicon, an indirect band-gap material, which means that the emission or absorption of a photon is also required for photon energies near that of the band gap. When more energy is converted in the PV cell, less energy is available for external use.

5.14.1 TEMPERATURE

Studies have yielded empirical results that prove there is an approximately linear decrease in V_{oc} with increasing temperature (~–2.3 mV/°C; Talbot 2007). The ideal fill factor depends on the value of V_{oc} normalized to $\frac{K_B T}{q}$ so that the fill factor will also decrease with an increase in temperature.

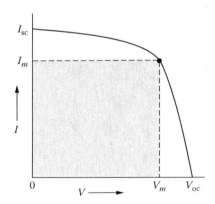

FIGURE 5.14 I-V curve of a solar cell.

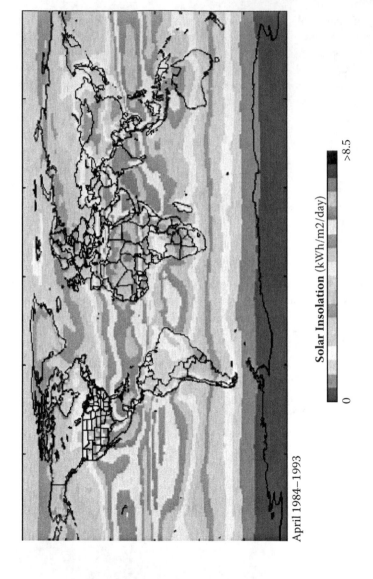

April 1984–1993

Solar Insolation (kWh/m2/day)

0   >8.5

FIGURE 5.15   Best reported PV cell efficiencies (Courtesy of National Renewable Energy Laboratory, 2008).

5.14.2 LIGHT

Light affects primarily the short circuit current. Effects to consider in order to increase efficiency are antireflective coatings (bare silicon is very reflective), minimization of the surface grid (shadowing will reduce I_o), light trapping, and the thickness of the semiconductor.

5.14.3 TYPE AND PURITY OF MATERIAL

Solar cells for terrestrial applications are typically made from silicon as single-crystal, polycrystalline, or amorphous solids. Single-crystal silicon is the most efficient because the crystal is free of grain boundaries, which are defects in the crystal structure caused by variations in the lattice that tend to decrease the electrical and thermal conductivity of the material. They can be thought of as barriers to electron flow. Polycrystalline silicon has obvious grain boundaries; the portions of single crystals are visible to the naked eye. Amorphous silicon (a-Si) is the noncrystalline form of silicon where the atoms are arranged in a relatively haphazard way. Due to the disordered nature of the material, some atoms have a dangling bond that disrupts the flow of electrons. A dangling bond occurs when an atom is missing a neighbor to which it would be able to bind. Amorphous silicon has lowest power conversion efficiencies of the three types, but is the least expensive to produce. Figure 5.16 depicts these types of solids pictorially.

Monocrystalline silicon. Cells are made from an ingot of a single crystal of silicon, grown in high-tech labs, sliced, then doped and etched. For commercial terrestrial modules, efficiencies typically range from about 15–20%. Modules made of this type of cell are the most mature on the market. Reliable manufacturers of this type of PV module offer guarantees of up to 20–25 years at 80% of nameplate rating.

Polycrystalline silicon. These cells are made up of various silicon crystals formed from an ingot. They are also sliced and then doped and etched. They demonstrate conversion efficiencies slightly lower than those of monocrystalline cells, generally from 13 to 15%. Reliable manufacturers typically guarantee polycrystalline PV modules for 20 years.

Amorphous silicon. The term *amorphous* refers to the lack of any geometric cell structure. Amorphous modules do not have the ordered pattern characteristic of crystals as in the case of crystalline silicon. Commercial modules typically have conversion efficiencies from 5 to 10%. Most product guarantees are for 10 years, depending on the manufacturer. The technology has yet to gain widespread acceptance for larger power applications largely due to shorter lifetimes from accelerated cell degradation in sunlight (degradation to 80% of original output in most cases). However, amorphous PV has found wide appeal for use in consumer devices (e.g., watches and calculators). It does have the advantage for some grid-tied or water-pumping systems in that higher voltage modules can be produced more cheaply than their crystalline counterparts. Another limiting factor to efficiency is that of

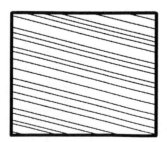

FIGURE 5.16 Monocrystalline, polycrystalline, and amorphous solids.

traps in the material. Traps are semiconductor material impurities in the depletion region and can greatly increase recombination of electrons and holes. Recombination reduces V_{oc}, in turn reducing the fill factor and efficiency.

5.14.4   PARASITIC RESISTANCES

The solar cell can be schematically depicted as shown in the equivalent circuit of Figure 5.17. Note that the ideal solar cell is given by full lines. Most industry models use dotted line components as well. The model shows the parasitic series and shunt resistance, depicted as Rs and Rsh, respectively. Major components of series resistance are the bulk resistance of the semiconductor material, bulk resistance of the metallic contacts and interconnections, and the contact resistance between the metallic contacts and the semiconductor. Shunt resistance is caused by leakage across the p–n junction caused by crystal defects and foreign impurities in the junction region.

5.15   CURRENT RESEARCH

Improvements to solar cell efficiency and cost reduction are critical factors in demonstrating the viability and marketability of these devices. Some methods of improving efficiency currently under investigation include efforts to reduce the size of the frontal grid and the thickness of a single cell. Current research in cell types includes concentrator and tandem cells as well as thin film technologies. Additional research to improve efficiencies and/or drive down cost is being conducted in the areas of surface metallic grid reduction, various antireflective coatings, and amorphous silicon cells. Recent laboratory research has introduced solar cells that incorporate the use of carbon nanotubes (Talbot 2007) (peak efficiency is 7% but using indirect illumination), quantum dots (Lee, 2008), and photonic crystals (Bullis 2007). Each type of cell will be briefly discussed.

5.15.1   CONCENTRATING SOLAR CELLS

Concentrator cells are smaller than standard PV cells (<1 cm²) and are manufactured specifically to absorb direct sunlight normal to the surface. They are highly efficient and therefore reduce the surface area needed to produce electricity. Note that the effective concentration level (C) is defined as

$$C = \frac{I_{sc(concentration)}}{I_{sc(sun)}}$$

(5.17)

As discovered in Chapter 2, the irradiance of 1,000 W/m² is usually referred to as "one sun." The problems inherent to concentrator cells or systems of concentrator cells are cooling, light tracking, and series resistance. These considerations have been studied and systems developed using Fresnel

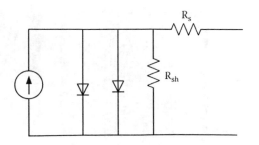

FIGURE 5.17   Equivalent circuit of a solar cell.

lens and back cooling; however, to date, the cost trade-offs have not proven positive. Due to these factors, optimum size has been determined to be in the range of 1–2 mm^2.

5.15.2 Tandem Cells

These more highly efficient heterojunction cells use a series of varying band-gap cells stacked on top of one another with the widest band-gap material uppermost. High-energy photons are absorbed in the wider band-gap material while lower energy photons pass through the stack until they reach a cell of low enough band gap to utilize them. In order to reduce circuit complexity, tandem cells are typically connected in series. Given that both voltage and current may vary with each cell, the current is therefore limited by the lowest value. Efficiencies of tandem cells are commonly claimed to be approximately 30% when they are made of layers such as GaInP/GaAs/Ge and are a vast improvement over the standards for polycrystalline silicon.

A heterojunction is the interface that occurs between two layers or regions of dissimilar crystalline semiconductors. These semiconducting materials have unequal band gaps as opposed to homojunctions, which are made from the same semiconductor material.

In addition to standard (Ge/GaAs/InGaP) tandem PV cells, preliminary laboratory research has been proposed by Stanford University, the National Renewable Energy Laboratory, and the University of New South Wales to investigate silicon tandem cells utilizing quantum dots. It has been proposed in order to "develop an innovative PV device based on integrating low-cost polycrystalline silicon thin films with higher band-gap semiconducting materials synthesized using silicon quantum dots embedded in a matrix of silicon oxide, nitride, or carbide to produce two- or three-cell tandem stacks" (Green and Conibeer 2006).

5.15.3 Thin Film Technologies

Semiconductor materials are applied in a thin film to a substrate—typically glass or ceramics. Rather than growing a crystal, the material is sputtered on to the substrate, enabling devices made this way to be highly portable. Devices made in this manner may not be as efficient as the cell types already mentioned, but they are effective in reducing costs by reducing the amount of material used and reducing labor in wiring the cells together in modules. This method of manufacturing is sometimes referred to as "second generation."

Copper indium gallium diselenide devices—CIGS (Cu/In/Ga/Se$_2$)—as indicated by the name, are heterojunction type devices. While he was working at Boeing, Billy Stanbury discovered the CIGS phenomenon: nanostructured domains acting as p–n junctions with a tendency toward high electron generating efficiency; this was subsequently called the Stanbury model. CIGS are direct band-gap devices and can be used in building integrated Photovoltaics (BIPV) such as integral shingles and windows or even interior window curtains. CIGS show great promise in both conversion efficiency and manufacturing.

5.15.4 Quantum Dots

Quantum dots are essentially tiny crystals of semiconductors just a few nanometers wide (Figure 5.18). The electron–hole pairs in quantum dots are confined in three dimensions and therefore exhibit properties different from those of standard semiconductors. Adding quantum dots to a solar cell increases the cell's ability to respond to a particular wavelength of light. Specifically, instead of releasing one electron for one photon of light, two or more electrons are released by a single photon, thus increasing the electrical current and therefore the efficiency. Because quantum dots can be made with relatively simple chemical reactions, adding them to solar cells may reduce cost; however, most of this technology is still in the research phase.

FIGURE 5.18 Colloidal quantum dots irradiated with UV light (From http://en.wikipedia.org/wiki/Quantum_dot, 2008).

5.16 CELL APPLICATIONS

Various methodologies to utilize PV devices have been proposed. The most common uses include production of electrical power for remote homesites and power generation for spacecraft. Other applications include utility grid connection to offset peak demand in large urban areas. In addition, various products are manufactured that incorporate solar power (e.g., calculators).

5.16.1 UTILITY POWER GENERATION

Some corporations are opting for "green power" (the use of renewable technology to generate power), perhaps in an effort to boost their public appeal. Google may not necessarily need to boost its public appeal, but it is revamping its headquarters to reflect this trend. Google converted part of its Googleplex headquarters in Silicon Valley to solar power with the installation of more than 9,000 solar panels, which produces enough electricity to power 1,000 homes. The project is now in place. In addition, the U.S. government built a 14 MW 140 acres in December 2007 at Nellis Air Force Base with panels of silicon wafers that rotate to follow the sun across the sky. Other countries are also using PV technology in an effort to offset utility costs. An April 2007 issue of *Technology Review* states:

> One of the largest solar power plants in the world went on line this winter in the sunny pastures of Serpa, a town in southern Portugal. The plant is owned by General Electric and operated by PowerLight of Berkeley, CA. At its peak, around noon on a sunny day, the solar park can generate 11 megaWatts of electricity—enough to power 8,000 homes. (Bourzac 2007)

5.16.2 SPACE SYSTEMS

Solar cells have been used in space application since 1958. Initially, they were single-crystal arrays that had an efficiency of only 10%. In the last decades, single-crystal silicon cell efficiencies have climbed to 18% and they have been used for "space missions that do not strictly require III-V cells with both higher efficiency and better radiation stability" (Talbot 2007). Currently, the most exotic of PV devices are generally used in mission-critical spacecraft. This is largely due to the cost of research and manufacturing. The devices are inclusive of triple tandem heterojunction cells consisting of III-V materials. Indium phosphide (InP)-based cells are being considered as well due to the fact that InP has a higher radiation resistance than GaAs.

5.16.3 SOLAR-POWERED PRODUCTS

Many modern products incorporate PV cells in order to operate independently of other electrical supplies. Indoor products include calculators, watches and clocks, battery chargers, smoke alarms, and units for rotating plants and shop window displays. Outdoor products include path and accent lighting, aquatic products such as fountains, small-animal garden deterrents, greenhouse vents, vehicle air vents, and radios. Amorphous silicon impregnated into backpacks can be used to charge cell phones that have GPS service in order to provide an added safety feature for long hiking trips. In addition, solar panels that appear similar to tinted glass are architecturally aesthetic.

In summary, PV device technology has been proven over the past few decades. Advances have shown the feasibility of many applications; however, solar cell efficiencies are still lower than desired. In order to increase efficiency, necessary research and subsequent manufacturing must be completed within well defined budgets. A delicate balance must be maintained and trade-offs may need to be made in both research and manufacturing that will increase efficiencies and drive down cost.

REFERENCES

Bourzac, K. 2007. Good day sunshine. *Technology Review*, March/April 2007 (journal online). Available from http://www.technologyreview.com/player/07/03/MagPhotoEssay/1.aspx

Bullis, K. 2007. Cheaper, more efficient solar cells. *Technology Review,* March 21, 2007 (journal online). Available from http://www.technologyreview.com/Energy/18415/

Green, M. A., and G. Conibeer.2006. Nanostructured silicon based tandem solar cells. Stanford University, Global Climate and Energy Project (online). Available from http://gcep.stanford.edu/pdfs/QeJ5maLQQr ugiSYMF3ATDA/2.2.6.green_06.pdf http://en.wikipedia.org/wiki/Quantum_dot.

Lee, H. Mighty small dots. Lawrence Livermore National Laboratories (journal online). Available from https://www.llnl.gov/str/Lee.html

Talbot, D. 2007. Solar cells that work all day. *Technology Review,* April 17, 2007 (journal online). Available from http://www.techreview.com/Energy/18539/

PROBLEMS

5.1   Explain what a p–n junction is.
5.2   Draw an I-V curve of a PV cell.
5.3   Given an I-V curve, find I_{mp}, V_{mp}, I_{sc}, and V_{oc}. Calculate the fill factor (FF).
5.4   Given P_{in}, calculate the conversion efficiency of the preceding problem.

6 Photovoltaic Conversion Systems

6.1 SOLAR BENEFITS

Solar energy represents an inexhaustible clean energy source that allows for local energy independence, and photovoltaics (PV) is the one technology that makes electric power available to anyone virtually anywhere on the planet. Solar is indeed the energy force that sustains life on Earth for all plants, animals, and people. The Earth is situated at the perfect distance and orbit from the Sun to make the miracle of life possible and is essentially a giant solar collector that receives radiant energy from the Sun in the form of electromagnetic radiation. As noted in Chapter 2, the Sun's power flow reaching the Earth is typically about 1,000 W/m^2, although availability varies with location and time of year.

Solar energy can be converted through chemical (e.g., photosynthesis), thermal, or electrical (i.e., PV) processes. Capturing solar energy typically requires equipment with a relatively high initial capital cost. However, over the lifetime of the solar equipment, these systems can prove to be cost competitive, especially because there are no recurring fuel costs, as compared to conventional energy technologies.

Solar electric power, or PV systems, is a cost-effective and viable solution to supply electricity for locations off the conventional electrical grid. PV power systems have been utilized almost everywhere, literally from the poles to the equator. However, the higher capital cost of PV means it is most cost effective for remote sites where other, more conventional options are not competitive. There are often misperceptions regarding what constitutes a good candidate PV application and site; thus, careful site consideration is necessary to eliminate unsuitable locations. For instance, projects that require large amounts of power are generally nonstarters for PV consideration. PV systems have both advantages and disadvantages that should be carefully considered by the project implementer and the end user (see Table 6.1).

PV is used from very small items such as calculators and watches to large installations for electric utilities. Even though PV systems are expensive, in a number of applications they are cost effective, especially for stand-alone systems some distance from the utility grid. Small power applications next to the utility grid can also be cost effective because the cost for a transformer is more than that of the PV system. An example is the flashing light for school lane crossings.

PV project success is directly related to a clear knowledge of site conditions and resources, as well as an understanding of PV capabilities and limitations. What makes a site adequate for solar energy? What differences are there in resources from site to site? What is the approximate system cost? These are a few of the questions that a project developer should answer. This text provides some basic tools necessary to help answer these questions. However, the PV industry is rapidly evolving and no book can be a replacement for consulting with a professional with actual field experience.

Carrying out a solar energy project requires time and resources. The initial investment is relatively high, so the project should be well thought out and designed to avoid possible future disappointments. The following basic points should be considered when developing a project:

- the availability of other sources of energy, such as electricity from the grid, gasoline, diesel, wind, etc.;

TABLE 6.1
Advantages and Disadvantages of Solar Energy Systems

Advantages	Disadvantages
High reliability (good system design)	High initial capital investment
Low operating costs	Modular energy storage increases costs
Local fuel (not imported fossil fuels)	Lack of infrastructure and limited access to
Long, useful life (from 20+ years)	technical services in remote areas
Clean energy	Variable energy production based on changing
Dry-weather production maximized	meteorological conditions
No on-site operator required	
Low operation and maintenance costs	

- how the energy will be used—for example, to pump water, refrigerate food, power lights, communications, etc.;
- project sustainability, such as who is responsible for the system (including both the PV system and ancillary systems);
- realistic energy requirements and anticipated usage; and
- availability of the solar resource on site.

6.1.1 ENERGY ALTERNATIVES

The availability of other energy sources is the first factor that should be considered. For example, the distance to the electric grid and availability of internal combustion engines should be researched because it might be more cost effective to extend the grid line to a nearby site or use an available internal combustion engine if energy requirements are large. In the case of grid extension, an immediate question arises: How far should the grid line be from the site to ensure the cost effectiveness of its extension? The answer varies. In relatively flat terrain locations, grid extension can cost about $10,000+ per kilometer (~$16,000+ per mile), while in more rugged terrain areas the cost can go as high as $20,000+ per kilometer (~$32,000+ per mile). To cross a gully, mountain, or other difficult terrain, costs more. However, actual costs vary by country. Normally, the solar option is considered feasible for most small and medium energy projects where the grid is more than a kilometer distant.

For *village power systems,* communities should be at least a couple of kilometers from the nearest electrical service; the more distant they are, the more competitive the solar option will be. If electrical service is nearby, it usually is a lower cost option to extend the electrical line and purchase a transformer. This might also allow for community electrification as a whole. However, this criterion is sometimes distorted. For example, if development funds are made available but only for renewable energy options, the conventional electricity may well be the best choice, but the technology selection is forced in favor of PV.

The dynamic of mandated rather than best-option technology choice should generally be avoided in project design. Also, communities are often inclined toward PV, even in electrified areas, because this arrangement leaves them free of any future power bills. This may not be the beautiful solution that it appears to be: The cost burden is merely loaded onto the initial project cost borne by the donating agency, and reducing the community inputs over time can run counter to sustainability principles. Communities need ongoing expenses (of manageable proportions) to motivate and solidify the pattern of regular payment of electrical tariffs.

6.2 BASIC MODULE ELECTRICAL CONCEPTS

As discussed in Chapter 5, electric fields, *E,* are created by charged particles, and if a charge particle is placed in an external electric field, there is a force on it that will make it move. Then energy or work is available. The electric potential, *V,* is the energy/charge. Current (I), voltage (V), power (W), and electrical energy (Wh) are key simple electrical concepts needed to understand PV systems. *Electrical current* is akin to a flow and is defined as the number of electrons that flow through a material. Current is measured in Amperes. *Electrical voltage* is the work that an external force must do on the electrons within the material to produce current and is measured in volts. As discussed in Chapter 5:

Electric potential (V) = energy/charge: $V = E/Q$, where Volt = Joule (J)/coulomb (C), or energy $= V * Q$.

The flow of electrons, current (I), is defined as $I = dq/dt$, the number of charges moving past a point in 1 s, and is measured in Amperes (A), where 1 A = 1 C/s.

The resistance to the flow of electrons in a wire is defined as resistance $(R) = VI$, where Ohm = Volt/Ampere.

Power (P) = $V * I$; Watt = Volt * Ampere. Electrical power (P) is that which is generated or consumed in any given instant and is the product of current and voltage.

Power is measured in Watts, where Watt = Volt * Ampere. The unit of power is the Watt (1 W = 1 V × 1 A). Electrical energy (E) is the power generated or consumed during a period of time (t) and is defined as $E = W \times t$.

The time period of consumption is given in hours; then the unit of energy is the Watt-hour (Wh).

6.2.1 PV ELECTRICAL CHARACTERISTICS

In order to generate usable power, PV cells are connected together in series and parallel electrical arrangements to provide the required current or voltage to operate electrical loads. PV cells are connected in series, grouped, laminated, and packaged between sheets of plastic and glass, thus forming a PV module. The module has a frame (usually aluminum) that gives it rigidity and allows for ease of handling and installation. Junction boxes, where conductor connections are made to transfer power from the modules to loads, are found on the backs of the PV modules.

The number of cells in a module depends on the application for which it is intended. Terrestrial solar modules were originally designed for charging 12 V lead-acid batteries; thus, many modules are nominally rated at 12 V. These PV modules typically have 36 series-connected cells, but there are also self-regulating modules with fewer cells. These modules produce a voltage output that is sufficient to charge 12 V batteries plus compensate for voltage drops in the electrical circuits and in the energy control and management systems. With the increased growth of grid-tied PV in recent years, there is a growing assortment of larger modules (e.g., 300 W_p) for these applications with more cells and higher voltages.

All PV modules produce direct current (DC) power. For AC applications, it is important to match the array voltage to that of the inverter under real-world operating conditions rather than standard test conditions (STCs). There are also some "AC modules" on the market, but in reality the inverter is built into the back of the module junction box; the PV cells themselves always produce DC power.

The most common solar cells are basically large p–n (thought of as positive–negative) junction diodes that use light energy (photons) to produce DC electricity. No voltage is applied across the junction; rather, a current is produced in the connected load when the cells are illuminated. The electrical behavior of PV modules is normally represented by a current versus voltage curve (*I-V curve*). Likewise, a power curve is generated by multiplying current and voltage at each point on the I-V curve. However, the only point desired to operate on this curve is the maximum power point. Figure 6.1 shows the I-V and PV power curves of a typical photovoltaic module.

6.2.2 COMMON PV TERMINOLOGY

Solar cell. The PV cell is the component responsible for converting light to electricity. Some materials (silicon is the most common) produce a PV effect, where sunlight frees electrons striking the silicon material. The freed electrons cannot return to the positively charged sites ("holes") without flowing through an external circuit, thus generating current. Solar cells are designed to absorb as much light as possible and are interconnected in series and parallel electrical connections to produce desired voltages and currents.

PV module. A PV module is composed of interconnected solar cells that are encapsulated between a glass cover and weatherproof backing. The modules are typically framed in aluminum frames suitable for mounting. Modules are rated to UL1703 and IEC 1215 standards.

PV array. PV modules are connected in series and parallel to form an array of modules, thus increasing total available power output to the needed voltage and current for a particular application.

Peak Watt (W_p). PV modules are rated by their total power output, or peak Watts. A peak Watt is the amount of power output a PV module produces at STC of a module operating temperature of 25°C in full noontime sunshine (irradiance) of 1,000 W/m². Keep in mind that modules often operate at much hotter temperatures than 25°C in all but cold climates, thus reducing crystalline module operating voltage and power by about 0.5% for every 1°C hotter. Therefore, a 100 W module operating at 45°C (20° hotter than STC, yielding a 10% power drop) would actually produce about 90 W. Amorphous modules do not have this effect.

6.2.3 I-V CURVES

Current–voltage relationships are used to measure the electrical characteristics of PV devices and are depicted by curves. The current–voltage, or I-V, curve plots current versus voltage from short circuit current I_{sc} through loading to open circuit voltage V_{oc}. The curves are used to obtain performance

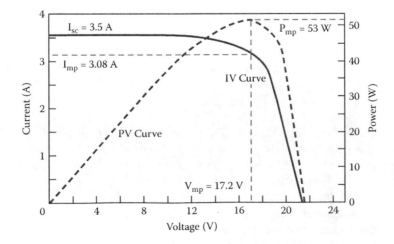

FIGURE 6.1 Typical I-V and power curves for a crystalline PV module operating at 1,000 W/m² (STC).

levels of PV systems (cells, modules, arrays). Strict standards for equipment and procedures are essential in achieving high-quality, consistent results. The I-V curve is obtained experimentally by exposing the PV cell or module to a constant level of irradiance while maintaining a constant cell temperature, varying the load resistance, and measuring the current produced. The horizontal and vertical axes measure voltage and current, respectively.

The I-V curve typically passes through the two end points: the short-circuit current, I_{sc}, and the open-circuit voltage, V_{oc}. The I_{sc} is the current produced with the positive and negative terminals of the cell shorted; the voltage between the terminals is zero, corresponding to zero load resistance. The V_{oc} is the voltage across the positive and negative terminals under open-circuit conditions with no current, corresponding to infinite load resistance. I-V curves can show the peak power point located on the farthest upper right corner of where the rectangular area is greatest under the curve.

The PV cell may be operated over a wide range of voltages and currents. By simply varying the load resistance from zero (a short circuit) to infinity (an open circuit), it is possible to determine the highest efficiency as the point where the cell delivers maximum power. Because power is the product of voltage times current, the maximum-power point (P_m) occurs on the I-V curve where the product of current (I_{mp}) times voltage (V_{mp}) is a maximum. No power is produced at the short-circuit current with no voltage or at open-circuit voltage with no current, so maximum power generation can be expected to be somewhere between these points. Note that maximum power is generated at only one point on the power curve; this occurs at the knee of the curve. This point represents the maximum efficiency of the device in converting sunlight into electricity.

Each I-V curve has a set of distinctive operation points that should be understood in order to appropriately install and troubleshoot PV power systems:

Short-circuit current (I_{sc}) is the maximum current generated by a cell or module and is measured when an external circuit with no resistance is connected (i.e., the cell is shorted). Its value depends on the cell's surface area and the amount of solar radiation incident upon the surface. It is specified in Amperes and, because it is the maximum current generated by a cell, I_{sc} is normally used for all electrical ampacity design calculations.

Nameplate current production is given for a PV cell or module at *standard reporting condition (SRC)* as specified by ASTM. The SRC commonly used by the PV industry is for a solar irradiance of 1,000 W/m², a PV cell temperature of 25°C, and a standardized solar spectrum referred to as an air mass 1.5 spectrum (AM = 1.5). This condition is also more commonly referred to as *standard test condition (STC)*. However, in reality, unless one is using PV in a relatively cold climate, the cells operate at a much hotter temperature (often 50°C or more), which reduces their power performance. The temperature effect is much greater for crystalline cells as compared to amorphous cells.

Maximum power operating current (I_{mp}) is the maximum current specified in Amperes and generated by a cell or module corresponding to the maximum power point on the array's current–voltage (I-V) curve.

Open-circuit voltage (V_{oc}) is the maximum voltage generated by the cell. This voltage is measured when no external circuit is connected to the cell.

Rated maximum power voltage (V_{mp}) corresponds to the maximum power point on the array's current–voltage (I-V) curve.

Maximum power (P_{mp}) is the maximum power available from a PV cell or module and occurs at the maximum power point on the I-V curve. It is the product of the PV current (I_{mp}) and voltage (V_{mp}). This is referred to as the maximum power point. If a module operates outside its maximum power value, the amount of power delivered is reduced and represents needless energy losses. Thus, this is the desired point of operation for any PV module.

The peak voltage (V_p) of the majority of nominal 12 V modules varies from 15 V (30 cells in series) to 17.5 V (36 cells in series). Each module has on its back side a decal placed by the manufacturer that shows the electrical specifications. For example, the decal on the back of a BP polycrystalline VLX-53 module whose characteristics are mentioned are provided in Table 6.2.

The power produced by a crystalline PV module is affected by two key factors: solar irradiance and module temperature. Figure 6.2 shows how the I-V curve is affected at different irradiance levels. The lower the solar irradiance is, the lower is the current output and thus the lower is the peak power point. Voltage essentially remains constant. The amount of current produced is directly proportional to increases in solar radiation intensity. Basically, V_{oc} does not change; its behavior is essentially constant even as solar-radiation intensity is changing.

Figure 6.3 shows the effect that temperature has on the power production capabilities of a module. As module operating temperature increases, module voltage drops while current essentially holds steady. PV module operating voltage is reduced on average for crystalline modules approximately 0.5% for every degree Celsius above STC (i.e., 25°). Thus, a 100 W_p crystalline module under STC now operating at a more realistic 55°C with no change in solar irradiance will lose about 15% of its power rating and provide about 85 W of useful power. In general, when sizing terrestrial PV systems, one should expect a 15–20% drop in module power from STC. This is important to remember when calculating daily actual energy production.

One may ask why the industry does not use a more realistic operating temperature for defining STC conditions—indeed, many module manufacturers will provide a more realistic 45°C or other rating.

TABLE 6.2
Example Manufacturer's Specifications for a 53-Wp PV Module

Operating point	Model BP VLX-53
P_{mp}	53 W_p (peak Watts)
V_{mp}	17.2 V
I_{mp}	3.08 A
V_{oc}	21.5 V
I_{sc}	3.5 A
Standard test conditions (STCs)	1,000 W/m², 25°C

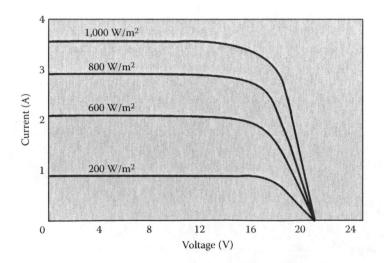

FIGURE 6.2   PV module current diminishes with decreasing solar irradiance.

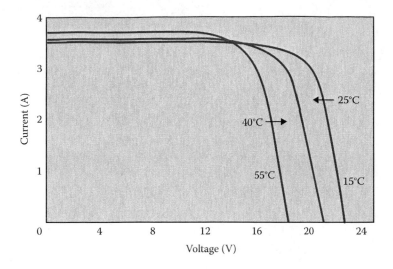

FIGURE 6.3   PV module voltage drops with temperature, as does power.

Historically, the NASA Jet Propulsion Laboratory (JPL) defined PV cell conditions for extraterrestrial applications and the notation stuck. Just remember that real-world operating conditions will see a derating in module performance due to the temperature effect for crystalline modules. Conversely, in very cold climes, a module operating under 25°C will produce more power than rated.

The PV module is typically the most reliable component of any PV system. The quality of installation and other components, such as the wiring connections between the modules, motors, etc., will ultimately determine the reliability of the PV system as a whole. But only a small fraction (less than 1%) of PV systems in the field have failed due to module failures.

6.3   PV ARRAYS

A PV array is a group of modules that are electrically connected either in series or in parallel. The electrical characteristics of the array are analogous to those of individual modules, with the power, current, and voltage modified according to the number of modules connected in series or parallel.

6.3.1   INCREASING VOLTAGE

PV modules are connected in series to obtain higher output voltages. Output voltage, V_o, of modules connected in series is given by the sum of the voltages generated by each module:

$$V_o = V_1 + V_2 + V_3 + \ldots \qquad (6.1)$$

An easy way to understand the concept of series-connected systems is through the analogy between a hydraulic system and an electrical system shown in Figure 6.4. As can be observed in the hydraulic system (left side), the water that falls from four times the 12 m height produces four times the pressure of water falling from the first level. This is analogous to the 48 V that the electrical system (right side) reaches after passing a current of 2 A through four modules connected in series. The current can be compared to the flow because both remain constant within their respective circuits, and the voltage is analogous to the role of pressure in the hydraulic system.

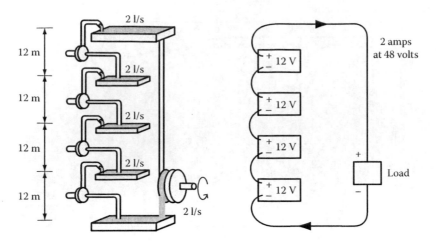

FIGURE 6.4   Analogy of a series connection using a hydraulic and an electrical system.

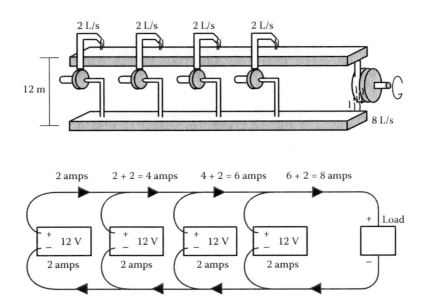

FIGURE 6.5   Hydraulic analogy of a parallel electrical connection, which is analogous to increasing flow of electrons.

6.3.2   INCREASING CURRENT

PV modules are connected in parallel to obtain greater current. The voltage of the parallel-connected modules is the same as the voltage of a single module, but the output current, I_o, is the sum of the currents from each unit connected in parallel:

$$I_o = I_1 + I_2 + I_3 + \ldots \tag{6.2}$$

In a manner similar to that of systems connected in series, systems connected in parallel can also be compared to a hydraulic system, like the one shown in Figure 6.5. In the hydraulic system (top), water that falls from the same height gives the same pressure as each individual pump, but the flow is equal to the total flow from all of the pumps. In the electrical system, then, the voltage remains constant and the output current of the four modules is added, producing 8 A of current and 12 V.

Figure 6.6 provides an example of modules connected in both series and parallel. The positions of blocking and bypass diodes are also shown. Diode sizes should by determined taking into consideration the maximum current generated by the PV array under short-circuit conditions. The electrical code stipulation used internationally requires that the current value supported by the diode should be at least 1.56 times the short-circuit current value of the array.

Finally, the nominal power of the array is the sum of the nominal-power values of each module, irrespective of how the modules are wired in series or in parallel.

Example 6.1

Sixteen PV modules like the one shown in Table 6.2 have been interconnected to operate a water pumping system. The array consists of eight modules in series and two strings of these in parallel (8s × 2p). The I-V and PV curves that describe the behavior of the array will have the same shape as those shown in Figure 6.2, but with the following parameters: $I_p = 3.08 \times 2 = 6.16$ A, $V_p = 17.2 \times 8 = 137.6$ V, $P_p = 53 \times 16 = 848$ W$_p$; maximum array current (I_{sc}) = 3.5 × 2 = 7.0 A; maximum array voltage (V_{oc}) = 21.5 × 8 = 172 V. These values correspond to the electrical characteristics under standard measurement conditions—AM 1.5, irradiance = 1.0 kW/m^2—and the operating temperature (T) of each module is 25°C. In the real world, expect array output to drop by 15–20% depending on ambient temperature.

6.4 PV ARRAY TILT

Maximum energy is obtained when the Sun's rays strike the receiving surface perpendicularly. In the case of PV arrays, perpendicularity between the Sun's rays and the modules can be achieved only if the modules' mounting structure can follow the movements of the Sun (i.e., track the Sun).

Mounting structures that automatically adjust for azimuth and elevation do exist. These types of structures are called trackers. Usually, the angle of elevation of the array is fixed. In some cases, azimuth-adjusting trackers are used. Depending on the latitude of the site, azimuth-adjusting trackers can increase the annual average insolation received up to 25% in temperate climates.

For the case in which a tracker is not used, the array is mounted on a fixed structure as is shown in Figures 6.7 and 6.8. This structure has the advantage of simplicity. Because the angle of elevation of the Sun changes during the year, the fixed-tilt angle of the array should be chosen so that maximum energy production is guaranteed. In the Northern Hemisphere, the Sun tracks

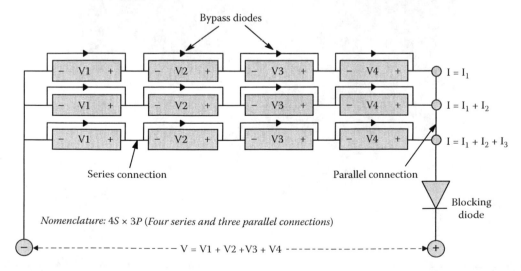

FIGURE 6.6 Connection of PV modules in series and parallel, increasing both voltage and current.

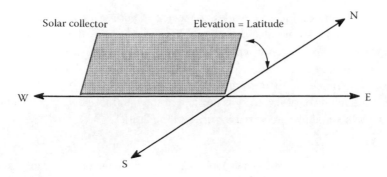

FIGURE 6.7 Module orientation for maximum solar gain year round in the Northern Hemisphere is to tilt the modules south at latitude tilt; the converse holds true for Southern Hemisphere sites.

primarily across the southern sky; for this reason, fixed PV arrays should be inclined (from the horizontal) to face south.

The angle of inclination of the array is selected so as to satisfy the energy demand for the critical design month. If producing the maximum energy over the course of the year is the desired goal, the value of the tilt angle of the array should be equal to the latitude of the site. Wintertime production can be maximized by tilting the array 10–15° more than latitude. Likewise, summertime production can be maximized by tilting the array 10–15° less than latitude.

6.5 PV BALANCE OF SYSTEMS

PV systems are made up of a variety of components, which may include arrays, wires, fuses, controls, batteries, trackers, and inverters. Components will vary somewhat depending on the type of application. PV systems are modular by nature; thus, systems can be readily expanded and components easily repaired or replaced if needed. PV systems are cost effective today for many remote power applications, as well as for small stand-alone power applications in proximity to the exist-

FIGURE 6.8 PV modules tilted to maximize annual energy production for utility interactive residential installation in Las Cruces, New Mexico.

ing electric grid. These systems should use good electrical design practices, such as the National Electrical Code (NEC) or its equivalent (Wiles, 2005; *IEEE* 1374–1998, 929).

Energy that flows through a power system necessarily runs through a variety of devices and wires between the system components. In a PV system, *balance of system* (*BOS*) refers to all of the system components except the PV modules. These components can account for half of the system cost and most of the system maintenance. The BOS components may include fuses and disconnect switches to protect the systems, structures, enclosures, wire connectors to link different hardware components, switch gear, fuses, ground fault detectors, charge controllers, general controllers, batteries, inverters, and dials and meters to monitor the performance and status of the systems. The selection of good BOS components is as important as the selection of PV modules. Low-quality BOS is often responsible for many avoidable maintenance problems for PV systems in remote areas and can lead to premature failure and disuse of the whole system. The PV industry goal is to provide PV systems with operational life spans of 25 or more years. Despite this, inexperienced system designers and installers still improperly select connectors, cable, etc., for PV systems, with predictable results. The National Fire Protection Association requires minimal safety standards for PV installations using the NEC (NFPA, 2009).

6.5.1 ENERGY STORAGE

Energy storage for PV systems commonly consists of batteries to store and discharge electrical energy as needed. However, each time a battery is charged or discharged, some energy is lost from the system. Batteries vary by type, depth of discharge, rate of charge, and lifetime (in PV applications). The most common types of batteries used with PV systems are lead-acid, but other more exotic and expensive batteries are sometimes used, such as nickel metal hydride. A new area of PV battery applications is emerging in which the PV battery is used for backup power when the utility grid fails for grid-tied PV systems. This application has unique battery charging and maintenance requirements. Batteries are usually installed in well-ventilated locations such as garages, utility rooms, and outbuildings to minimize the potential for capturing explosive concentrations of hydrogen gas and to minimize possible hazards from electrolyte spills. A complete analysis of battery storage systems is provided in Chapter 11.

6.5.2 CHARGE CONTROLLERS

Charge controllers manage the flow of electricity among the array, battery, and loads. The appropriate charge control algorithms and charging currents should be matched for the batteries used in the system; no one size controller fits all batteries. Better quality charge controllers allow for adjustable regulation voltages, multiple stage charge control, temperature compensation, and equalization charges at specified intervals for flooded batteries. The main purpose of a charge controller is to protect batteries from damage from excessive overcharging or discharging. A complete analysis of charge controllers is also provided in Chapter 11.

6.5.3 INVERTERS AND CONVERTERS

Inverters accept an electrical current in one form and output the current in another form. An inverter converts DC into AC, whereas a rectifier converts AC into DC. There are also DC–DC converters, which step up or step down the voltage of a DC current. Inverters convert DC power from the batteries or solar array into 60 or 50 Hz AC power. Inverters can be transformer based or high-frequency switching types. Inverters can stand alone, be utility interconnected, or be a combination of both.

As with all power system components, the use of inverters results in energy losses due to inefficiencies. Typical inverter efficiency is around 90%; however, inverters that are poorly matched to array and loads can operate at considerably less efficiency. Inverters are an interesting option due to the great variety of low-cost appliances that run on AC.

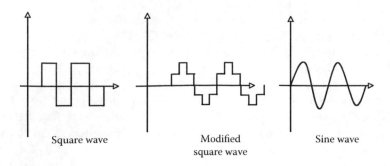

Square wave   Modified   Sine wave
                   square wave

FIGURE 6.9 Inverter wave outputs.

Inverters are a key component to most PV systems installed in grid-connected or distributed applications. Aside from the modules themselves, inverters are often the most expensive component of an installed PV system, and frequently are the critical factor in terms of overall system reliability and operation. Utility-interactive PV systems installed in residences and commercial buildings will become a small, but important, source of electric generation over the next 50 years. This is a new concept in utility power production—a change from large-scale central generation to small-scale dispersed generation. The basic system is simple, utilizing a PV array producing DC power that is converted to AC power via an inverter to the grid—very simple, yet elegant.

The AC produced by inverters can have square, modified-sine, or quasi-sine waves and pure sine wave outputs. The pure sine wave is high cost, high efficiency, and has the best power quality. Modified sine wave is mid-range cost, quality, and efficiency. Square wave is low cost and low efficiency, and it has poor power quality that is useful for some applications. Square wave signals can be harmful to some electronic appliances due to the high-voltage harmonic distortion. All inverters emit electromagnetic noise. This noise can cause interference with sound and video equipment. One method of attenuating this electromagnetic noise in some cases is by grounding the inverter case, which is also a code requirement for safety reasons (NEC, 2008).

The harmonic frequencies and their magnitudes that appear on a system are governed by the shape of the distorted wave. The output capacity of an inverter is expressed in volt-amperes (VA). Two output capacity specifications are generally given: continuous output and starting (or surge) output. Continuous output must be enough to operate all the AC loads at the same time.

During start-up, devices such as motors require a VA power input several times greater than continuous power. This demand exists for only a brief period of time. Motor starting current is from two to six times the steady state; induction motors like compressors and pumps that start under load are the toughest to start and, for capacitor start motors (drill press, band saw), one can expect to start only up to 1 hp. Most motors use 20% more power and run hotter with modified sine wave than with pure sine wave.

Inverters typically have starting outputs a couple of times greater than their continuous output. If at any time the output capacity is exceeded, inverters typically protect themselves by disconnecting the loads. Usually, a manual reset or fuse replacement is needed for the inverter to work again.

Maximum output power is the maximum number of Watts the inverter can produce continuously.

Surge power is the number of Watts the inverter can handle when a reactive load is turned on (1–5 s).

Efficiency is 92–98% modified sine wave and 80–95% sine wave, rated at a specified wattage.

Harmonic distortion is distortion of the output waveform (2–35%).

DC voltage limits are 10.5–15 V for a 12 V model.

Stand-alone inverters are designed to work for off-grid systems. Key design parameters include load compatibility, power rating, power quality, and maintaining battery health. Because inverters in stand-alone systems are connected directly to batteries, an overcurrent protection device (such as a fuse or automatic breaker) needs to be installed between the batteries and the inverter. Other distributed energy sources, such as fuel cells and microturbines, use inverters as well. Most inverters for this application are sized for a few kiloWatts. Very small loads may not keep the inverter running because it has a minimum threshold to start up (may cycle the inverter). For stand-alone inverters, a separate low-voltage disconnect is not necessary because the inverter disconnects the load to protect batteries from overdischarge. Inverter bases low voltage disconnect (LVD) on battery voltage, current, and the capacity that is entered.

Stand-alone inverters are power conversion devices installed in compliance with the requirements of the electric code, which generally requires the use of fixed input and output wiring methods. A stand-alone PV inverter should have provisions for hardwiring at least the DC input/outputs and possibly the AC inputs/outputs, although sometimes these are just plug-in connections, depending on the inverter size and design.

In nearly all stand-alone installations (no utility connection), the AC outputs of stand-alone inverters are connected to an AC load center (either a set of circuit breakers in a PV power center or a standard AC load center—panel board—in a residential or commercial building). The bond between AC neutral and the grounding system is normally made in these panel boards or in a related area such as the main disconnect enclosure when the main disconnect is not collocated with the load circuit breakers, as in a mobile home.

Stand-alone inverters are connected to batteries through the load center. The cables between the inverter and the batteries are generally kept as short as possible to minimize voltage drop and keep the number of disconnects and overcurrent devices to a minimum.

Grid-tied inverters are widely used in Europe, Japan, and the United States to inter-tie PV systems with the electric utility grid. These inverters convert the DC power to AC power in synchronization with the electric grid (UL 1741). When the grid goes down, the inverters go off by design. PV system utility interconnection considerations include safety, anti-islanding, and power quality. *PV system islanding* is the condition present when the utility power grid fails and the inverter attempts to power the grid. An inverter that is *islanding protected* senses the loss of AC power from the grid and does not back-feed into the grid system. All AC grid-tied inverters are designed to be anti-islanding and the voltage on the inverter side must reduce to zero within 2 seconds of the grid going down. The inverter should be correctly wired according to the manufacturer's instructions and have proper wire sizes, fusing, and breaker sizes and types. PV system anti-islanding protection methods include grid shorted, grid open, anti-islanding inversion synchronization, over or under frequency, and over or under voltage.

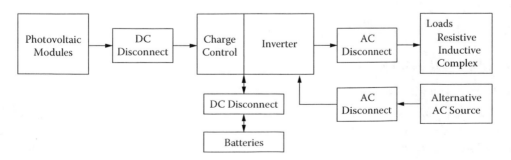

FIGURE 6.10 PV system schematic incorporating a stand-alone inverter to meet AC loads.

FIGURE 6.11   Example of a stand-alone inverter setup (Xantrex/Trace) as part of DC load center.

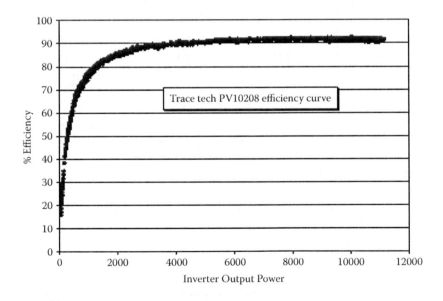

FIGURE 6.12   Measured efficiency curve for an inverter.

6.6   PV SYSTEM UTILITY

6.6.1   GROUNDING AND BONDING DC AND AC CIRCUITS

Grounding and bonding is important to maintain system integrity. PV systems that operate under 50 V are not required to be grounded according to the electric code, although chassis grounds are required for all hardware, even that operating under 50 V. The U.S. electrical code requires that one

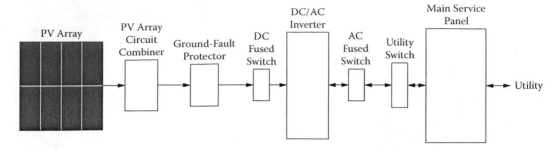

FIGURE 6.13 Typical utility interactive PV system components.

FIGURE 6.14 Utility interactive PV system installation in Austin, Texas (2008). Note the DC and AC disconnects, electric meters, and Fronius inverter.

and only one bond be made in a power system (AC or DC) between the grounded conductor and the grounding system. In a stand-alone system where all AC power is derived from the inverter, most residential and commercial buildings that are wired according to NEC requirements have the bond made in the first panel board or at the first disconnect in the system.

In a utility connected PV system with battery backup, power may be supplied to the loads from the load center (when the utility is available), the battery/inverter system coupled to a subpanel for some loads upon loss of utility power, or the subpanel where utility power is supplied through the inverter in a bypass mode.

If the inverter has a common neutral for all AC inputs and outputs, then a bond anywhere in the external AC system or a bond in the inverter would meet the requirements for a single bond under all operating modes. Because the building will necessarily have a bonding point in or near the main panel (to provide grounded conductors for all circuits not associated with the inverter), neither the inverter nor the subpanel may have a second bonding point. If the inverter has isolated neutrals between AC inputs and outputs, then the inverter must have some sort of relay-controlled neutral switching or ground bond switching.

6.6.2 Net Metering

Net metering or net energy billing is where the utility meter, which runs forward and backward, is read at the end of a specified time period. The time period can vary from a month to a year. The customer pays the utility for net energy purchased and if more energy is produced than is used, then the utility pays the customer the avoided cost. When energy produced on site is not being used at that specific time but offsets energy from the utility company at a later time, that energy is worth the retail rate. If there is net metering by year, then the customer receives the retail rate during seasons of low renewable energy production. Net metering is allowed in a majority of U.S. states. It permits customer-connected power-generating equipment without changing the existing meter. The meter may rotate *backward* when power generated *by* the customer is being delivered *to* the utility. The customer should wire the PV system and utility interconnections safely according to NEC and local codes; install equipment that conforms to applicable IEEE standards of design and operation; and install equipment that is UL listed for safety.

Before connecting to the utility system, be sure to

- request permission from the local electric utility;
- request technical rules for interconnecting;
- request rates for installation of metering equipment and cost of service;
- be prepared to provide full details and documentation of the proposed PV installation, including sell-back methods if desired; and
- learn the details *before* installing the system.

6.7 PV SYSTEM SAFETY

Finally, when working with PV systems, please be careful. Never work on a PV system alone. Have proper knowledge of the PV system. Be careful accessing roofs and ladders. Be careful with batteries and be sure to have bicarbonate, etc., to neutralize battery acid. Dress appropriately. Have an alert mind, a skeptic instinct, and a slow hand. The goal is to avoid accidents and injuries. This requires the following:

- good work habits;
- awareness of potential hazards;
- proper tools and hardware;
- safe PV systems; and
- working in pairs (buddy system).

Table 6.3 lists some electric shock hazards.

6.8 PV SYSTEM TESTING RULES

- Remove all jewelry.
- Visually inspect the system and take notes of risks and problems.
- Be aware of telephone and first-aid equipment locations.
- Be careful climbing up and down ladders and roofs.
- Identify and locate disconnects.
- Measure the open-circuit voltage.
- Measure the voltage of each conductor.

TABLE 6.3
Electric Shock Hazards

	Current	
Reaction	AC	DC
Perception: tingle, warmth	1 mA	6 mA
Shock: retain muscle control; reflex may cause injury	2 mA	9 mA
Severe shock: lose muscle control; cannot let go; burns; asphyxia	20 mA	90 mA
Ventricular fibrillation: probable death	100 mA	500 mA
Heart frozen: body temperature rises; death will occur in minutes	>1 A	<1 A

Source: Sandia National Laboratories, 1990.

REFERENCES

ANSI/IEEE Std 1374-1998. 1998. IEEE guide for terrestrial photovoltaic power system safety. New York: ANSI/IEE.

ANSI/IEEE Std 929-1999. 1999. IEEE recommended practice for utility interface of photovoltaic (PV) systems. New York: ANSI/IEE.

IEC 1215-1993. 1993. Crystalline silicon terrestrial photovoltaic modules: Design qualification and type approval. Geneva, Switzerland: International Electrotechnical Commission.

National Electrical Code, National Fire Protection Association, Quincy, Massachusetts, 2008.

National Fire Protection Association. http://www.nfpa.org, Quincy, Massachusetts, 2009.

Sandia National Laboratories, Design Assistance Center. 1990. Working safely with photovoltaic systems. Albuquerque, NM.

UL/ANSI Std 1703-1993. 1993. Standard for flat-plate photovoltaic modules and panels. Northbrook, IL: Underwriters Laboratories.

UL/ANSI Std 1741-1999. 1999. Static inverters and charge controllers for use in photovoltaic systems. Northbrook, IL: Underwriters Laboratories.

Wiles, J. 2005. Photovoltaic power systems and the National Electrical Code: Suggested practices. Albuquerque, NM: Sandia National Laboratories.

PROBLEMS

6.1 Using the PV module specifications from Table 6.2, determine the nominal current and voltage outputs for a PV system that is wired 5s × 3p.

6.2 For a PV module operating in a hot desert climate at 65°C, what would be the approximate system percentage temperature derate from STC?

6.3 Draw a general I-V curve of a PV module indicating where the P_{mp}, I_{sc}, and V_{oc} points are located.

6.4 Current output from a solar module is proportional to what variable?

6.5 PV modules produce what kind of electrical current?

6.6 Describe the temperature effect on crystalline PV modules and how it affects power output.

6.7 Draw an electrical schematic of major components for a stand-alone PV system with battery storage and both AC and DC power distribution.

7 Photovoltaic System Sizing and Design

7.1 INTRODUCTION

In order to size and design a solar energy system, it is necessary to conduct a reasonable assessment of the energy requirements that the system will have to meet. With this information, a reasonable estimate for the size of a PV system required to supply the energy needed can be made. The following section outlines the typical design and installation process for PV power systems.

In PV systems, energy demand is specified on a per-day basis; this leads to the next factor for consideration: the proposed use of the energy. Is the energy for a regional telecommunications system that will operate 24 hours/day, 7 days/week? Is it for lights only at night? Perhaps it is for a water-pumping system that will mostly be needed during the hot summer. Typical and viable applications of PV systems are those in which the power demand is relatively small, such as in providing drinking water for cattle and water for human consumption. Flood irrigation of farmland is usually not cost effective due to the high water demand and low value of harvested crops. One size does not fit all when it comes to solar energy system sizing and design. The key is first to minimize energy consumption by using the most efficient equipment and then to design a solar power system around the energy-efficient system.

7.2 SOLAR RESOURCE SIZING CONSIDERATIONS

To predict solar energy system size accurately requires understanding the local solar resource. These resources can vary tremendously depending on location. The solar resource is available almost everywhere on the planet and more than adequate in most temperate and tropical locations to be utilized successfully. Locations where complete cloud cover occurs continuously during weeks at a time (e.g., tropical mountain rain forests) can present challenges and PV systems will have to be larger to meet energy needs. Some power can be generated even under overcast conditions, but it is just a fraction (<10%) of what is available during sunny, clear-sky conditions. Concentrating solar energy systems only work where direct sunlight is available.

Vegetation, such as that found in a dry, arid region (e.g., cacti tend to grow where it is sunny and dry), can be a useful indicator of solar resource. Regions within the tropics have a less variable solar resource over the course of a year as compared to higher latitude temperate regions with long summer days and short winter days.

As discussed in Chapter 2, maps and tables available that indicate average monthly *solar insolation* (i.e., energy) are available for many different geographic areas. The common unit used for insolation is kiloWatt-hour per square meter (kWh/m^2), also called a sun-hour. Appendix A contains a table with insolation values for various geographic regions. Several adequate Web sites, such as those by NREL or NASA, also provide solar insolation data. The insolation value that most closely fits the project location should be used. When in doubt, it is a good idea to be conservative (i.e., use fewer sun-hours).

The solar energy system should be designed to fit the need with the seasonal solar resource. For an off-grid home, one may want to design a PV system for the winter season when there is less sunlight.

Water pumping needs often drop greatly during cool or cold winter months and increase during the hot and sunny summer months. This a natural correlation that bodes well for PV water pumping, when systems can be designed for best tilt to maximize usage for when the water is needed.

As discussed in Chapter 2, radiation striking the Earth's surface can be classified as direct or diffuse. *Direct radiation* is radiation that has reached the surface of the Earth without being reflected. *Diffuse radiation* is received after sunlight has changed direction due to reflection or refraction that occurred in passage through the Earth's atmosphere. A surface receiving solar energy "sees" the radiation as if it arrived as both components (direct and diffuse), which can lead to irradiance values even greater than $1,000$ W/m^2. On clear days, a surface receiving solar energy will capture mainly direct radiation; on cloudy days, the surface will receive mostly diffuse radiation because direct radiation has been obstructed by the clouds.

Throughout the day and under stable atmospheric conditions, irradiance will vary, with minimum values at dawn and dusk and maximum values at mid-day. For example, on a clear day, the irradiance value at 9:00 a.m. will be less than the irradiance value at noon. This is explained by the Earth's rotation about its axis, which causes the distance traveled by sunlight through the Earth's atmosphere to be at a minimum at solar noon. At this hour, the Sun's rays are striking a surface perpendicularly and through the least atmosphere (exactly 1 atm).

For practical solar energy system design and sizing, we consider the average energy available over a day; this is the *insolation* and corresponds to the accumulated irradiance over time. Insolation is typically provided in units of kiloWatt-hours per square meter. Normally, this value is reported as an accumulation of energy over a day. Insolation is also expressed in terms of *peak sun-hours*. A peak sun-hour is $1,000$ W/m2. The *energy* produced by a PV array is directly proportional to the amount of insolation received (see Figure 7.1).

7.3 SOLAR TRAJECTORY

In addition to atmospheric conditions, another parameter affects the incident solar radiation and system sizing: the apparent movement of the Sun throughout the day and throughout the year as discussed in Chapter 2. The Earth's tilted axis results in a day-by-day variation of the angle between the Earth–Sun line and the Earth's equatorial plane, called the solar declination (angle varies with the date). The daily change in the declination is the primary reason for the changing seasons, with their variation in the distribution of solar radiation over the Earth's surface and the varying number of hours of daylight and darkness. Solar energy systems must be sized to the critical season for their use.

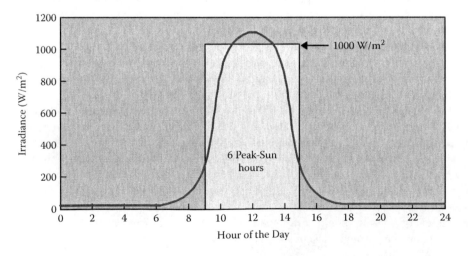

FIGURE 7.1 Irradiance and insolation expressed as peak solar hours (i.e., 6 sun-hours = 6 kWh/m^2).

The position of the Sun can be defined in terms of its altitude above the horizon and its azimuth, measured as an angle in the horizontal plane. Because the Earth's daily rotation and its annual orbit around the Sun are regular and predictable, the solar altitude and azimuth can be calculated for any time of day when the latitude, longitude, and date (declination) are known. Because the Sun appears to move at the rate of 360° in 24 hours, its apparent rate of motion is 4' per degree of longitude. At solar noon, the Sun is exactly on the meridian, which contains the south–north line. Consequently, the solar azimuth, f, is 0°. The angle between the line normal to the irradiated surface and the Earth–Sun line is called the incident angle. It is important in solar technology because it affects the intensity of the direct component of the solar radiation striking a surface. For practical purposes related to PV system design, it is only necessary to consider the total daily sun-hours for calculating daily energy production; however, it is important to understand season variations due to the Sun's apparent movement across the sky dome.

A PV array receives the maximum insolation when it is kept pointing directly normal (perpendicular) toward the Sun. In order to accomplish this, the Sun must be followed throughout the day and throughout the year, requiring the constant adjustment of two angles: the azimuth, to track the daily movement of the Sun from East to West, and the angle of elevation, to track the north–south trajectory of the Sun through the seasons. In order for a PV array to follow the Sun in this manner, tracking structural mounts designed for this purpose (either single or dual axes) are needed.

7.4   SOLAR ENERGY SYSTEM SIZING CONSIDERATIONS

Insolation is the key parameter for solar energy system design. The main factors affecting the amount of insolation incident upon a solar surface are orientation, mounting angle with respect to horizontal, and climatic conditions. In areas where cloudy days are relatively frequent, average insolation is less. At latitudes of greater than the tropics (i.e., >20°), winter days are significantly shorter than summer days. This results in a greater average insolation during the summer. For instance, in rainy areas of the tropics near the equator such as the rainforests of southern Mexico, insolation upon the horizontal plane reaches 4 kWh/m² per day in the winter, 5.2 kWh/m² per day in the summer, and 4.5 kWh/m² per day as an annual average. In the dry areas of northern Mexico, insolation upon the horizontal plane reaches 5 kWh/m² per day in the winter, 8 kWh/m² per day in the summer, and 6.5 kWh/m² per day as an annual average. This difference results from a combination of the longer summer days of the higher latitudes in northern Mexico and overall less cloudy conditions. One-size solar energy system does not fit all, and it depends greatly as to latitude and corresponding solar resource.

Because the insolation received by a solar surface depends on orientation and inclination with respect to the apparent position of the Sun, the *solar resource* of a designated site is specified as the amount of insolation measured upon the horizontal plane. Drawing on data for insolation upon the horizontal plane, insolation values can be estimated for surfaces set at specific azimuths and angles of elevation. Maps and tables are available from various sources that give horizontal-plane insolation values for numerous regions and times of the year. Appendix A contains insolation values for selected cities.

It should be noted that the maximum solar energy available for any fixed array at a location is for latitude tilt. However, the tilt of the array can be maximized for adjusted energy production during the critical design period, such as a more horizontal tilt to maximize summer water pumping, if needs warrant.

Example 7.1: Solar Resource

A 848 Watt PV array has been installed on a family farm near Aldama, Chihuahua, Mexico. The array is pointed true south and has a tilt angle equal to the latitude (30°). An azimuth-adjusting tracker is not used. The real capacity of the array, operating at a cell temperature of 55° C, is 0.85 × 0.848 = 0.72 kW. According to the table in Appendix A, expected insolation is 6.1 kWh/m2 per day in the first third of the year. The energy that can be expected from the array is approximately 6.1 × 0.72 = 4.4 kWh per day in the first third of the year and 6.6 × 0.72 = 4.8 kWh per day in the last third of the year. If the array were installed at a tilt angle of 15° (latitude minus 15°), the estimated insolation is 5.7 kWh/m² per day in the first third of the year and 6.9 kWh/m² per day in the last third of the year. In this case, the expected electrical energy for the system is 4.1 kWh and 5.0 kWh per day in the first and last thirds of the year, respectively.

7.5 SOLAR ENERGY SYSTEM SIZING

As previously discussed, the amount of energy delivered by a PV array or module depends on irradiance and temperature. It is possible to estimate the electrical energy (in kiloWatt-hours/day) expected of an array with known nominal power by using the following approximations:

- PV modules installed on structures anchored to the ground operate at approximately 55°C during the day during the summer; some desert climates may be hotter yet. This is 30° above standard test conditions (25°C). This means that the real capacity of the array is approximately 15% less than the nominal power rating. The effective capacity, then, is 85% of the nominal capacity.
- Expected electrical energy (kiloWatt-hours) is the product of the real capacity of the array (kiloWatt-hours) and insolation (peak sun hours) at the angle of elevation of the array. Generated PV energy varies seasonally, as do the levels of insolation.
- If an azimuth-adjusting tracker is used, annual energy production can increase up to ~25% in temperate climates.

A multitude of PV-sizing methodologies, spreadsheets, and computer programs exist. However, the hardest part in sizing a system is anticipating the expected end user loads; this drives solar energy system design. Some users are energy wise, while others (e.g., teenagers) have no concept of, and little concern about, energy use.

One of the simplest and most effective methods for solar energy system sizing is to take a look at solar energy system efficiencies (Table 7.1). Thus, the energy required from a solar energy system is roughly halved from array nameplate rating for off-grid systems with battery storage and drops one-quarter for grid-tied systems. Grid-tied systems are more forgiving in that even if the solar energy system is undersized, the user will never notice because the grid will be supplying any power short-fall. For off-grid systems, the user will be forced to live within the confines for the energy produced by his solar energy system.

7.5.1 Example of Simple PV DC System Sizing

Table 7.1 assumes some basic loads for a small off-grid residence in Oaxaca, Mexico. The critical design month is assumed to be in winter with 5.4 sun-hours available in December. The battery

TABLE 7.1
Average PV System Component Efficiencies

PV array	80–85%
Inverter	80–90%
Wire	97–98%
Disconnects, fuses	98–99%
Total grid-tied PV system efficiency	60–75%
New batteries (roundtrip efficiency)	65–75%
Total off-grid PV system efficiency (AC)	40–56%
Total off-grid PV system efficiency (DC)	49–62%

bank should be designed for 3 days' autonomy, not to exceed 45% depth of discharge (see Chapter 11). The loads are assumed to be 1,040 Wh/day.

Load	Hours operating/day (hours)	Power (W)	Daily energy (Wh)
Four fluorescent lamps	4	30	480
One refrigerator (DC)	5	80	400
One laptop	2	50	100
One stereo (teenager-free home!)	2	30	60

Total energy required:	1,040 Wh
Critical month of use (winter):	5.4 sun-hours
PV module STC rating:	50 W each
Battery size:	105 Ah (ampere-hours) each, 3 days' storage
PV system size required:	1,040 Wh/(5.4 peak sun-hours * 50 W/module * 50% system efficiency) = 7.7 modules required (which means buy 8 modules)
battery bank size required:	1,040 Wh * 3 days = 3,120 Wh storage

3,120 Wh/(45% DOD * 75% battery efficiency * 105 Ah * 12 V) = 7.3 batteries in parallel (which means buy seven batteries)

7.5.2 SIZING INVERTERS

The inverter for a PV lighting system is an important benefit in running specific AC appliances. Modern inverters are extremely reliable and there are several hundred types, sizes, brands, and models. Choosing the best from such a long list can be a chore and there is no "best" inverter for all purposes. Power output is usually the main factor. An inverter needs to meet two needs: peak, or surge, power and continuous power:

Surge is the maximum power that the inverter can supply, usually for only a short time. Some appliances, particularly those with electric motors, need a much higher power level at start-up than they do when running. Pumps are another common example, as well as refrigerators (compressors).

Continuous is the power that the inverter has to supply on a steady basis. This is usually much lower than surge power. For example, this would be what a refrigerator pulls after the first few seconds it takes for the motor to start or what it takes to run the microwave or the sum of all combined loads.

Inverters are rated by their continuous wattage output. The larger they are, the more they cost. For example, assume sizing for loads where there is a 19-inch TV (80 W), blender (350 W), and one fluorescent light at 20 W, and two fluorescent lights at 11 W each, for a total of 472 W. An inverter that can supply a least 472 W continuously will be chosen. The only concern here would be the blender's initial surge requirement. Normally, a small motor like the one the blender has will surge for a split second at twice the amount of power it normally uses—in this case, 350 doubled equals about a 700 W surge. Some existing loads may need to be turned off to help meet the surge if the inverter is already continuously loaded.

Suppose that an inverter is selected at 500 W with a surge capacity rating of 1,000 W, which is more than sufficient to handle the blender surge. The following information was considered for selecting an inverter for the PV lighting application example mentioned earlier.

7.5.2.1   Technical Specifications

- Nominal system operating voltage (input V):   12 V DC (10.8–15.5 V)
- Output voltage: 120 V AC, 60 Hz sinewave
- Continuous output: 500 W
- Surge capacity power: 1,000 W
- Standby power: 3 W
- Average efficiency: 90% at full rated power
- Recommended input fuse: 75 A (500 W/10.8 V * 0.8 *1.25)
- DC wire size minimum: 8 AWG
- Availability of system status (light indicator):   yes

7.5.2.2   Load Estimation

Item:	Watts:	Hours of Use:	Calculated Consumption:
one lamp	20	3/day	20 W * 4 hrs/day = 80 Wh/day
two lamps	11	3/day	3(11 W*4 hrs/day) = 132 Wh/day
one 19-in TV (AC)	80	4/day	80 W * 4 hrs/day =320 Wh * (1.1) = 352 Wh/day
one blender (AC)	350	0.5/day	350 W 0.5 hrs/day = 175 Wh * (1.1) = 193 Wh/day

Total Watt-hour/day requirement: (80 + 132 + 352 + 193) Wh = 757 Wh/day

7.5.2.3   Battery Storage Requirement

Battey size is a design variable and is generally based on the desired autonomy period, depth of discharge (e.g., 50%), and derating for round-trip efficiency (e.g., 75%). Here, the autonomy desired is 3 days of storage; maximum allowable depth of discharge (DOD) for deep cycle battery is 50%. See Chapter 11 for definitions. Calculations are as follows:

Total daily load Ah requirement = daily energy Watt-hours/system nominal voltage. This is a 12 V DC for lighting systems:

$$757 \text{ Wh/12 V} = 63 \text{ Ah/day}$$

Required battery bank capacity = (days of autonomy * daily load Ampere-hours)/(battery DOD)

$$= (3 \text{ days} * 63 \text{ Ah/day})/(0.75 * 0.5)$$
$$= 504 \text{ Ah}$$

Average daily depth of discharge = (total daily load Ampere-hours)/(total battery bank capacity)

$$= 63 \text{ Ah/295 Ah}$$
$$= 12.5\% \text{ daily}$$

For lead-acid batteries, generally one wants to design for 10–15% daily DOD. Battery selection would comprise 12 V DC at 100 Ah capacity each. Series/parallel configuration for the battery bank would be

number of batteries in series = 12 V/12 V = one battery; and
number of batteries in parallel = 504 Ah/100 Ah = five batteries.

This would yield a configuration total of five batteries at 500 Ah capacity.

7.5.2.4 Array Estimation

Estimating the size of the PV array for a PV lighting system is based on providing adequate energy to meet the load during the period with the highest average daily load and lowest solar insolation on the surface.

- Design month is December at 5.4 h (Oaxaca).
- Assume PV array temperature derate averages 15% of daily requirement.
- Assume inverter losses at 10% of daily requirement.
- Assume fuses/disconnect losses at 1%.
- Assume wiring losses at 3%.
- Assume battery losses at 25%.
- Total system losses are then 0.85 * 0.90 *0.99 * 0.97 * 0.75 = 55%
- Adjusted system load requirement = 63 Ah/day/0.55 = 114 Ah/day.

Selection of the PV module for use in design (the module derating factor is usually 80–90%) is for a rated peak current of 3.55 A and rated peak voltage of 16.9 V at 60 W. The number of parallel modules = the adjusted load Ampere-hour requirement/module peak current output * peak sun-hours. Note that we have already adjusted for module derate in the load requirement: = 114 Ah/day/(3.55A * 5.4 h) = 5.95, rounded to six modules in parallel. The number of series modules = nominal system voltage/module peak voltage output * voltage temperature derate = 12 V/(16.9 V * 85%) = 0.84, rounded to one module in series. Thus, the total PV array is 1s × 6p, totaling 360 W.

7.5.2.5 System Summary

Total Watt-hour requirement:	63 Ah/day
PV array size:	360 W—six 60-W modules
Nominal system voltage:	12 V DC
Battery bank capacity:	five batteries totaling 500 Ah
Battery type:	deep cycle 12 V at 100 Ah each
Autonomy:	3 days
Average daily depth of discharge:	12.5%
Inverter:	12 V DC nominal (10.8–15.5V); 500 W continuous output; 1,000 W surge capacity power; 90% efficiency

7.6 SOLAR WATER PUMPING SYSTEM SIZING

Chapter 8 discusses solar water pumping applications in detail.

A solar pump can be chosen based on its peak flow rate when the daily water requirement is known. In other words, for X liters per day, X liters per second must be pumped. There are three methods of choosing a flow rate to meet daily water requirements:

1. If the pump manufacturer has created a pump selection table based on *daily volume,* the flow rate is calculated for the consumer. The manufacturer's calculation is probably more precise than the following two methods and it based on measured performance.
2. Required liters per day can be divided by *peak sun-hours* to determine liters per hour. This is a logical but overly simplified method that is usually optimistic. The reasons will become apparent in the next section.
3. Required liters per day can be divided by *effective peak sun-hours* based on the following method. This uses general rules to define how the performance varies throughout the day.

This is a function of the type of pump (centrifugal or positive displacement) and the design of the solar array (fixed mount or sun tracking) and oversize factor.

7.6.1 General Method of Sizing a Solar Pump

This method, developed by Windy Dankoff (Dankoff, 2003), is reasonably accurate when one is given a pump's flow and power draw at a given head, based only on its peak (full-sun) performance. The graph in Figure 7.2 illustrates four ways to obtain the same daily volume of water (also see Table 7.2). The variables are (1) centrifugal pump (C), positive displacement pump (P), (3) fixed-mount (F) PV array, or (4) tracking (T) PV array. In the figure, curve CF represents high flow with poor efficiency early and late in the day. Curve PT represents a low-flow pump utilizing the full length of the solar day.

Of the four types of systems represented, the most economical will usually be that which produces the *lowest* flow rate for the *longest* duration—that is, the lowest and widest curve. It will use a smaller PV array that makes better use of the available solar resource. It also results in

1. reduced pump size and weight (more portable, easier to install and remove by hand);
2. reduced pipe and electrical wire sizes (lighter weight; more economical, especially if very long; easier to install and remove by hand);
3. reduced solar array size; and
4. reduced costs of freight and transportation, storage, and installation.

The next steps are to

1. determine which system is commercially available (a high-flow system may be available only with a centrifugal pump);
2. determine whether solar tracking is economical and desirable; and
3. compare the total delivered and installed cost of each system, if the choice is still not clear.

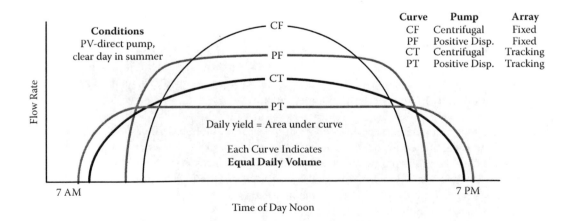

FIGURE 7.2 Solar pump performance (Developed by Dankoff Solar, 2003).

TABLE 7.2
How to Estimate Effective Peak Solar Day for Clear Weather

Type of pump	Type of array	Pump factor		Array factor	Array oversize factor	Net multiplier factor
Centrifugal	Fixed	0.7	X	1.0	X	=
Centrifugal	Tracking	0.9	X	1.7	X	=
Positive displ.	Fixed	1.0	X	1.0	X	=
Positive displ.	Tracking	1.5	X	1.4	X	=

Notes: For a cloudy climate, oversizing the array is more effective than tracking. Array oversize is another variable that can enhance the effective solar day. Array oversize = PV array Watts divided by minimum array size recommended for the pump. Effective peak solar day = peak solar day hours from solar insolation data map × net multiplication factor. Peak flow required (liters per hour) = daily water demand/effective peak solar day (Dankoff, 2003).

7.7 GENERIC WATER PUMP SIZING METHODOLOGY

Spreadsheets can also be used to determine the size of the PV array for solar pumping generically. Although this procedure is not as exact as using pump curves, it can serve to determine preliminary technical parameters for a generic pump, assuming a certain efficiency. The following three spreadsheets contain boxes that should be filled in the order in which they are presented (Sandia, 1995; Foster, 1997). They have the following format:

1. Water volume needed (l/day): Make note of the daily water requirement desired by the user. Choose the month of the year that requires the greatest water-pumping flow rate. Table 7.3 shows sample calculations using data from a particular site. Simply substitute the numbers from the proposed site to arrive at the critical month. Make note of the daily demand and divide this number by the number of hours of peak solar insolation to obtain the flow rate. Insolation values can be found in the appendix. Use estimated or available solar data closest to the project site.
2. Site insolation (kWh/m²/day): Make note of the peak hours of insolation per day corresponding to the critical-pumping month.
3. Pumping regime (l/day): Calculate this value using the previous information. This value should not surpass the well recharge capacity. If this is the case, consider reducing the daily demand, using battery storage for 24-hour pumping, or drilling additional wells.
4. Static level (m): This is the vertical distance measured from ground level to the water level when the pump is not operating.
5. Drawdown (m): This is the vertical distance measured from the static level to the water level when the pump is operating.
6. Height of discharge (m): This is the vertical distance measured from ground level to point at which the water is discharged.
7. Static head (m): Calculate the vertical distance traveled by the water, from the point of drawdown to the point of discharge, using the values in list items 4, 5, and 6.
8. Additional pipe length (m): This is the remainder of pipe not included in the static head calculation. Take into account the vertical distance from the drawdown to the position of the pump, as well as any horizontal distance traveled by the tubing.
9. Total pipe length (m): This is the combined pipe length of the entire system. Calculate this value using list items 7 and 8.

TABLE 7.3
Critical-Month Calculation

Month	Daily demand (l/day)		Insolation (peak h/day)		Flow rate (l/h)
January	8,000	÷	5.8	=	1,379
February	8,000	÷	6.4	=	1,250
March	10,000	÷	6.8	=	1,471
April	10,000	÷	6.9	=	1,449
May	10,000	÷	6.9	=	1,449
June*	12,500	÷	6.4	=	1,953
July*	12,500	÷	6.4	=	1,953
August	12,500	÷	6.5	=	1,923
September	12,500	÷	6.8	=	1,838
October	10,000	÷	6.8	=	1,471
November	10,000	÷	6.0	=	1,667
December	8,000	÷	5.2	=	1,538

* Critical design month: Month when the maximum load occurs during the lowest insolation month.

10. Friction factor (decimal): This is caused by the frictional forces that result when water passes through a pipe (typically should be less than 5% by selecting an appropriate diameter pipe).
11. Frictional head (m): This accounts for the frictional losses that occur when water is in contact with the "rough" pipe wall; it is expressed in meters.
12. Static head (m): Write in the value calculated in list item 7.
13. Total dynamic head (m): Calculate this value using list items 11 and 12. It is the sum of the frictional head and the static head. Note: With the information obtained through this item, it is possible to select an adequate pump. Consult the literature provided by the manufacturer. Fill in the box "Pump and Motor Information" and then continue with the next list item.
14. Volume of water needed (l/day): Note the value from list item 1.
15. Total dynamic head (m): Write down the value from list item 13.
16. Conversion factor: The factor, 367 l-m/Wh, is used to determine the hydraulic pumping energy required in Watt-hours needed to lift 1 liter of water a distance of 1 m. This value is based on Newton's law and is a physical constant.
17. Hydraulic energy (Wh/day): Calculate the energy needed to lift the water using list items 14–16.
18. Pump efficiency (decimal): This is the percentage of electrical energy transformed into hydraulic energy. Daily production will vary according to total dynamic head (TDH), solar

TABLE 7.4
Generic Values for Pumping-System Efficiencies

Total dynamic head (meters)	Pumping system type	Efficiency (%)
5	Surface centrifugal	20–25
20	Surface centrifugal	10–15
20	Submersible centrifugal	20–25
20–100	Multistage centrifugal	25–35
50–100	Positive displacement	25–35
>100	Piston-type positive displacement	35–45

insolation, and pump type. Look for specific information in publications provided by manufacturers. If this information is not available, the generic values presented in Table 7.4 can be used for an estimate. Actual pump performance will vary depending on the pump manufacturer.

19. Energy provided by the PV array (Wh/day): Calculate the energy needed to operate the system using list items 17 and 18.

20. Nominal system voltage (V): Write the voltage that the system should be running at during the day. This is the array input voltage to the inverter or controller.

21. Array current (Ah/day): Calculate the production of the PV array expressed in Ampere-hours per day using the values from list items 19 and 20.

22. Array current (Ah/day): Write in the value from list item 21.

23. Electrical conductor efficiency factor (decimal): Appropriately sized electrical conductors have an approximate efficiency of 95% or better.

24. Corrected current (Ah/day): Amperage required to satisfy the daily load, after considering the losses noted in the previous item.

25. Insolation (kWh/m^2/day): Write in the value from list item 2.

26. Project current (A): Calculate the current necessary to satisfy the system load for the design month using the values in list items 24 and 25.

27. Project current (A): Write in the value from list item 26.

28. Module reduction factor (decimal): PV modules suffer a derate due to temperature. PV modules' nominal ratings are based on an operating temperature of 25°C, but will typically run much hotter than this (>55°C) unless in a cold climate. Assume an average 80% average operating efficiency for crystalline PV modules. Amorphous PV modules have much fewer voltage losses due to temperature effects (≥95% efficiency).

29. Adjusted project current (A): Calculate the minimum array current necessary to activate the pumping system using the previous items.

30. Maximum power current (Imp) of the module (A): Write in the Imp indicated by the manufacturer. Note: Select a PV module and note its specifications in the boxes found in the *PV module information* block.

31. Modules in parallel: This calculation will determine the number of modules to be connected in parallel. If the calculated value is not a whole number, round up to the next highest whole number. Another option is to look for a module with a different Imp and repeat the process beginning with the previous item.

32. Nominal system voltage (V): Write in the value from list item 20.

33. Maximum power voltage (Vmp) of the module (V): Find the Vmp of the module from the information provided by the manufacturer.

34. Modules in series: Calculate the number of modules connected in series necessary to produce the system voltage. If the determined value is not a whole number, round to the next highest whole number.

35. Modules in parallel: Write in the value from list item 31.

36. Total number of modules: Calculate the total number of modules in the array, which is the product of the number of modules in parallel and the number of modules in series. Be sure that the calculated value is a multiple of the number of modules in parallel.

37. Maximum power current (Imp) of the module (A): Write in the value from list item 30.

38. Maximum power voltage (Vmp) of the module (V): Write in the value from list item 33.

39. Size of the PV array (W): Calculate the power of the PV array using the three previous list items.

40. Modules in parallel: Write in the value from list item 31.

41. Maximum power current (Imp) of the module (A): Write in the value from list item 30.

42. Nominal system voltage (V): Write in the value from list item 20.

43. System efficiency factor (decimal): Write in the value from list item 18.

44. Conversion factor: Write in the value from list item 16.
45. Site insolation (peak hours/day): Write in the value from list item 2.
46. Module reduction factor (decimal): Write in the value from list item 28.
47. Total dynamic head (m): Write in the value from list item 13.
48. Amount of water pumped (l/day): This is the number of liters pumped per day using this design. Calculate this value using the values from list items 40–47.
49. Amount of water pumped (l/day): Write in the value from list item 48.
50. Site insolation (peak hours/day): Write in the value from list item 2.
51. Pumping regime (l/h): Calculate the water-pumping regime and compare it to the water-source capacity and the value obtained from the critical-pumping-month tables.

(See worksheets next page)

The example given before indicates that the array necessary for this system consists of 14 modules of 53 peak Watts each—connected two in parallel and seven in series—thus giving a nominal power of 742 peak Watts. When sizing took place, another module or pump could have been selected.

After filling in list item 10, it was found that the total dynamic head was close to 40 m. This information was used to select the pump. All manufacturers publish tables and graphs that assist in the selection of an appropriate pump. Some include recommendations on the approximate size of the PV array necessary to power the pump.

Solar pump manufacturers publish pump graphs that relate daily water volume, TDH, insolation, and PV-array size. These graphs are pump production curves that are absolutely essential in selecting a pump. The best method to size a specific pump is to use the manufacturers' technical specification sheets.

7.8 ELECTRICAL CODES FOR PV SYSTEM DESIGN

Globally, there is a lack of standards for PV system safety and quality. The National Electrical Code (NEC) has been in use for over a century in the United States, while in Europe the IEC codes are widely applied. Japan, with some of the highest quality PV installations, has its own set of simple electric codes (discussed further in Chapter 8). The electrical codes apply to all building electrical systems, including the PV power systems that continue to be installed in ever increasing numbers around the world. PV power systems, like all other electrical systems, should be designed, specified, and installed to ensure their safety and reliability while complying with appropriate codes and standards, such as the NEC. (Wiles, 2003).

PV systems have current-limited generating PV arrays that are energized when exposed to light. They may use electrochemical energy storage, which can be quite hazardous. Unfortunately, many PV installations around the globe continue to have problems related to poor design and installation that cause safety concerns and diminish system reliability. Some common PV system problems include:

- improper conductors;
- unsafe wiring;
- improper overcurrent protection;
- unsafe batteries;
- lack of grounding;
- use of nonlisted components; and
- improper use of listed components.

Note that "listed components" refers to equipment and materials in a list published by an organization that is acceptable to the authority having jurisdiction and concerned with evaluation of products or services that meet appropriate designated standards (e.g., Underwriter's Laboratories,

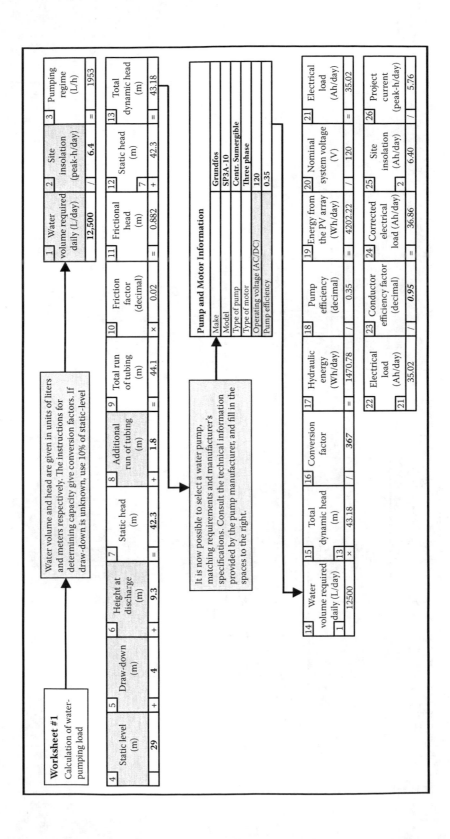

Worksheet #1
Calculation of water-pumping load

Water volume and head are given in units of liters and meters respectively. The instructions for determining capacity give conversion factors. If draw-down is unknown, use 10% of static-level

1 Water volume required daily (L/day)	2 Site insolation (peak-h/day)	3 Pumping regime (L/h)
12,500	/ 6.4	= 1953

4 Static level (m)	5 Draw-down (m)	6 Height at discharge (m)	7 Static head (m)
29	+ 4	+ 9.3	= 42.3

8 Additional run of tubing (m)	9 Total run of tubing (m)	10 Friction factor (decimal)	11 Frictional head (m)	12 Static head (m)	13 Total dynamic head (m)
1.8	= 44.1	× 0.02	= 0.882	+ 42.3	= 43.18

It is now possible to select a water pump, matching requirements and manufacturer's specifications. Consult the technical information provided by the pump manufacturer, and fill in the spaces to the right.

Pump and Motor Information

Make	Grundfos
Model	SP3A-10
Type of pump	Centr. Sumergible
Type of motor	Three phase
Operating voltage (AC/DC)	120
Pump efficiency	0.35

14 Water volume required daily (L/day)	15 Total dynamic head (m)	16 Conversion factor	17 Hydraulic energy (Wh/day)
12500	13 × 43.18	/ 367	= 1470.78

18 Pump efficiency (decimal)	19 Energy from the PV array (Wh/day)	20 Nominal system voltage (V)	21 Electrical load (Ah/day)
/ 0.35	= 4202.22	/ 120	= 35.02

22 Electrical load (Ah/day)	23 Conductor efficiency factor (decimal)	24 Corrected electrical load (Ah/day)	25 Site insolation (Ah/day)	26 Project current (peak-h/day)
21 35.02	/ 0.95	= 36.86	2 6.40	/ 5.76

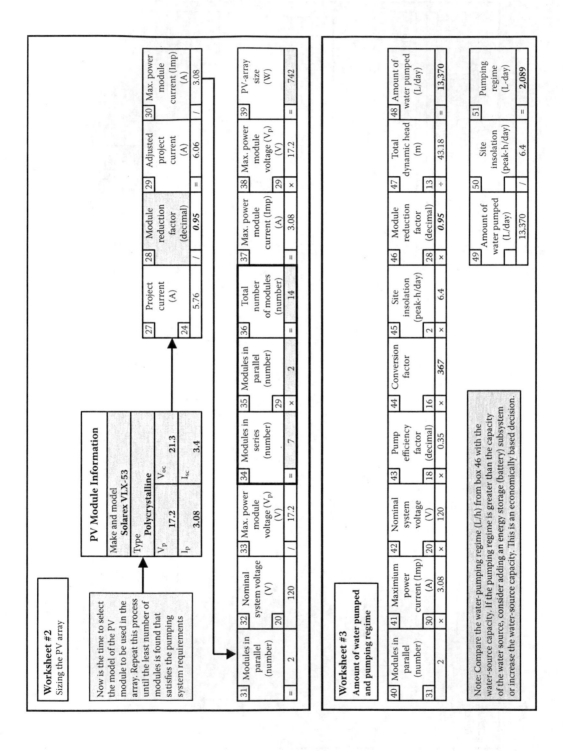

Japan Electrical Safety and Environment Technologies, Canadian Standards Association, and Environmental Testing Laboratories listings).

It is surprising, however—given the millions of Third World PV systems installed by poorly trained personnel (often with little or no overcurrent protection)—that remarkably few fires, injuries, or deaths have been reported from PV; the very few reported cases are usually associated with large utility interactive systems or with someone falling off a roof.

In the United States, the NEC was originally developed in 1897 as a fire-safety code and has been under the auspices of the National Fire Protection Association (NFPA) since 1911. NFPA 70 covers the installation of electrical systems and should be adhered to when electrical systems are designed and installed. The NEC is recognized as a legal criterion for safe electrical design and installation in 48 states. Compliance with the provisions of the NEC can help minimize fire and accident hazards in any electrical design.

Article 690 ("Solar Photovoltaic Systems") was added to the code in 1984 and addresses safety standards for solar PV electrical energy systems including array circuits, power conditioning units, and controllers. Likewise, many of the other sections of the code contain pertinent sections for PV installations, such as wiring, grounding, overcurrent protection, etc. These practices and requirements are applicable to most PV installations in the United States. The NEC is updated every 3 years by the NFPA; the latest version of the code is the 2008 NEC.

NEC-compliant PV systems generally have better performance and reliability than noncompliant systems. Likewise, NEC-compliant installations are safer; even small PV systems can present fire hazards because a single, deep-cycle storage battery (12 V, 100 Ah) can discharge over 6,000 A into a short circuit. Batteries have additional risks due to hydrogen gas explosive potential and acid or caustic burns. Larger PV systems with voltages greater than 50 V also present shock hazards. Of course, even NEC code-compliant systems can still fail and there is no substitute for a long-term maintenance regime.

The NEC addresses nearly all PV power installations for both stand-alone and grid-connected systems. The code deals with any PV system that produces power and has external wiring or electrical components accessible by the public. The NEC is organized into nine chapters (NEC, 2008):

Chapter 1—general: definitions and general requirements for electrical installations;

Chapter 2—wiring and protection: load calculations and circuit sizes, overcurrent protection, and grounding;

Chapter 3—wiring methods and materials: wiring methods and conductor ampacity tables;

Chapter 4—equipment for general use: types of electrical equipment and portable cords and how they are connected and used;

Chapter 5—special occupancies: hazardous locations, health-care facilities, recreational vehicles, mobile homes, motion picture theaters, and other commercial buildings;

Chapter 6—special equipment: electric vehicle chargers, audio systems, fuel cells, swimming pools, x-ray equipment, electric welders, and PV systems;

    Article 690 specifically applies to PV installations. When Article 690 requirements differ from other requirements in the NEC, Article 690 takes precedence.

Chapter 7—special conditions: emergency systems, standby systems, utility-interactive systems, fiber optic systems, and power-limited systems;

Chapter 8—communications systems: radio and TV transmitting and broadband communication systems; and

Chapter 9—tables: conductor properties, sizing, raceways, and conduit.

The NEC requires that available equipment be rated for safety and listed by an acceptable independent testing laboratory such as Underwriter's Laboratories (UL), Environmental Testing Laboratories (ETL), Factory Mutual Research (FM), Asociación de Normalización y Certificación (ANCE) in Mexico (see Figure 7.3), or Canadian Standards Association (CSA). Thus, PV modules

FIGURE 7.3 Mexico's first *NEC*-compliant PV system installed at Montes Azules Biosphere Reserve in Chiapas (1998) by WWF/USAID, Condumex, Sandia, and NMSU.

should be listed under the UL 1703 or IEC 1215 standard for NEC compliance. Some equipment, such as batteries or lightning surge arrestors, is not listed (Ellis, 2001).

The local inspector ultimately has the final interpretation of the NEC and approval of electrical installations. Local inspection authorities sometimes have regional electrical codes, but most jurisdictions use the NEC. The United States has a sophisticated electrical inspection system with about 65,000 electrical inspectors operating in 42,000 local jurisdictions (county, city, etc.). Foreign countries that have adopted the NEC typically have only a handful of electrical inspectors operating in the largest urban centers.

Unfortunately, even in the United States, where the NEC has existed for over a century, a majority of PV systems installed are not yet NEC compliant. Many U.S. inspectors are not familiar with PV systems and rarely come across them in their work. Likewise, many PV installers are unlicensed and may have little familiarity with the code.

Although numerous countries have adopted the NEC, they do not update the code every 3 years and are often working with outdated versions. This is a more serious concern for newer NEC technologies like PV, where many changes are still occurring during each code modification cycle as the technology rapidly evolves.

Gradually, the NEC, IEC, and UL are coming together to coordinate standards. A great deal of work is needed in educating the PV industry globally on how to design and install code-compliant systems. There should be a focus on PV developers to require code compliance until such time as the individual countries have an established inspection mechanism in place. In general, project developers want good-quality and safe PV systems and are willing to specify codes once they learn about them.

Similarly, the Global Approval Program for Photovoltaics (PV GAP) was founded in Geneva, Switzerland, as a nonprofit organization focused on developing global PV markets, particularly in the rural sector. The World Bank and the United Nations Development Program (UNDP) initially funded and helped organize PV GAP. A variety of PV manufacturers belong to PV GAP, including BP, Isofoton, Photowatt, and Schott). PV GAP is administered by the IEC's System for Conformity Testing and Certification of Electrical Equipment (IECEE), which is responsible for the certification program. PV GAP has developed a reference manual for PV manufacturers, first published in 1998. PV GAP maintains a list of recognized PV testing laboratories worldwide, as shown in Table 7.5 (PVGAP, 2005).

TABLE 7.5
PV GAP Recognized PV Testing Labs

Institution	Location	Country
Arizona State University: PV Test Lab	Mesa, AZ	United States
CIEMAT-PVlabDER	Madrid	Spain
European Solar Test Installation: Joint Research Center	Ispra	Italy
Japan Electrical Safety and Environment Technology Lab: Yokohama Industries	Kanagawa	Japan
Kema Quality B.V.	Arnhem	Netherlands
TÜV Rhineland Product Safety GmbH	Koln	Germany
UL Northbrook Office	Northbrook, IL	United States
VDE Testing and Certification Institute	Offenbach	Germany

A PV Quality Mark was established for PV components and a PV Quality Seal was established for systems. The mark and seal are licensed to companies and products that achieve approval under the PV GAP Program. The World Bank has developed the Quality Program for Photovoltaics (QuaP-PV), which promotes PV GAP in its technical specifications, especially for the Asia region (World Bank, 2009).

7.9   STAND-ALONE PV LIGHTING DESIGN EXAMPLE

The following simple example shows the steps for designing a PV system to be code compliant. To follow along, it is helpful to consult the NEC, especially Article 690 on PV and Article 310 on wire ampacities (Wiles, 2003).

General system specifications:
Array size—4 modules, Isc = 4.0 A
Voc = 21.3 V, 64 W module

System electrical configuration—24 Vdc (2s × 2p)
Battery size—200 Ah at 24 Vdc
Load size—60 W at 24 Vdc
Lamp—metal-halide, electronic ballast

Goals:
- determine design current to use for wire and fuse selection;
- size wiring for the array at anticipated temperatures (insulation type is a determining factor);
- determine the suitability of the overcurrent device terminals (can be dependent on selection of wire size and insulation type);
- determine the suitability of the overcurrent device;
- use open-circuit voltage times 125% as design voltage (Figure 7.4).

All ungrounded conductors should have overcurrent protection.

Use signs and labels to indicate operating parameters.

Design current equals correction factor times array short-circuit current at 1,000 W/m and 25°C

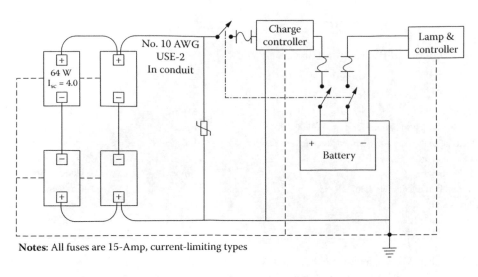

Notes: All fuses are 15-Amp, current-limiting types

FIGURE 7.4   PV lighting system example electrical schematic (DC).

- irradiance correction factor = 125%
      1,000 W/m = 1,200 W/m
- terminal correction factor = 125%
      reduce connector stress.

Design current = array short circuit (156%).

Design array current.
array short-circuit current 8 A (2p × 4A Isc each)

terminal correction factor: 1.25 × 8 A = 10 A

irradiance correction factor: 1.25 × 10 A = 12.5 A

cable run = 6 m

Design load current:
- load units to Amperes 60 W/24 V = 2.5 A
- terminal correction factor: 1.25 × 2.5 A = 3.1 A

From NEC wire tables, number 10 AWG USE-2, 90°C insulation:
- module interconnections
- modules to control box
      ampacity 40 A at 30°C ambient temperature in conduit
      temperature correction factors for 90° insulation:

ambient temp.	correction factor
21–25°C	1.04
26–30°C	1.00
31–35°C	0.96
61–70°C	0.58

(40 A) (0.58) = 23.2 A at 70°C ambient temperature

design array current = 12.5 A

Check overcurrent device terminal requirements:
- fuse holder terminal 60°C

Ampacity of cable:
- #10 AWG USE-2, 40 A at 30°C ambient temperature

Normal operation ratings:
- Fuse size determines ampacity
- terminal rating 60–75°C at 40°C ambient temperature

Determine cable temperature at full current and compare to terminal rating
selected cable = #10 AWG USE-2 90°C insulation
determine whether temperature of #10 AWG USE-2 wire at 40°C ambient temperature and 12.5 A is <60°C to rate wire temperature rating
   90°C insulation maximum values:

ambient temp.	correction factor	Amperes
26–30°	1.00	40.0
36–40°	0.91	36.4
61–70°	0.58	23.2

Wire temperature versus current—#10 AWG USE-2 90°C insulation at 40°C ambient temperature:

current	wire temperature
0	40°C
36.4	90°C
12.5	??? °C

$$I = \sqrt{\frac{TC - (TA + DELTA\,TD)}{RDC(1 + YC)RCA}}$$

(7.1)

where

I = corrected current in Amperes
TC = conductor temperature in degrees Celsius
TA = ambient temperature in degrees Celsius
DELTA TD = dielectric loss temperature rise
RDC = DC resistance of conductor at temperature TC
YC = component AC resistance resulting from skin effect and proximity effect
RCA = effective thermal resistance between conductor and surrounding ambient temperature

For 10 AWG, 60°C insulation:

ambient temp.	current (Amperes)
26–30°C	30.0
31–35°C	27.3
36–40°C	24.6

41–45°C   21.3

this application: 12.5 A < 24.6 A

derated ampacity of 10 AWG USE-2 cable is 23.2 A and required current is 12.5 A; use a 15 A DC fuse or circuit breaker

Design currents for array are 156% × Isc at 1,000 W/m2 and 25°C, 12.5 A

#10 AWG USE-2, 90°C insulation is OK

at 70°C, module back temperature has 23.2 A ampacity and design current is 12.5 A

fuse terminal will operate below its 60°C rating at 40°C ambient temperature

a 15 A fuse will protect the wiring of this system (ampacity of wiring is 23.2 A)

Voltage drop of wires was not a determining factor. This example was provided to show a safety guideline requirement only. DC losses are typically designed to be less than 3% of voltage losses (Ellis 2001).

REFERENCES

2008 National Electrical Code. Quincy, MA: National Fire Protection Association.
Dankoff, W. 2003. Sizing a solar water pump to meet daily water requirements. Santa Fe, NM.
Ellis, A., R. Foster, A. Sánchez, A. Romero, and C. Flores. 2001. Sistemas Fotovoltaicos de Acuerdo a las Normas de Seguridad [PV systems safety codes]. XXV Semana Nacional de Energía Solar, ANES, San Luis Potosí, San Luis Potosí, Mexico, October 1–5, 2001, 115 pp.
Foster, R. E., A. Ellis, O. Carrillo, and G. Cisneros. 1997. Guía para el Desarrollo de Proyectos de Bombeo de Agua con Energía Renovable [Renewable energy water pumping guide]. Programa de Cooperación FIRCO-Sandia, Sandia, USDOE, FIRCO, USAID, San Luis Potosí, August 13–15, 1997.
Sandia National Laboratories, Stand-Alone Photovoltaic Systems, 1995. SAND87–7023, Albuquerque, New Mexico.
Wiles, J., A. Ellis, and R. Foster. 2003. Sistemas de Energía Fotovoltaica de Acuerdo a las Normas de Seguridad: Prácticas Recomendadas para CentroAmérica [PV energy systems and standard safety practices: Recommendations for Central America]. Guatemala City, Guatemala, August 2003.
World Bank. 2009. http://www.pvgap.org
World Bank. 2009. http://www.worldbank.org/astae/quappv/

PROBLEMS

7.1   Assuming a total load of 1,200 Wh/day and a solar resource of 6 peak sun-hours/day, what is the size PV array required to meet this load? Assuming a 105 Ah battery at 12 V, how many batteries are required for 3 days of autonomy?

7.2   Calculate the PV system size required to pump 10,000 l/day at a total dynamic head of 25 m. Assume a pumping system efficiency of 35% and an average solar resource of 5 peak sun-hours/day.

7.3   Consulting Article 310 and related tables in the NEC, what is the allowable ampacity for a single insulated copper conductor AWG#12 THHN in free air, based on a temperature rating of conductor at 90°C?

8 Photovoltaic (PV) Applications

8.1 INTRODUCTION

PV systems can be used for a wide variety of applications, from small stand-alone systems to large utility grid-tied installations of a few megaWatts. Due to its modular and small-scale nature, PV is ideal for decentralized applications. At the start of the twenty-first century, over one-quarter of the world's population did not have access to electricity, and this is where PV can have its greatest impact. PV power is already beginning to help fill this gap in remote regions, with literally millions of small residential PV systems installed on homes around the world, most commonly as small stand-alone PV systems, but also increasingly as larger on-grid systems in some industrialized regions (notably Japan, Germany, and California). Ironically, the wealthy, who want to demonstrate that they are "green," or often impoverished remote power users, who need electricity and have limited options, form the majority of PV users.

8.2 GRID-TIED PV

Decentralized PV power production promises to be a widely applicable renewable energy source for future clean energy production. Because most of the electric power supply in industrialized countries is via a centralized electric grid, the widespread use of PV in industrialized countries will be in the form of distributed power generation interconnected with the grid. Indeed, since 2000, the fastest growing market segment for PV has been in the grid-tied sector. Utility-interactive PV power systems mounted on homes and buildings are becoming an accepted source of electric generation. This tremendous growth has been due to government incentives and policies encouraging clean energy out of concern for the environmental impacts, especially global warming, of conventional electric generation technologies (especially coal). Growth has been particularly phenomenal in Europe, Japan, and California.

Grid-tied PV represents a change from large-scale central generation to small-scale distributed generation. The on-grid PV system is really the simplest PV system. No energy storage is required and the system merely back-feeds into the existing electrical grid. This growth has also had unintended consequences for the off-grid market, in that many module manufacturers have ceased production of their smaller, battery-charging PV modules in favor of larger, higher voltage modules made for on-grid inverters.

Utility-interactive PV systems are simple yet elegant, consisting of a PV array (which provides DC power), an inverter, other balance of systems (such as wiring, fuses, and mounting structure), and a means of connecting to the electric grid (by back-feeding through the main electric service distribution panel). During the daytime, DC electricity from the PV modules is converted to AC by the inverter and fed into the building power distribution system, where it supplies building loads. Any excess solar power is exported back to the utility power grid. When there is no solar power, building loads are supplied through the conventional utility grid. Grid-tied PV systems have some advantages over off-grid systems:

- *Lower costs.* Grid-tied PV systems are fairly simple and connect to the standard AC wiring. Only two components are required: the PV modules and the inverter (with associated wiring and overcurrent protection).

- *No energy storage.* Because the utility grid provides power when the PV system is off-line, no energy storage is required. The grid effectively is the energy-storage bank, receiving energy when a surplus is generated and delivering energy when the load exceeds on-site generation.
- *Peak shaving.* Typically, sunlight and thus PV peak power production coincide with utility afternoon peak loading periods; the utility gains from solar peak shaving. Even better, during the summer cooling season when the sun is out and hottest, this is exactly when the PV system will be producing maximum power. With grid-tied PV systems, daytime peaking utilities gain a reduction in peak load while not impacting off-peak energy sales. The customer benefits by having lower utility bills while helping the utility reduce peaking loads.

Utility-interactive PV systems cost about \$6–\$8/watt peak (W_p) when installed. Existing rooftops are the lowest cost siting option because both the real-estate and mounting structures are provided at no cost. The system cost includes about \$3–\$4/W_p for the PV modules, about \$0.60/W_p for power conditioning, and from \$2 to \$3/W_p for mounting and labor. Thus, a turn-key 2 kW_p PV residential system will cost about \$12,000–\$16,000.

For a location receiving an average of 5 sun-hours/day (for example, Atlanta, Oklahoma City, or Orlando), a 2 kW_p system after system losses will produce about 2,700 kWh/year. At a value of \$0.10/kWh, this energy is worth a little over \$270/year. Assuming that the system cost about \$12,000 to install, simple payback for a grid-tied PV system is over 40 years. Grid-tie PV life-cycle costs are typically over \$0.20/kWh, assuming a relatively good solar resource and amortizing over a couple of decades. Although PV system prices can be expected gradually to decrease, it will still be a couple of decades before they are competitive with the grid in the United States. However, in places like Japan or Germany, where grid power is already more than double the cost in the United States, PV has achieved basic parity with grid-tied power on a life-cycle cost basis, as discussed in Chapter 9.

There are also no real issues with PV systems endangering line workers; indeed, many knowledgeable utilities no longer require an outside disconnect. A PV inverter behaves very differently than a conventional rotating-type generator that powers the grid. A rotating generator acts as a voltage source that can generate independently of the grid and is synchronized with it. A PV inverter acts as a sinusoidal current source that is only capable of feeding the utility line by synching up with it when voltage and frequency are within standard limits. Thus, islanding (independent operation of the PV inverter) is for all practical purposes impossible because line voltage is not maintained by PV inverters. Also, under fault conditions, a rotating generator can deliver most of its spinning energy into the fault. A PV inverter, which is a controlled-current device, will naturally limit the current into a fault to little more than normal operating current. The PV cells themselves act as current-limited devices (because output current is proportional to sunlight).

Modern PV inverters use pulse-width modulation (PWM) to generate high-quality sinusoidal currents, so harmonic distortion is not a problem. Modern PWM inverters also generate power at unity power factor (i.e., the output current is exactly in phase with the utility voltage). Grid-tied PV inverters are designed with internal current-limiting circuitry, so output circuit conductors are inherently protected against overcurrent from the PV system. The overcurrent protection between the inverter and the grid is designed to protect the AC and DC wiring from currents from the grid during faults in the PV system wiring. PV inverters are available in a range of sizes, typically 1–6 kW with a variety of single phase voltage outputs including 120, 208, 240, and 277 V. The interconnection from the inverter to the grid is typically made by back-feeding an appropriately sized circuit breaker on the distribution panel. Larger inverters, typically above 20 kW, usually are designed to feed a 480 V three-phase supply.

Typically, PV power producers enter into an interconnection agreement with the local utility for buying and selling power and the necessary metering scheme to support this arrangement. The basic options include:

- for net metering, a single bidirectional meter;
- for separate buy and sell rates, two individual ratcheted meters to determine the energy consumed and generated; and
- other arrangements that take advantage of time-of-use rates. These may require additional meters capable of time-of-use recording, which is particularly advantageous for PV power producers because PV power production normally coincides with peak rate periods.

Grid-tied PV power systems have proven to be a reliable method of generating electricity. Some of the largest grid-tied PV power installations and highest concentrations of PV residences in the world can be found in Japan (Figure 8.1). A closer look at what the Japanese have accomplished will give a good idea as to where the rest of the world will be going over the next couple of decades.

8.3 JAPANESE PV DEVELOPMENT AND APPLICATIONS

Japan has one of the most advanced and successful PV industries in the world, which warrants a closer look. Japan became the first country to install a cumulative gigawatt of PV back in 2004. Through aggressive government policies beginning with the SunShine Program launched in 1974 and then more recent subsidies promoting deployments, Japan has become a global PV production and industry leader. PV-powered homes are now a common site throughout Japan. Japan used to provide half of global production, but now provides one-fifth of global PV production as the rest of the world ramps up. Japan's Sharp was the second largest global producer in 2007 with 370 MW (but it has produced more than that in the past and was supply constrained) and Kyocera ranked fourth with 200 MW. Other key producers in 2007 included the world's largest producer in Germany (QCells with 400 MW) and China's Suntech, which is ranked third with 300 MW in 2007 (Renewable Energy World 2008).

The Japanese government is making solar energy an important part of its overall energy mix, with a goal of 10% electricity production from PV by 2030. It seeks to reduce PV costs to be on par with conventionally generated electricity. Likewise, Japan is a signatory to the Kyoto Protocol and sees solar power as a viable part of the solution to meeting CO_2 reduction targets. Japan became a global PV leader for three key reasons:

FIGURE 8.1 Ohta City has the highest concentration of grid-tie PV homes, with over 500 homes installed in this neighborhood.

- aggressive government policies promoting PV to help meet Kyoto Protocol goals;
- tight research and development (R&D) collaboration among industry, government, and academia; and
- majority overseas exports helping to drive down PV in-country manufacturing costs.

Individual homeowners are the most common PV buyers in Japan, comprising nearly 90% of the market. In Japan, there is a twofold reason for buying PV. First, the Japanese consider it "good to be green" and have ties to nature that are culturally embedded. Second, the retail price of residential electricity in Japan is the highest in the world, at ~¥23/kWh (~US$0.21/kWh). Thus, over a 20-year lifetime, grid-tied PV power is actually cost effective. Initially, the government offered substantial rebates on PV installations (50% in the mid-1990s), but these rebates were dramatically reduced and phased out as PV prices dropped. Japan's budget for development and promotion of PV systems has more than halved since its peak in 2002. This has been possible as PV prices have decreased, and homeowners without rebates today are paying approximately the same price they paid a decade ago with rebates. Some local city and county governments do continue to offer incentives for PV installations.

The costs of PV systems in Japan are among the lowest in the world and were down about ¥670/W_p (or about US$6/$W_p$) installed for residential installations by 2004. Japan is able to achieve lower costs through simplified balance of systems, including transformerless inverters. All equipment used is manufactured in country. Japan also has a customized mass production technique and some housing manufacturers offer PV options on homes. Likewise, regulations are simple and nonprescriptive for PV installations. There are no special PV installers; rather, electricians are trained by industry to install PV systems. Installations are self-inspected. The Japanese electric code for PV is simple (one page) and not prescriptive. The Japanese rely on the industry to self-police and do a good job out of a cultural honor tradition. If there is a problem, the homeowner can make claims against the warranty and company. Most Japanese companies are very responsive if there is a problem because it is a matter of honor and pride for them to do a good job. Indeed, Japan has among the best installed PV systems anywhere.

PV systems are also made easy for homeowners to use and understand. Simple graphical displays are used so that homeowners can easily see how their PV systems are doing on a real-time and cumulative basis. This generates interest and participation from the homeowner, who in turns shows off his system to his friends and learns to conserve electricity. Systems are metered and the homeowner sees a reduction in his monthly electric bill by using a PV system.

Overall, PV technology deployment in Japan is mature and there are few reported failures. The government has put most of its funding into deployment and determining how to maximize power from clustered PV systems. Basic research is shifting toward thin film technologies and the Japanese are leading the world on how to recycle PV modules.

Japan is a global PV manufacturing leader and also has the most mature PV market in the world. The Japanese market represents about one-twentieth of global PV sales, and the country exports over 60% of its PV module production overseas. The rapidly changing Japanese market and experience hold a number of lessons learned that are pertinent for other countries interested in large-scale PV deployment. Numerous technology and policy insights can be gained from the Japanese experience.

The Japanese government has been developing a self-sustaining residential PV market free of incentives. There has been a successive annual decline in government subsidies that were phased out by 2006. The reason for this is that PV prices have declined over 30% in the last decade, and PV is now competitive in Japan, especially because domestic grid power costs about US$0.21/kWh. PV is now an attractive and economically competitive electricity option for many homeowners.

Few nondomestic companies operate in the Japanese market. Although there are no particular trade barriers for other companies to sell product in Japan, the national Japanese market is so competitive that most foreign manufacturers find it difficult to enter. The Japanese PV manufacturers should continue to lead global PV production in the future. They have learned how to make it cheaper and better through mass commercialization.

The Japanese culture has always had strong ties to nature, exemplified through the country's famous gardens, poetry, etc. Likewise, the Japanese culture has always had a unique relationship with the sun, reflected on its national flag as the "Land of the Rising Sun." Thus, many Japanese view the use of solar energy as in keeping with their cultural traditions. With the signing of the Kyoto Protocol on Global Warming, the Japanese also see it as a matter of national pride for Japan to meet its share of the protocol's objectives on limiting CO_2 emissions. Thus, again, solar energy is seen as an important part of the solution to achieving these objectives. This attitude permeates all levels of the society, from homeowners to schools, government, and industry. Most want to use solar energy on their buildings and help the country become "solarized."

Countering the effects of global warming is a mainstay of Japanese government policy. Economics for PV plays a secondary role as compared to national goals of meeting the Kyoto Protocol. The prime minister's residence, as well as the Japanese Parliament and many key government buildings, all have 30- to 50-kW PV arrays mounted on their rooftops (Figure 8.2). There is nearly a megaWatt installed on key government buildings in downtown Tokyo. A total commitment to making Japan a solar nation exists from the government officials and planners, industry leaders, and the public. Japan has an integrated solar development approach. Also, there is a sense of need for energy independence. Because grid electric costs are the highest in the world in Japan, there is also an economic return for residential PV.

The Japanese also feel that the expansion of PV power generation systems in Japan will greatly contribute to creating new jobs and industries in the coming decades. This meets the goals of energy and industrial policies that the Japanese government is pursuing.

Most of the Japanese PV systems are installed on single-family residences belonging to average homeowners. These are typically middle-aged Japanese parents with a couple of children. The typical household income in Japan is ¥6.02 million per year (MHLW 2002). Most of the Japanese PV systems (about three-quarters) are installed as retrofits on existing homes. Typical household electricity consumption in Japan is 290 kWh/month (JAERO 2004); this is more than half that of the United States. In Japan, a 1 kW_p PV system annually generates about 1,050 kWh/kW_p on average.

FIGURE 8.2   Installed grid-tied PV array on Japanese prime minister's official residence (Sishokante) signals to the country the government's deep commitment to solar energy.

Although the majority of PV systems are installed as retrofits on existing homes, some prefabricated homes also offer PV as part of a package deal. There is no standardized specification, and manufacturers are free to partner with the PV companies that offer them the best deals. More and more of the prefabricated homes will offer a PV option in the future.

Close cooperation among government, industry, and academia has made Japan a leading producer of solar cells in the world, with about 16% of global production (previously Japan had over 40% of global production as did the U.S. before that, but other countries like China and Germany have greatly increased production). Of the installed systems in Japan, about 92% are for grid-connected distributed applications such as residences and public buildings. Total PV production in Japan for 2006 was 927 MW. Sharp is the largest PV module producer, with about 370 MW of production in 2007 (Renewable Energy World 2008).

Japan sets the global standard for residential PV installation programs in terms of size and cost. The country currently installs 50,000–60,000 PV homes per year in cooperation with the large cell manufacturers and the home builders. Japan has more PV homes than any other country; total number of residential PV systems will surpass a half million by 2010. Given the large PV manufacturing base in Japan, PV systems are more inexpensive there than in the rest of the world. The balance of systems (BOS) is also cheaper due to simplified electrical code requirements. The average residential PV system cost is about ¥650/W_p (<US$6/$W_p$) (Kaizuma, 2005/2007).

8.3.1 JAPANESE GOVERNMENT'S APPROACH

The Japanese government supports PV development at every step, from the prime minister and Parliament down to the different implementing agencies. The Ministry of Economy, Trade and Industry (METI) began a subsidy program for residential PV systems (PV modules, BOS, and installation) in 1994. At first, the subsidy covered 50% of the cost. The program was open to participants from residential homes, housing complexes, and collective applications. By 1997, METI grew the program to encourage mass production of PV systems (Figures 8.3 and 8.4). After achieving their price goals, the Japanese government rolled back the subsidy program in 2003 and had largely phased it out by 2006. The Japanese government has now shifted focus to larger commercial- and utility-scale systems (e.g., water plants for backup power; see Figure 8.5).

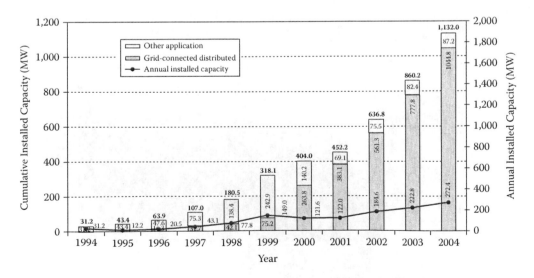

FIGURE 8.3   Growth of Japanese PV installations from 1994 to 2004 (IKKI, 2005).

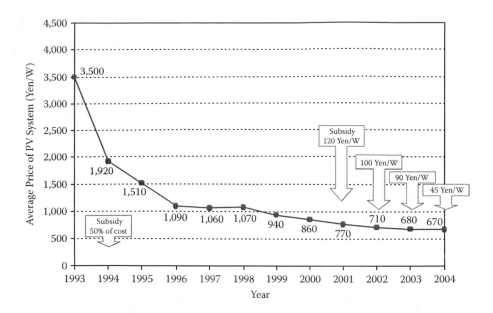

FIGURE 8.4   Average PV system price in Japan from 1994 to 2004 and corresponding national government subsidy, which was phased out by 2006 (Ikki, 2005).

FIGURE 8.5   Project designer Mr. Ohashi and Osaka Waterworks Kunijima Treatment Plant with a 150 kW$_p$ PV plant (Kyocera), one of over a dozen such PV water plants in Japan.

8.3.2 JAPANESE PV UTILITIES

The electrical sector in Japan is deregulated. There are five electric utilities in Japan, all of which are investor owned. Generation, transmission, and distribution are vertically integrated. Some independent power producers also generate electricity. The electric generation industry is regulated by the Agency for Natural Resources and Energy (ANRE) of METI.

The distribution network for electricity in Japan is single-phase, three-line, 100/200 V AC. The western part (e.g., Osaka) of the country uses 60 Hz, while eastern (e.g., Tokyo) Japan uses 50 Hz power. This fact also is an advantage for the Japanese inverter industry, which designs inverters for both 50 and 60 Hz for its own market and thus has ready-made products for the European and U.S. markets.

Typical metering arrangements and tariff structures for electricity consumers are 30-minute interval readings. A time-of-use tariff is available. Utilities are responsible for their side of the grid. The PV installation is done by the PV and contractors' industry. There are some big utility PV installations, but over 90% of PV is installed on residences. Normally, a separate meter monitors PV system performance. Japan has about a half million PV-powered homes (Figure 8.6).

8.3.3 JAPANESE MARKETING

PV plays an important role within Japan's overall energy strategy. The government has raised public awareness on climate and energy matters and on how solar PV can bring global and personal benefits. Ongoing government publicity campaigns from both national and local governments discuss the benefits of PV related to environmental issues. PV technology is promoted through a range of media from newspaper to television.

The Japanese PV industry conducts marketing activities for its own PV products. In Japan, solar energy is a popular idea with the public, so industry sales need to differentiate themselves from their competitors rather than selling the public on the solar energy system concept. Most are systems sold to homeowners who have a profound understanding of the ecological impacts of their purchases and are not as concerned about the decades' long payback for the system.

PV commercials are aired on television. One classic solar commercial in Japan by Kyocera shows a young Japanese woman homeowner proudly viewing the energy production of her Kyocera PV system with the Kyocera graphical display meter inside her home. Then there is a clap of thunder and rain, and she is sad that her system is not producing power. The shot cuts away to the PV system and explanation. Soon, the sun comes out again and the birds are singing and the PV system owner is once again pleased about producing energy. Likewise, Sharp has a commercial promoting the ecological aspects of solar energy and exhorts viewers to "change all the roofs in Japan into PV plants."

The Japanese PV industry has also made it easy for consumers to understand the performance of their PV systems, which also figures prominently in advertisements. Instrumentation on installations comes from industry. Graphical meters are simple to read so that homeowners can easily follow their system's performance (Figure 8.7).

FIGURE 8.6 PV system grid inter-tie (note 2 meters) in Ohta City. Inverter and battery bank are housed in the large boxes on the right.

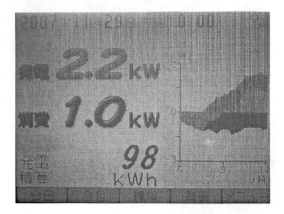

FIGURE 8.7 Consumer-friendly Kyocera residential PV meter display

Overall, Japanese PV systems are professionally installed and exhibit excellent workmanship with dedication to detail. The image of PV in Japan is a positive one that the technology works. Overall, the industry is not highly regulated and the Japanese companies are entrusted to design and install PV systems. Some general guidelines for grid-tied installations have been recommended by JET; although these are not law, they are generally followed by the industry (Jet, 2002).

8.3.4 JAPANESE PV ELECTRICAL CODE

The Japanese Industrial Standards (JIS) specify the standards used for industrial activities in Japan. The standardization process is coordinated by the Japanese Industrial Standards Committee (JISC) and published by the Japanese Standards Association (JSA). The objective of the JSA is "to educate the public regarding the standardization and unification of industrial standards, and thereby to contribute to the improvement of technology and the enhancement of production efficiency (JSA, 2007)." The Japanese have a well established electric code developed after 1945, known as the Technical Standard for Electric Facilities. This simple, technical approach has proved to be very effective and safe in Japan for installing high-quality PV systems. Engineers do not get lost over detailed nonsensical discussions about "how many angels can fit on the head of a pin"—unlike some other industrialized countries with prescriptive electric codes that inhibit growth and innovation of PV systems design.

Japan has among the highest quality PV installations in the world, while maintaining some of the simplest regulations. The equivalent to the U.S. NEC Article 690 for PV in Japan is Section 50 in the Japanese code. It is essentially a simple one-page checklist. Unlike the U.S. NEC, the Japanese code is not prescriptive, but rather more of a handbook. Individual manufacturers are responsible for following the code on their installations. In Japan, the work ethic is such that companies take pride in their work and want to do quality installations. The code does not require use of listed modules, inverters, etc.; however, the manufacturers take pride in getting their equipment listed with JET, and installers will want to use listed equipment. The main points of the Japanese electric code related to PV installations are simple and straightforward (Kadenko, 2004):

- Charging parts should not be exposed.
- PV modules should have a disconnect located near the array.
- Overcurrent protection should be installed for PV modules.
- The minimum size wire used for module installations should be 1.6 mm^2 and follow existing wiring codes.
- Interior installations should follow all other codes (Sections 177, 178, 180, 187, and 189).

- Outside installations should follow all other wiring codes (Sections 177, 178, 180, 188, 189, and 211).
- Wires should be connected using terminal connectors and the connections should have appropriate strain relief.

Japanese PV systems are installed in compliance with the Japanese electrical code. In eastern Japan, systems use a European standard of 50 Hz AC, while western Japan uses a U.S. standard of 60 Hz. Japanese electrical codes are somewhat similar to European electrical codes, with PV systems ungrounded on the DC side and grounded on the AC side. A chassis ground is always used (AC and DC sides; Tepco, 2004).

8.3.5 Japanese PV Design

PV companies and electrical contractors design PV systems in Japan. Utilities sometimes may get involved in the design of a few large-scale systems, but typically not for the smaller residential systems. Residential PV systems generally range from about 3–4 kW$_p$ and average about 3.6 kW$_p$ (Kaizuma, 2005/2007).

PV arrays are often mounted directly onto reinforced corrugated metal roofs (no roof penetrations). Most roofs in Japan are metal or a traditional style ceramic for high-end roofs. There is a great deal of concern in Japan that PV systems be able to withstand typhoon (hurricane)-force winds, which are common during the late summer months. Often, commercial PV installations in Japan are not optimally tilted for solar energy production but are tilted in favor of better wind survivability (typhoons). System profiles are installed low to the roof to reduce wind loading (Figure 8.8). Local codes typically call for PV systems to withstand winds of 36 m/s in Tokyo, 46 m/s in Okinawa, and even 60 m/s in some places, such a Kanazawa City.

One unique aspect for some Japanese PV installations is that many systems are installed with PV arrays facing south, east, and west on the same roof. This is due to the limited roof space of smaller Japanese homes. The west and east arrays typically produce about 80% of the energy of a south-facing array. Some inverters (e.g., Sharp) are designed to max power point track three different subarrays independently for this reason.

Japanese PV systems are not grounded on the DC side (although they all have a chassis ground). Only the AC side is grounded. Operating voltage is 200/100 V AC. The distribution network for electricity in Japan is single-phase 100/200 V AC. The western part of the country uses 60 Hz (e.g., Osaka), while eastern Japan uses 50 Hz power (e.g., Tokyo).

Crystalline PV modules are by far the most popular in Japan, representing over 80% of PV modules produced and installed in the country. Modules normally carry a guarantee on performance

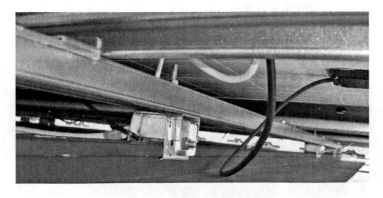

FIGURE 8.8 Underside of typical Japanese PV array clamp mounting on metal corrugated roof (no roof penetration) designed to withstand typhoon force winds.

from 10 to 25 years, depending on the manufacturer (those active in U.S. markets will have a superior warranty). Thin film modules are slowly gaining in popularity, but still greatly lag sales of crystalline modules. Cadmium telluride (CdTe) modules will never be found in Japan due to the society's disdain for using toxic materials. In Japan, a lot of thought has been given to how to recycle a PV module; thus, toxic materials are quickly eliminated from consideration of use in PV modules.

In Japan, there are about two dozen residential PV inverter manufacturers. Most Japanese inverters do not use transformers. There are over 100 listed residential PV inverters in Japan. Inverters are single phase and three wire (100 and 200 V). Inverter warranties vary by manufacturer (typically 1–3 years).

Several Japanese PV producers also make their own inverters, such as Kyocera, Sanyo, and Sharp. Sharp and Daihen are developing inverters jointly for large-scale PV systems installed by commercial users and electric utilities. Daihen is responsible for manufacturing the solar inverters, and Sharp focuses on PV modules targeting electric utilities. In the future, it can be expected that Japanese inverters will become as prevalent as Japanese PV modules around the globe. Some of the major inverter manufacturers include Sharp/Daihen, Omron, Toshiba, Mitsubishi, Sanyo, GS, Matsushita, and Kyocera (Figure 8.9). Inverters in Japan are a mature technology. One very interesting application in Japan is that the industry is looking at how large clusters of inverters work together and how to improve performance, such as the Ohta City project with over 500 PV homes (Figure 8.10).

PV installations in Japan exhibit excellent workmanship and are done by certified electricians. There are no independent certified installers (e.g., no North American Board of Certified Energy Practioners (NABCEP) equivalent). Industry is responsible for training its installers and maintaining quality standards. Some module manufacturers, like Kyocera, will also install PV systems; others rely on electrical contractors. In new homes, often the same electricians that install a home's wiring system also install the PV system.

Overall installation costs for PV systems in Japan are generally less than in the United States because systems have simpler BOS requirements and more streamlined installation procedures (e.g., no roof penetrations). Systems for 3 or 4 kW_p can be installed efficiently in only a couple of days. Electrical crews generally consist of two or three electricians/assistants. PV installations are normally completed within 2 or 3 days. No on-site QA records are maintained, and it is up to the installer to do a good job. If there is a failure, the installer will be held responsible. Generally, in Japanese culture, the installer and also manufacturers will want to fix any problems with their products. It is a matter of cultural honor for them to have satisfied customers.

FIGURE 8.9   Four-kiloWatt transformerless Omron inverter on AIST PV parking structure.

FIGURE 8.10   Excellent workmanship typifies Japanese installations, such as with this PV system breaker box with monitoring transducers at Ohta City clustered systems project.

8.3.6   JAPANESE PV SYSTEM GUARANTEES

Japanese PV systems and components are warranted against defects in product or workmanship. A normal PV system installation warranty is for 3 years. Of course, additional module warranties vary by manufacturer (10–25 years). Some in the industry believe that a 10-year module warranty is sufficient (e.g., a car has only a 3-year warranty but everyone knows it will last longer).

There are no requirements for using listed equipment in Japan. It is strictly voluntary to have listed modules and inverters. However, most manufacturers will seek a voluntary listing from JET to be more competitive. Japanese installers are left on their own to do the right job (this is akin to how the Japanese automobile industry operates). It is a matter of cultural and professional pride for Japanese industry to install quality PV systems.

8.3.7   JAPANESE PV DEVELOPMENT

Japan is a global leader when it comes to PV manufacturing and innovation. Residential system needs have helped promote higher cell efficiencies and smaller sizes. Larger commercial systems have led to innovation in PV for building integration that requires flexible, lightweight, light-transmitting, or bifacial products for facades and large-area installations. A number of office buildings now have see-through PV on their south-facing windows (Figure 8.11). Some prefab homes use PV,

FIGURE 8.11   Building integrated see-through PV modules (Sanyo) at the Ohta City government office complex.

but only 25% of installed residential systems are on new construction. Research and development on expanding the use of PV on prefab construction continues. The factory will offer a PV system packages for delivery. Most assembly is still done in the field.

Japan is also shifting home construction toward a "mass customization" approach. A future home-owner is given a wide menu of standardized options to customize his or her prefab home design (e.g., a dozen different stairway designs, windows, etc.). Customized modifications can be significant on homes and gets the homeowner involved with the home design. PV manufacturers do offer standardized systems, but these vary from manufacturer to manufacturer.

The Japanese industry forms the backbone of the global PV industry. The government research program has been tightly coordinated with Japanese industry and academia. There are 13 major PV module manufacturers in Japan; these include some of the world's leading PV companies, such as Sharp, Sanyo (Figure 8.14), Kyocera, Mitsubishi, and Kaneka. Japanese industry continues to strive for cost reductions in PV manufacturing while maintaining a healthy profit, especially for those companies well established in the sector. Residential PV installations are the driving application for the domestic PV market in Japan (Figure 8.12).

PV growth in Japan has also nurtured peripheral industries, such as production of silicon feed-stock, ingots and wafers, inverters, and reinforced aluminum frames. Sharp is the number one PV manufacturer, followed by Kyocera and Sanyo. Japan overtook the United States in terms of manufacturing in 1999 and their current market share of overall worldwide PV production is about 15% (Figure 8.13).

8.3.8   Japanese PV Module Certification

As a METI-designated testing body and independent and impartial certification institution with a proven track record, Japan Electrical Safety and Environment Technologies (JET) provides product certifications by use of a symbol that represents "safety and authority" to manufacturers and import-ers as well as to consumers. JET receives a range of requests from government agencies, including requests to conduct tests on electrical products purchased in the market, to harmonize domestic standards with IEC standards, and to conduct research and development on technologies for assessing solar power electric generation systems.

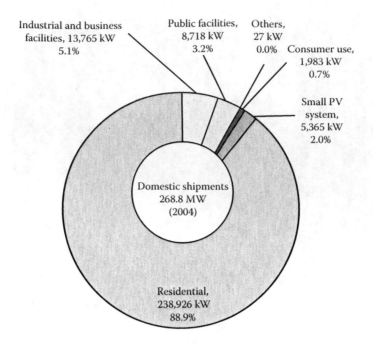

FIGURE 8.12 Japanese installations by sector type in 2004, dominated by residential (OITDA).

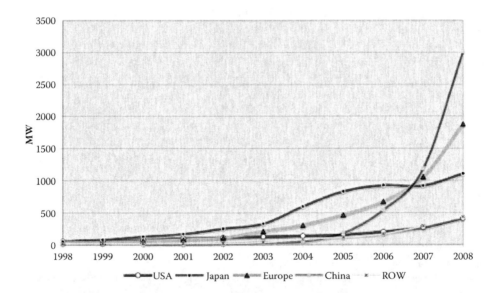

FIGURE 8.13 Japan led annual global PV production until 2007, when China became the global leader exporting 98% of its production (Approximated from sources: Worldwatch, Maycock, Kaizuka, Marketbuzz, and Wicht).

With regard to PV generation systems, in 1993, JET began registration of system interconnection devices linking PV modules installed on residential homes with electric power company systems. In addition, JET began calibration service for PV modules in April 2002 and began certification of PV modules in 2003. The JET PVM Certification Scheme is a voluntary program operated by JET and certified to IEC standards 61215 and 61646. The main objectives are to verify the safety and reliability of PV modules. Certificates are granted to modules after successful completion of applicable

FIGURE 8.14 Sanyo corporate headquarters in Tokyo with BIPV on the south, east, and west sides of this office complex.

tests based on IEC module test standards. In 2006, JET also began testing for PV module safety in accordance with IEC 61730 (JET, 1998, 2002).

Likewise, JET certifies inverters for PV systems. In Japan, there is no requirement to use JET-listed inverters and modules. However, most manufacturers want to participate in the JET program so that their modules are viewed as independently certified and thus be more competitive in the marketplace.

8.4 FUTURE JAPANESE TRENDS

In Japanese society, the use of PV is seen as important and necessary from a social, cultural, and ecological perspective. Likewise, Japanese leaders and industry see PV as a revolutionary technology that can make significant contributions to the electric power sector while making good business sense. A combination of R&D support and installation subsidies support has proven an effective strategy to promote PV technology development.

Government involvement has been important at the initial stage of technology introduction. Market subsidies help create initial markets. The Japanese PV system market will continue to benefit and expand even as government subsidies for the residential sector are eliminated. The leading market sector will continue to be residential installations for the near future. However, there will be greater emphasis on PV systems growth in the public, industrial, and business facilities sectors.

The Japanese government and industry view the next 25 years as a critical period for the creation of a full-scale PV market. A cumulative capacity of 83 GW of PVs in Japan is seen as achievable by 2030, by which time PV could meet 50% of residential power needs. This is equivalent to about 10% of Japan's entire electricity supply.

The PV price targets to be achieved by means of R&D, large-scale deployment, and export sales are ¥23/kWh by 2010, ¥14/kWh by 2020, and ¥7/kWh by 2030. Future PV cost goals were chosen based on making PV competitive with conventional energy rather than on any type of technology feasibility study. Thus, the goal of ¥23/kWh by 2010 corresponds to the current residential electric rate, 14¥/kWh by 2020 corresponds to the current commercial rate, and ¥7/kWh by 2030 corresponds to the current industrial rate. All price goals are defined in terms of 2002 yen.

As PV systems grow across the world, Japan has placed itself as a global leader to meet future PV demand. The Japanese industry model is outwardly focused toward export markets and the majority of Japanese-produced PV product is exported. Japanese industry has set up overseas manufacturing operations in Europe, the United States, Asia, and Mexico.

8.5 STAND-ALONE PV APPLICATIONS

Over the past quarter century, the developing world has adopted stand-alone PV technologies in earnest for social and economic development. PV is a viable alternative to traditional large-scale rural grid systems. With the advent of PV as a dependable modern technology alternative and more private participation and choices made available to the general public, PV systems have become attractive throughout the less developed parts of the world. The challenge is to develop financing strategies that are affordable to potential clients.

Off-grid markets represent the natural market for PV technology, which does not require any government subsidies to be competitive or successful. The technology fills a real-world niche and is especially useful for developing countries, where often the national electric grid is lacking coverage. The use of PV systems in rural regions of the developing world has increased dramatically from an initial concept pioneered by a few visionaries over 25 years ago to many thriving businesses throughout the developing world today.

PV is a viable alternative to traditional large-scale rural grid systems. With the advent of PV as a dependable modern technology alternative and more private participation and choices made available to the general public, PV systems have become attractive all over the globe, with literally millions of rural households electrified via PVs. Indeed, the most common PV system on the planet is the small ~50 W_p solar home system providing basic electricity for a few lights, radio, and maybe a small TV. Even smaller solar lanterns and flashlights incorporating LCDs are more popular. The challenge is to develop financing strategies that are affordable to potential rural clients, who often have incomes dependent on crop harvest cycles.

8.5.1 PV SOLAR HOME LIGHTING SYSTEMS

PV first served space and remote communication needs, but quickly became popular for basic domestic electricity needs for residences in rural regions in the United States and then throughout Latin America, Africa, and Asia. During the mid-1980s, solar energy pioneers began to disseminate PV technologies in rural Latin America as a solution for providing basic electricity services for populations without electricity. Some of the first pilot projects in the world were undertaken by non-government organizations (NGOs), such Enersol Associates in the Dominican Republic beginning in 1984 (Figure 8.15). Gradually throughout the developing world, small solar companies began to form as key module manufacturers of the time, such as Solarex and Arco, sought out distributors for off-grid rural markets. By the mid-1990s, these activities were followed by large-scale solar electrification activities sponsored by government agencies in Mexico, Brazil, South Africa, etc.

Many of these early large-scale PV government electrification efforts faced sustainability issues as planners attempted to force large-scale rural solar electrification projects onto unknowledgeable rural users. Common problems included use of inappropriate battery technologies, substandard charge controllers, unscrupulous sales personnel, and poor-quality and unsupervised installations. Often these were giveaway programs, so there was no sense of ownership from the recipients, which can often lead to a lack of responsibility to care for systems. Despite these hurdles, only rarely did PV modules themselves ever fail; in fact, they continued to be the most reliable part of any installed system.

In response to early system failures, implementing agencies gradually began to adopt basic technical specifications that observed international standards that improved the quality and reliability of PV systems. Rural users mostly want a PV system that works to provide basic electric light and entertainment with radio and TV. PV users are not interested in the finer points of technical operation and maintenance. They want a simple and functional system that is easy to maintain.

FIGURE 8.15 Latin America's first PV training center established by Richard Hansen (far left) of Enersol Associates in the Dominican Republic, training both local technicians and Peace Corps volunteers (1985).

Think sustainability. All paths should lead to this and institutions applying solar energy systems must have a true commitment for long-term sustainability. Government agencies face particularly difficult challenges because the parties in power often change. The ultimate goal is to have a well designed and installed solar energy system that will provide many years of reliable and satisfactory service. The past quarter century has set the stage for future solar development, which is growing exponentially.

One good example of a PV lighting system (PVLS) for the home was deployed in Chihuahua, Mexico, by Sandia Labs/NMSU for the USAID/DOE Mexico Renewable Energy Program in the late 1990s with the state of Chihuahua. The program installed a Solisto PVLS designed by Sunwize Technologies to meet the Mexican electric code requirements (i.e., NEC). This is a prepackaged control unit specifically engineered for small-scale rural electrification and long life. Key characteristics of this system were that both the positive and negative legs were fused (an ungrounded 12 V system) and proper disconnects were used. The system employed a sealed maintenance-free lead-acid battery and a solid-state UL listed charge controller that uses fuzzy logic to help determine battery state of charge.

A total of 145 systems were installed in the municipality of Moris, located about 250 km west of Chihuahua City. The terrain consists of steep mountains and 1,000 m deep canyons in the midst of pine forests. The steep topography makes electric grid access difficult and indeed there is no interconnection with the national electric grid and no paved roads. Over three-fourths of Moris residents do not have access to electricity, and the few that do are mostly on diesel-powered minigrids.

The Moris PV systems consist of one 50 W Siemens SR50 module, which was the first deployment of these modules specifically developed for the rural lighting market. The PV modules are mounted on top of a 4 m galvanized steel pole capable of withstanding high winds. The module charges a nominal 12 V sealed gel VRLA battery (Concorde Sun-Xtender, 105 Ah at C/20 rate for 25°C; Figure 8.16). These are sealed, absorbed glass mat (AGM) and never require watering. The immobilized electrolyte wicks around in the absorbed glass mat, which helps the hydrogen and oxygen that form when the battery is charged to recombine within the sealed cells.

The thick calcium plates are compressed within a microfibrous silica glass mat envelope that provides good electrolyte absorption and retention with greater contact surface to plates than gel batteries. The Concorde batteries are in compliance with UL924 and UL1989 standards as a recognized system component. These batteries meet U.S. Navy specification MIL-B-8565J for limited hydrogen production below 3.5% during overcharging (less than 1% in Sun-Xtender's case), which

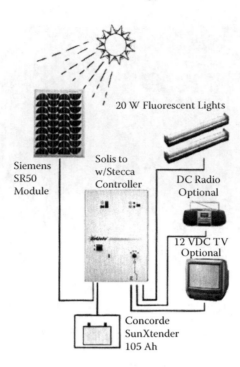

FIGURE 8.16 Residential Solisto PV system used in Chihuahua, Mexico.

means they are safe for use in living spaces. All batteries were installed inside a spill-proof, child-proof, heavy plastic battery case strapped shut.

Control is maintained through the Solisto power center via a UL-listed Stecca charge controller with a 10 A fuse. The system has a DC disconnect and six other DC fuses protecting different circuits. The Stecca controller uses fuzzy logic to monitor battery charging to avoid under- or over-charging the battery and is equipped with an LED lighted display to indicate state of charge. The Solisto power center is still available on the commercial market; Chihuahua marked the first use of these power centers in the world.

The PV system powers three compact fluorescent lamps with electronic ballasts (20 W each). It also has a SOLSUM DC–DC voltage converter (3, 4.5, 6, 7.5, and 9 V options) and plug to allow for use of different types of appliances, such as radio and TV. For an extra US$200, end-users could also elect to install a Tumbler Technologies Genius 200 W inverter; although few chose to do so, several users did install satellite TV service, which comfortably allowed them about 3 hours of color TV viewing in the evenings. The design of the Solisto SHS assumed that a household using the full set of three fluorescent lamps for an average of 2 hours a day would consume about 120 Wh/day on average. Given that Chihuahua averages about 6 sun-hours/day and assuming an overall PV system efficiency of 60% for this fairly well designed system (i.e., including battery efficiency losses, module temperature derate, line losses, etc.), the user could expect on average to have about 180 Wh/day of available power.

Of course, there are seasonal variations and double or more power could be extracted from the battery on any single day, but could not be sustained long term. As is typical for solar energy users, the Mexican users quickly learned to live within finite energy system bounds and to ration energy use during extended cloudy periods, which are relatively rare in Chihuahua.

Also of particular interest was an additional innovative financing component representing the first financing of PV systems anywhere in Mexico. The financing activities of this program were conducted by the State Trust Fund for Productive Activities in Chihuahua (FIDEAPECH). This state trust fund provides direct loans and guarantees, primarily based on direct lending (e.g., to

FIGURE 8.17   Solisto 50 W_p PV lighting system installed in Talayotes, Moris County, Chihuahua.

farmers for tractors). FIDEAPECH designed and implemented the revolving fund in which the municipality paid 33% of the total cost of PV home-lighting systems up front, end users provided a down payment of 33%, and the remaining 34% was financed for 1 year by FIDEAPECH. The municipal government provided the loan guarantee and eventual repayment to FIDEAPECH. The total installed cost of each quality-code-compliant PV home lighting system was about US$1,200. Other 50 W_p PV systems had been installed at the same cost in this region, at considerably worse quality and performance (e.g., with some failures reported in less than a month) (Figure 8.17).

Since October 1999, the performance of a Solisto PV lighting system has been continuously monitored at the Southwest Region Solar Experiment Station of New Mexico State University (NMSU) in Las Cruces, New Mexico, simulating usage of about 171 Wh/day. The long-term monitoring provides a reasonable base case with which to compare fielded systems. The monitored system was still functional in 2008. The Stecca charge controller successfully protected the battery from severe abuse from overcharging and deep discharging during prolonged cloudy periods. Charge regulation using pulse-width modulation charging and fuzzy logic to determine battery state of charge has performed very well for the sealed batteries, providing good lifetime. The nominally regulated voltage on the battery averaged 12.9 VDC each day, with the lowest battery voltages observed as 11.9 VDC after cloudy periods. The average daily depth of discharge (DOD) was about 13.5%. The Sun-Xtender battery manufacturer claims that the 105 Ah battery should have a cycle life of approximately 1,600 cycles for 40% DOD at 25°C and 5,200 cycles at 10% DOD.

NMSU also had the opportunity to monitor the systems in the field after 5 years. Performance was assessed through electrical measurements, visual inspection, and an end-user survey to determine user satisfaction. A total of 35 evaluations were performed. The results showed that over 80% of the installed systems were operating correctly and as designed, 11% were in fair condition (most commonly, one of three lamps was no longer working), 6% were nonoperational, and 3% of systems had been dismantled (e.g., user moved). The high percentage of working PV lighting systems after 5 years demonstrates the potential reliability for PV home lighting systems. In the household survey, NMSU found that 94% of users expressed complete satisfaction with their PV lighting systems, 86% thought that PV was better than their previous gas lighting source, and 62% believed that the PV systems were reasonably priced for the service provided (Foster, 2004). The sealed battery lifetimes were good. PV modules proved to be one of the most reliable components, all modules were functional, and no module problems had been reported. New and expanded evening activities, such as sewing, watching TV, reading, and studying, were also reported.

The PV lighting systems in Moris Chihuahua performed well after 5 years and met original system design and life criteria (Figure 8.18). The PV systems saved an average of US$300 over 5 years

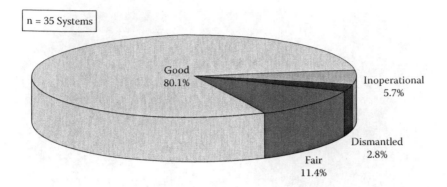

FIGURE 8.18   Over 90% of Solisto PV SHS were operational in Chihuahua after 5 years.

in lieu of previous gas and dry-cell battery options, while providing superior light and entertainment capabilities. The end-users have been very satisfied with the PV lighting systems. The Moris PV lighting systems demonstrate that, with proper diligence and detail to design and installation, PVLS can provide many years of useful service with little or no maintenance actions required.

8.5.2   PV Battery Charging Stations

The Nicaraguan National Energy Commission (CNE) with the World Bank implemented a large-scale solar rural energy initiative called the Renewable Energy for Rural Zones Program (PERZA—Proyecto de Electrificación Rural para Zonas Aisladas) during the mid- to late-2000s. Approximately 80% of Nicaragua's rural population does not have access to electricity. PV is a promising alternative for providing energy to rural areas there, either through individual PVLS for the home or centralized PV battery charging stations (PVBCSs). The project installed centralized PVBCSs in the Miskito region of northeast Nicaragua, which is one of the countries with the lowest electricity coverage in Latin America.

Both approaches charge batteries through charge controllers. Typical appliances powered by one battery per household are a few energy-efficient light bulbs, a radio, and perhaps a black-and-white TV. The main difference is that the batteries are charged centrally in the PVBCS (and then transported by the users). For PVLS, each household has its own small PV module, battery, and charge controller. The advantages of PVBCS are potential economies of scale in management and battery charging, as well as the potential to adapt payment schedules to local needs. The main advantages of PVLS are the increased convenience and the household charge controllers, which avoid deep discharging and increase battery lifetime over PVBCS.

These indigenous Miskito communities are located in the North Atlantic Autonomous Region (RAAN) of Nicaragua north of Puerto Cabezas in the Waspam area. The project financed seven PV battery charging stations that provide energy for approximately 300 homes that represent about three-quarters of the total population of the communities of Francia Sirpi, Butku, Sagnilaya, and Ilbara. These battery charging stations were installed in November 2005 in locations selected by the communities so as to facilitate access by the population. Each home has a complete "kit" that includes a battery, two fluorescent lamps, and a voltage regulator. All of the PV systems and kits have similar design and construction.

This project was subsidized entirely by the government of Nicaragua, due to the extreme poverty conditions of the Miskito indigenous communities. The users paid a small fee, calculated based on their payment capacity, to recharge their batteries. A typical PV battery charging station in the community of Francia Sirpi comprises a 2,400 W PV array with three subarrays that can charge up to 24 lead-acid batteries at the same time (Figure 8.19). Shell SQ80 80 W_p PV modules are used.

FIGURE 8.19   One of the three PV battery charging stations (NW system) at Francia Sirpi, Nicaragua.

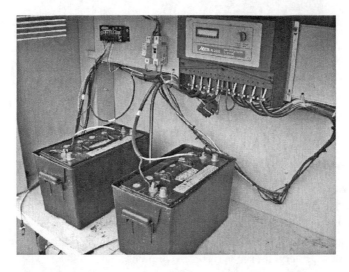

FIGURE 8.20   Battery charging at Francia Sirpi with Stecca controller capable of charging eight batteries simultaneously.

The complete system is composed of three PV 800 W_p substations with its own individual Stecca PL2085 controller capable of charging eight PV batteries per station simultaneously (Figure 8.20). The intelligent control unit in which the adjustment, operation, and display functions are carried out by a microprocessor serves as the brains of the battery charging station. The batteries are charged as quickly and efficiently as possible, in the order of priority according to when they are connected. In addition, an MPP-tracking system enables optimum use to be made of the energy available even if not all battery stations are fully utilized. No energy is wasted, even if all eight stations per subarray are not occupied (Ley 2006).

Approximately 150 residential household lighting packages were installed in the Francia Sirpi community. Residents were provided a PV lighting household kit with two or three 15 W fluorescent lamps. The lighting kit installed on each house had a small 6 A Morningstar SHS-6 charge controller used as a low-voltage disconnect for the 12 V, 105 Ah maintenance-free AGM battery

FIGURE 8.21 Nicaraguan home lighting kit with deep-cycle battery safeguarded in a battery box against the most curious PV system clients.

(Figure 8.21). No PV modules were installed on the individual homes. Instead, when the battery was low on energy, it was disconnected from the home lighting system and taken to the charging station site to be recharged. When fully charged, the battery was then brought back to the house and reconnected to the home lighting system.

The main concern for PVBCS is that, if the users overly deep-discharge their batteries (e.g., bypass the LVD), then battery lifetimes could be prematurely cut short. There were some early controller failures with the Stecca controller because, if the operator reversed polarity on the battery leads, the controller could fail because it was not polarity protected. These failed controllers were later replaced by Phocos controllers, which could only individually charge a battery. Some of the installed projects were also hit hard by a hurricane in October 2007, which hit the Miskito communities particularly hard.

The PERZA project essentially represents a "supply push" rather than a "demand pull" for off-grid PV applications. Off-grid rural energy services can be designed to be franchised and supplied through standardized distribution chains. The advantages of PVBCS are potential economies of scale in management and battery charging, as well as the potential to adopt payment schedules to local needs. The main advantages of PVLS are the increased convenience and the household charge controllers, which avoid deep discharging and increase battery lifetime over PVBCS. Typically, as seen by this project and others in Brazil and Bangladesh, PVLS is a more successful application.

8.5.3 PVLS HUMAN MOTIVATION: THE FINAL DRIVER OF SYSTEM SUCCESS [GUEST AUTHORS DEBORA LEY, UNIVERSITY OF OXFORD AND H. J. CORSAIR, THE JOHNS HOPKINS UNIVERSITY]

The community of Xenimajuyu is located in the highlands of Guatemala in the department of Chimaltenango. Although the electric grid ends fairly nearby, a confluence of factors including the mountainous terrain and the political fallout resulting from the division of this community from another larger community have made it very unlikely that the electric grid will be extended to Xenimajuyu in the foreseeable future.

One household in the community chooses to generate its own electricity. In addition to a small electric generator, the homeowner uses solar PV panels to meet his electricity demand for lighting and entertainment. The system has been in operation for over a decade, even though the panels are of poor quality, the system lacks a charge controller, and the automotive battery is inappropriate. This system uniquely illustrates the role of human motivation in the sustainability of rural solar PV systems: The individual decisions of the homeowner to keep the system operational have proved more powerful than the technical shortcomings of the system.

Guatemala has an overall electrification rate of 83.1% (CEPAL 2007a), although more than 40% of the rural population remains without electricity (Palma and Foster 2001). This is equivalent to approximately 2.2 million people or almost 440,000 homes (CEPAL 2007b) without access to the national electric grid. The so-called Franja Trasveral Norte, consisting of the Departments of Huehuetenango, Quiché, Alta Verapaz, Baja Verapaz, and Izabal, together with Petén, include the poorest departments with the most people without electricity in Guatemala. This population without electricity consists mainly of the 32% of the population that lives in extreme poverty, according to statistics from 2000 (Hammill 2007). The same statistics indicate that 56% of the population lived in poverty in 2000 (Hammill 2007). In addition to high rates of poverty and extreme poverty, these departments are characterized by communities with very difficult access and a high dispersion rate of houses (Arriaza 2005)—characteristics that make it economically infeasible for the grid to be extended. Because of this, renewable energy is often the best electrification option. This is especially true for Guatemala due to its solar resources.

According to various PV design guidebooks, the minimum solar resource that should exist before a project can be considered feasible is 300 W/m²/day. The *Solar and Wind Energy Resource Assessment* (SWERA), cofinanced by the Global Environmental Facility (GEF) and the United Nations Environment Program (UNEP), indicates good to excellent solar resources (400–600 W/ m²) in areas of Guatemala coincident with the most marginalized population of the country. Rural Guatemalan communities have been using isolated PV systems since the early 1990s in applications that vary from household and community lighting to productive uses to community services (CEPAL 2007a; Arriaza 2005; Palma and Foster 2001).

Although there is not an exhaustive list of installed PV systems in the country, the government, through the Ministry of Energy and Mines, has installed PV panels in approximately 80 communities serving nearly 3,435 families with 50 W systems. Some of these systems have been uninstalled and a subset of these relocated. In the 8 years leading up to 2001, other institutions installed nearly 5,000 household systems. These systems typically consist of a 50 W PV module, a 12 V deep-cycle battery, a charge controller, and three CF light bulbs, providing about 3 hours of illumination per night. This means over 220 kW of residential PV have been installed, generating over 400,000 kWh per year (Palma and Foster 2001).

Numerous lessons have been learned over the years and some of them have shaped more recent installations. Early PV projects focused on technical aspects while ignoring human and social needs (Palma and Foster 2001). Although technical shortcomings may be a cause of failure of PV projects, the case study in the next section illustrates that, despite these technical shortcomings, human motivations and convictions can lead to long-term project sustainability.

8.5.4   PV IN XENIMAJUYU: THE XOCOY FAMILY
[GUEST AUTHORS DEBORA LEY, UNIVERSITY OF OXFORD AND
H. J. CORSAIR, THE JOHNS HOPKINS UNIVERSITY]

The use of candles, kerosene, and ocote (a type of fuel wood) are common in rural unelectrified households in Guatemala. Richer families might be able to afford a car battery or a diesel generator to power light bulbs, radios, and televisions; however, most families do not have this option and burn three to five candles per night depending on the number of family members. With the cost of each candle at 1.5 quetzales, families can spend up to US$1 on lighting energy per day, representing a higher percentage of their income than lighting consumes in urban populations (UNDP 2005).

The Xocoy family lives in the community of Xenimajuyu. The terrain where the community is located is very mountainous, which can make the extension of the power grid prohibitively expensive even over short distances (Palma and Foster 2001; CEPAL 2007a). According to a local Peace Corps volunteer, the community of Xenimajuyu split off from Chuisac, a larger neighboring community, to form its own autonomous village. When the parent village got access to grid electricity in 2000, Xenimajuyu was not included in the plan. Because it is small and particularly difficult to access, the community does not anticipate grid extension in the foreseeable future. The Peace Corps proposed a community-wide photovoltaics electrical project, though it is in the very early stages and the timeframe for completion is not realistically known.

Estanislado Xocoy and his family currently meet their energy needs using three different sources: the traditional fuels often used in rural Guatemala, a gasoline-fired generator, and a small PV system. The family seems to prefer its small gas generator to more traditional lighting sources. Although the generator produces enough energy for the family's needs, it is not an ideal option because of the high and volatile price of fuel and the difficulty of transporting fuel from the gas station to the home.

The lighting energy source that the family claims to prefer is their solar PV system. The system is very simple and consists of solar panels, a battery, three compact fluorescent lamps, and a small black-and-white television.

The owner has two identical PV panels, neither of which has a module plate. The first one was purchased new almost 11 years ago, while the second was purchased used 3 years ago. The previous owner stopped using it when he got grid electricity. The first PV panel is mounted on the house's roof. It is a 50 W nominal amorphous thin-film silicon panel. The discoloration in the panel surface and the corrosion in the wires are evidence of significant degradation of the panel. The homeowner has had the panel for longer than the anticipated 10-year life of an amorphous panel, so this degradation is not unexpected. The degree of degradation could not be measured precisely because ambient conditions were not conducive to accurate measurement. It is installed on an eastward-facing roof slope, rather than facing south as recommended, and is mounted directly on the metal roof, rather than on a mounting structure that would allow air circulation to cool the panel and improve its performance. It is also installed with an inclination of less than $10°$; an inclination approximately equal to the latitude of the location (about $15°$ in this case) will produce optimal annual power output. The dirt on the panel reduces the amount of sun that hits the panel and therefore its output.

The second panel, which is even more severely degraded than the first, had been connected in parallel to the first panel for approximately 6 months, but was removed from service because the owner believed it was no longer providing any benefit. Measurements of output of this second panel showed that it was capable of producing some electricity, though the small quantity may or may not merit its being reinstalled to supplement the currently installed panel.

A charge controller can significantly improve system performance by preventing overcharging or overdischarging of the battery. This household system lacks a charge controller; thus the frequency with which the battery must be replaced and therefore the cost of the system are increased.

A deep-cycle battery designed for PV or marine use is best suited for solar applications. This system made use of an automotive battery, which degrades quickly under the deep discharge cycles demanded of this type of system. However, the battery is well maintained, without evident corrosion, overheating, or loose connections that can be problematic with batteries in solar home applications. The owner replaces the car battery every 2–3 years when the terminals "get humid," as he describes it. The system can only light one bulb for 1 hour instead the normal three bulbs for 3 hours, together with 3 hours of television or, alternatively, 5 hours of lighting with no television.

The Xocoy family's system almost entirely fails to meet the norms and standards expected of a robustly designed quality system: The panel is inappropriately mounted and in poor condition; the wiring is in poor condition; the battery is inappropriate to the application; important system components such as the charge controller are missing; and the system is installed without basic safety considerations such as electrical grounding or a compartment to protect family members from accidents with the battery. However, this household has kept this system successfully operational for over a decade, and even wants to expand the system by replacing the second faulty panel with a new one. The owner's conviction is that solar energy works: Photovoltaics represents a more attractive option to his family and his community than do fossil fuels or traditional energy sources, as well as a more realistic option than waiting for grid extension. Estanislado Xocoy may be an opinion leader and a technical resource who will be a powerful enabler of the project to electrify the community using solar energy.

8.6 PV FOR SCHOOLS

Thousands of rural schools in the developing world do not have access to electrical grid power. It is important to bridge this gap so that rural student populations living outside electricity grid services can also have the same opportunities as other students. An enhanced quality of education forms a foundation for increased productivity, leading to higher standards of living. Solar power offers a practical way to meet such power needs. Renewable energy technologies can be used to bring services, such as distance education and computer Internet access, to rural isolated communities, where the application of such technologies is appropriate and suitable. The high costs associated with fuel purchases, transportation of fuel, and engine maintenance, coupled with environmental costs that are difficult to quantify, make renewable energy an attractive alternative to conventional fuel-burning motor generators. PV systems are used to power televisions, DVDs, and computers to modernize the educational experience of rural schoolchildren (Figure 8.22).

Several programs in Mexico and Central America are using renewable energy to bring quality distance education programs to their rural populations. The Mexican Secretariat of Public Education is recognized for its distance learning programs that are based on satellite broadcast. Most of the schools in the programs are located on the electrical grid, but there is an increasing desire to extend the educational network to off-grid areas (Figure 8.23).

An ingredient often lacking in many programs is a clear understanding of what renewable energy technologies are, what equipment is available, and how they can best be used to meet energy needs. The technical expertise required to implement projects that incorporate the use of renewable energy is often overlooked and does not exist within implementing agencies. In-country partners need knowledge, experience, and engineering expertise to install and operate long-lived, quality systems.

PV systems are currently installed on more than 500 off-grid schools in Mexico and over 300 schools in Honduras and Guatemala. Some of the PV systems in use have been poorly designed and installed and thus are operating inefficiently. Most problematic PV systems have been identified to suffer from simple, resolvable problems such as

FIGURE 8.22   PV-powered COHCIT satellite telecenter with Internet connectivity using quality BOS components with master PV installer Ethel Enamorado in Sosoal, Lempira, Honduras.

FIGURE 8.23   A solar PV-powered one-room *Telesecundaria* school in Quintana Roo, Mexico.

- undersized battery cables, thus limiting battery recharge;
- improper orientation and location of the panels;
- incorrect types of batteries used for the application; and
- lack of end-user knowledge on proper operation and maintenance.

Public education agencies in Mexico and Central America had, in some measure, lost confidence in the use of renewable energy sources and technologies—thus diminishing the willingness to repair or replace existing systems and/or to purchase new systems for additional rural off-grid schools. However, with appropriate knowledge and institutional capacity, the majority of the problems are simple and resolvable. Around 2000, PV school installations in many parts of Latin America began to show great improvements as the industry matured and implementing agencies gained valuable training and experience. Successful large-scale rural school PV electrification programs have been implemented in Mexico, Guatemala, Cuba, Honduras, Peru, and Brazil. The PV systems are used to power televisions and computers to modernize the educational experience of rural schoolchildren.

8.7   PV FOR PROTECTED AREAS

Renewable energy technologies have been widely applied to support protected areas throughout Latin America, especially in Guatemala, Mexico, and Ecuador (Galapagos). Key environmental agencies such as the Nature Conservancy, World Wildlife Fund, and Conservation International have embraced PV technologies.

Use of solar energy in protected areas benefits the living conditions of researchers, technicians, and rangers, as well as providing energy for environmental training centers. The solar energy systems also have the advantage of providing power without the noise or pollution associated with conventional fossil-fueled generators, while reducing the risk of fuel spills in these sensitive biosphere reserves. As always, up-front design decisions, user operation, and long-term maintenance issues play an important role for overall system reliability.

Solar energy is an environmentally appropriate example to neighboring buffer communities (often without electricity) surrounding biosphere reserves, which can likewise benefit by replicating the protected areas' example. Solar energy systems also provide a useful example for visitors and tourists to take back home.

In addition, the remote protected-area facilities benefit economically from solar installations through reduced operation and maintenance costs associated with fossil fuel generators. Actual system life-cycle costs for any particular solar or wind energy system vary and are a function of design, usage, application, and maintenance. With proper system operation and maintenance, the expected solar energy system lifetime should exceed 25 or more years (with appropriate battery replacements, etc.).

One example of a PV-wind hybrid system application is found at Isla Contoy in Quintana Roo, Mexico (PNIC—Parque Nacional Isla Contoy) (Figure 8.24). It is informally known as Bird Island, due to the 151 bird species found on the island, surrounding which are over 5,000 frigates. It is also an important site for protecting marine turtles, crocodiles, 31 coral species, and 98 indigenous plant species.

The park was burning gasoline transported by boat from nearby Cancun for a 3.5 kW generator, with significant noise pollution that disturbed birds, as well as the constant threat of fuel spills. Eventually, $35,000 in funding was secured from USAID to support the installation of a hybrid renewable energy system. Of particular concern was the potential impact of wind turbines on the large bird sanctuary (i.e., threat of bird kills). Because small wind turbines spin very fast and are quite visible, it is unlikely that a bird would fly into the spinning blades. It was agreed that any wind turbines would not be installed on any of the key bird transit routes over the island (typically, right along the coastline) or in critical nesting areas (which are off limits to all visitors as well). Only two birds had been killed by the wind turbine after the first 5 years.

During the system design process, it was determined by Sandia Labs that a hybrid solar–wind energy system would be the best option for PNIC. The average annual wind speed was measured at 6.5 m/s. Loads were sized for an average daily usage of 5,000 Wh/day, mostly for lighting, communications, radio, fans, and TV/VCR, as well as an LCD projector (for workshops), shop equipment, kitchen appliances, and a water pump.

The architecture of the original hybrid system consisted of two 500 W wind turbines, a 256 W_p amorphous PV array, a 4,500 W Trace sine-wave inverter, and 19.2 kWh battery bank. The wind machines were originally installed on a tall dune on the east side of the small island. A 3-day training course was then conducted on renewable energy systems design, operation, and maintenance for 23 persons from area institutions, including PNIC. Also, individual training was provided to the three key PNIC maintenance personnel on appropriate RE system operation and maintenance (Romero-Paredes, 2003).

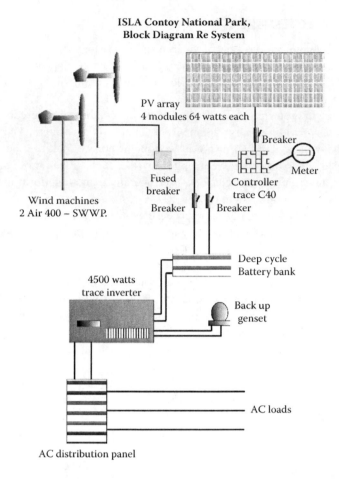

FIGURE 8.24 Isla Contoy National Park (Quintana Roo, Mexico) solar-wind hybrid system designed by Ecoturismo y Nuevas Technologias and funded by USAID with Sandia National Laboratories.

The hybrid system has evolved since installation, adjusting to expanding energy needs and operational conditions. After the first year of operation, the original wind machines had suffered from severe corrosion problems due to the salt spray environment and, under warranty, were replaced by the installer. Two Southwest WindPower (SWWP) Marine Air 400 W_p units (more corrosion tolerant) were installed at the top of the park's observation tower, and one H80 wind machine was left on the original dune site.

The PNIC station was completely remodeled in 2000 by SEMARNAT. A 40 kW diesel plant was installed to operate a reverse osmosis desalination unit, as well as vapor compression air-conditioning systems. However, the desalination plant was never operated and the diesel plant is used just a few hours per month. Subsequently, the battery bank and the PV array were further expanded thanks to a donation from the European Union (EU); the bank grew 300% in size (to 2,400 Ah) and an additional PV subarray was added for a total of 1.5 kW_p. This is one of the most complex renewable energy systems installed in Quntana Roo.

In September 2002, Hurricane Isidore caused substantial damage to the Isla Contoy hybrid system. The hurricane destroyed the dune-mounted H80 wind turbine due to a unique tower failure mode. The galvanized NRG tower had been guyed with stainless steel guys due to the severe corrosion of the area, so the guys and tower tubes did not fail; rather, the actual tower base failed. The base had corroded to the point that the high hurricane winds caused it to fail where the tower and base meet. The wind turbine itself did suffer some damage at its mounting base due to the fall. There was no damage to the actual rotor, but two of the three rotor blades were damaged when the tower collapsed. The smaller SWWP units had been lowered for the hurricane and did not suffer any damage.

The Trace inverter also failed during Hurricane Isidore due to rainwater entering the inverter through a conduit leak, which caused a circuit failure. It took about 4 weeks for the two technicians sent by the Ecovertice Company to repair the inverter failure. An additional floor fan is now necessary to provide cooling because the original inverter fans failed.

The battery bank is somewhat undersized given current loads and should be further expanded to about 4,000 Ah. This will reduce cycling and extend the battery lifetime. The expanded EU portion of the battery bank did not include spark arrestors, unlike the original USAID portion, and thus presents a safety hazard for such a large bank (Figure 8.25).

In addition to hurricanes, one problem the PNIC has had to contend with is the ever changing park technicians and maintenance personnel, who make various adaptations to the hybrid system configuration that are never documented. This complicates future maintenance actions by new personnel who have to spend a great deal of time trying to understand undocumented system changes; which exacerbates accident or failure potential if components are incorrectly connected due to poor system documentation.

In summary, the PNIC RE hybrid system has satisfactorily met park energy needs. Since the tower collapse of the H80 wind turbine, most of the energy is generated from the PV array. Despite system ups and downs, the PNIC staff have been able to maintain the solar/wind hybrid system successfully and it has survived several major storms and hurricanes over the past 5 years. The system clearly shows that an important component for successful application of RE technologies is the institutional aspects related to follow-up support and maintenance.

FIGURE 8.25 Isla Contoy solar and wind energy power center, inverter, controls, and batteries.

8.7.1 PV Ice-Making and Refrigeration

Another key market segment for PV technology is application for remote refrigeration or ice-making. This can be done with or without electrochemical battery storage. Battery-based PV refrigeration technology is relatively mature from the standpoint that DC compressors, batteries, and charge controllers have been in mass production for years, leading to lower cost manufacturing. The battery-free technology is newer and has a much lower level of production, so the manufacturing cost is still relatively high.

A PV direct–drive or "PV-direct" solar refrigerator uses thermal storage, and a direct connection is made between the vapor compression cooling system and the PV panel. This is accomplished by integrating a phase-change material into a well-insulated refrigerator cabinet and by developing a microprocessor-based control system that allows direct connection of a PV panel to a variable-speed DC compressor. This allows for peak power-point tracking and elimination of batteries. This new direct-drive approach with ice storage may revolutionize refrigeration in remote regions around the world.

Solar PV power system applications are increasing due to both technical and economic factors. Some of the most successful applications for solar energy, such as water heaters and PV water-pumping, benefit from built-in energy storage. This is now true for solar refrigerators that use ice thermal energy storage. Although past solar PV refrigerators used batteries to store electricity, the latest work focused on the "PV-direct" concept and thermal storage to eliminate electrical energy storage. PV-direct technology can be applied to freezers, air-conditioners, and larger scale refrigeration systems; however, initial efforts have focused on small-scale refrigerators, which are most appropriate for off-grid personal or small-scale commercial use.

The battery-free solar refrigerator stores thermal energy in a phase change material rather than storing electrical energy in a battery. To develop a practical thermal storage system that effectively replaces the batteries, a well insulated cabinet and a phase change material with a high latent heat of fusion is required. For the commercial application, a chest-style cabinet with standard insulation is used. For the phase change material, a nontoxic, low-cost, water-based solution that has good freezing properties is selected. Based on the heat-leak rate of the cabinet, a quantity of thermal storage material is calculated to provide 7 days of reserve cold storage for an assumed average ambient temperature of 29.5°C (85°F). This thermal storage reserve is intended to simulate approximately the electrical energy reserve of batteries used for solar refrigeration systems. For efficiency, it is also necessary to make good thermal contact between the thermal storage material and the refrigeration system evaporator. Poor contact reduces the refrigeration system efficiency as well as the cooling capacity of the compressor. The phase change material is stored in containers located against the cold inner wall of the refrigerator cabinet, behind a polyethylene liner that holds the containers in place and hides the thermal storage containers from view.

To drive the refrigeration system directly (and efficiently) from solar panels, a variable-speed DC compressor is used. The variable-speed feature allows the compressor to operate longer during the day and make better use of the variable solar resource. A fixed-speed compressor would not be able to begin cooling as early in the morning or as late in the afternoon and would waste power during solar noon (when the available power is more than the compressor needs to operate). A fixed-speed compressor can only utilize about 50% of the solar resource.

A variable-speed compressor uses about 75% of the available solar resource on a sunny day, because its speed can vary to match the available solar input. The speed is controlled by a microprocessor, which seeks to maximize the compressor speed for the available solar power. The control algorithm effectively maintains the PV array at its peak power point while the compressor is on. The microprocessor also performs load testing of the array before starting the compressor, temperature control of the cabinet, and additional speed control as required to keep the compressor power within the manufacturer's limits. Starting capacitors are also used to furnish the compressor with a short power burst during turn-on. A small DC cooling fan is used to improve condenser and compressor heat removal.

The SunDanzer direct-drive prototype refrigerator uses thermal storage, and a direct connection is made between the cooling system and the PV panel (Figure 8.26). This is accomplished by integrating a water–glycol mixture as a phase-change material into a well insulated refrigerator cabinet and by developing a microprocessor-based control system that allows direct connection of a PV panel to a variable-speed DC compressor. The refrigerator uses a more efficient variable-speed DC compressor.

The unit is designed to run on 90–150 W of PV power (needed for compressor start-up), but only draws about 55 W when cycling. During cloudy weather, internal thermal storage keeps products cold for a week, even in a tropical climate. The battery-free unit is designed to work optimally in locations with at least 4 sun-hours per day using a variable-speed compressor and peak power tracking. The unit offers the most economical method for on-site refrigeration for rural people. SunDanzer is an American success story and has now sold thousands of solar refrigerators (most using batteries) around the globe.

8.7.2 PV ICE-MAKING

PV ice-making has not been widely deployed yet, but there have been some attempts. The world's first automatic commercial PV ice-making system was installed in March 1999 to serve the inland fishing community of Chorreras in Chihuahua, Mexico (Figure 8.27). The system was designed and installed by SunWize and supported by the New York State Energy Research and Development Authority, which teamed with USAID, Sandia, the state of Chihuahua, and New Mexico State University.

The US$38,000 hybrid system produced an average of 8.9 kWh/day at 240 V to the ice maker. The system coefficient of performance (COP) was 0.65 and a total of 97% of the energy was supplied by the PV array; only 3% was supplied by the backup propane generator. Production of ice varied each month due to changes in insolation and ambient temperatures and averaged about 75 kg of ice/day (11.5 kg/sun-hour). About every 9 months, the ice-maker water lines would need to be cleaned to remove calcium deposits. With a fixed timer setting, the ice

FIGURE 8.26 SunDanzer PV-direct drive refrigerator piloted in the indigenous Mayan village in Quiché, Guatemala, by NASA and Fundación Solar.

FIGURE 8.27   World's first PV ice-maker developed by SunWize for fishermen in Chihuahua (1999).

maker operated daily for 3 hours with a dozen 15-minute cycles at night to make ice, except on Sundays when there is no fishing (Foster, 2001).

The ice maker performed well for the first few years of operation but eventually fell into disuse after about 4 years. Long-term commitment and follow-up by the Mexican project partners was necessary for continued project success. Unfortunately, there were state political changes and the area faced a severe drought. The lake receded over 2 km from the ice house by 2003 and the fishermen moved their catch out to the other end of the reservoir. The ice-making system was shut down and has not been operated for the past few years.

8.8   PV WATER-PUMPING

PV water-pumping is highly competitive compared to traditional energy technologies. PV power is often the least expensive alternative as compared to extending the power supply grid for applications in remote sites or where loads are small. PV is best suited for remote site applications that have small to moderate power requirements. Some typical cost-effective applications in addition to water-pumping include residential electrification, lighting, small-scale irrigation, refrigeration, and electric fences.

Pumping water is a universal need for agriculture and the use of PV power is a natural choice for this application. Agricultural watering needs are usually greatest during sunnier summer periods when more water can be pumped with a solar energy system. Arid regions, which have the greatest water needs, also have the greatest amount of sunlight available. PV-powered pumping systems can meet the range of needs between small hand pumps and large generator-driven irrigation pumps: drip/trickle, hose/basin, and some open channel irrigation, although flood or sprinkler irrigation are rarely used with photovoltaics. PV water-pumping systems are simple, reliable, and low maintenance. Tens of thousands of agricultural PV water-pumping systems are in the field today throughout the world. PV pumping systems' main advantages are that they are reliable and durable, no fuel is required, and little maintenance is needed. The principal disadvantage of a PV system is the relatively high initial capital cost.

A PV-powered water-pumping system is similar to any other pumping system, with the exception that the power source is solar energy. These systems have, as a minimum, a PV array, a motor, and a pump. PV water-pumping arrays are often mounted on passive trackers (which use no motors) to follow the sun throughout the day, which increases pumping time and water volume. AC and DC motors with centrifugal, displacement, or helical rotor pumps are commonly used with PV pumping systems. If absolutely needed, a battery bank can be used to store energy (e.g., some residential systems often use this approach), but water is typically much more cheaply and effectively stored in a tank.

The advantages of PV water-pumping are long-term lower costs when compared with other alternatives such as diesel- or gasoline-operated water pumps. PV pumping is never a least-cost option if a site is already on the existing conventional electric grid. PV water pumps do not require an on-site operator and have a low environmental impact (no water, air, or noise pollution). Another advantage is system modularity, which provides the owner with the ability to meet specific needs flexibly at any given moment and to increase system size as water-pumping needs grow. Well designed and installed systems are relatively simple to operate and maintain. In order to make a PV water-pumping project successful, it is best to understand basic concepts such as solar energy, PV, water hydraulics, pumps, motors, and other system requirements.

Solar water-pumping is one of the most simple yet elegant solar applications found today, often providing many years of reliable service.

8.8.1 HYDRAULIC WORKLOADS

The volume of water required daily is not adequate to determine the size and cost of a water-pumping system. The total dynamic head (TDH) should also be considered (pumping depth plus discharge height plus drawdown plus friction losses throughout the length of pipe.). For example, more energy is required to extract a cubic meter of water with a TDH of 10 m than with a TDH of 5 m.

A useful formula for quickly determining whether a given project is a good candidate for solar power pumping is to determine the hydraulic burden or duty (Figure 8.28). Multiply the daily volume of water that will be required (expressed in cubic meters) by the total dynamic head estimated for the pumping system (expressed in meters of height). This product is the *hydraulic workload* and provides an excellent indication of the power that will be required to meet the project's needs. If the result is less than 1,500 m^4, then the project is most likely feasible using PV. If it is between 1,500 and 2,000 m^4, it may or may not be feasible for solar pumping. If it is over 2,000 m^4, a technology other than solar options should generally be considered.

For example, 5 m^3 to be pumped with a TDH of 15 m gives a hydraulic workload of 75 m^4. Similarly, 15 m^3 to be pumped with a TDH of 5 m gives a hydraulic workload of 75 m^4. In both cases, the energy required is approximately the same and the cost of these systems is similar. When is the demand considered to be too great for solar water-pumping? Experience shows that a project

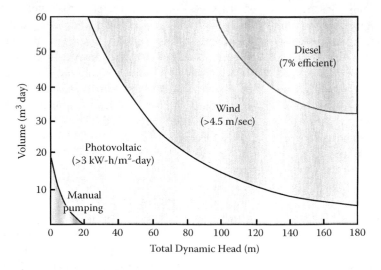

FIGURE 8.28 Water-pumping technology selection based on hydraulic workload.

is economically viable when the hydraulic workload is less than 1,500 m^4. Water-pumping systems powered by internal combustion engines or wind are more competitive when the hydraulic workload is greater than 2,000 m^4.

To obtain maximum benefit from a solar water-pumping system, the water pumped should be used for products of high value to the owner. The water should not be more expensive than the product. The hydraulic duty of any project is essential in determining the most appropriate technology. Figure 8.28 indicates the most appropriate technology, considering the daily volume and total dynamic head and assuming a minimal solar or wind energy resource. This assumes a daily solar insolation of greater than 3.0 kWh/m^2/day, an average annual wind resource greater than 4.5 m/s, and an average efficiency of 7% for diesel-powered systems (fuel at U.S. $1.50/gallon). Note that most diesel- and gasoline-powered pumps are often oversized and efficiencies are typically very poor, especially in the range competitive for PV.

8.8.2 OTHER CONSIDERATIONS

Other factors of significant importance are not easily quantifiable:

- *Experienced installers.* Ideally, PV water-pumping systems should be installed by professionals from the region, although this is not always easy for remote areas. In addition, it is important that the installer be easily located in case service should be required in the future (especially for the pump). The provider and installer should be able to demonstrate their experience, technical expertise, and integrity.
- *User acceptance.* Users should understand the abilities of solar energy systems, including their limitations, advantages, expected maintenance requirements, and principles of operation. Designers should involve users with general project design. This will allow them to grasp the technology better as well as feel a sense of buy-in to the project and its realistic outcome.
- *Security.* The nature and portability of solar water-pumping systems make them ideal for remote and isolated applications, but they also become vulnerable to theft and vandalism. They are best protected from theft if they are placed in areas that are not likely to be transited and seen by the general public.
- *Environmental benefits.* Solar energy technology helps maintain clean air and water quality. An added plus is that it pumps with little noise, unlike noisy diesel- or gasoline-powered pumps.
- *Batteries.* Batteries are a key part of PV systems in most applications, but are rarely used in stand-alone solar pumping systems. Batteries add cost and complexity to the system. It is far better to design a system where energy is stored in the form of additional pumped water available at the distribution tank instead of in electrochemical form with batteries. The only time batteries are commonly employed is for a household water pump with an existing battery bank supplying energy to other household loads as well.
- *Water needs.* In communities where water is easily available from traditional sources and the perceived benefit that the potential PV pumping project brings is mostly about an improvement in convenience, the attitudes regarding the real value of water and water conservation may be too cavalier to make a PV project feasible. This is a notable issue in some countries where many communities have multiple sources of spring water. The issue should be honestly examined with the community from two interrelated reference points. First, because PV is more expensive than other solutions, there should be some reflection of the higher cost in the tariff set. Communities unwilling to pay a higher price for water than the very low fees used in gravity flow systems are questionable prospects for PV. Second, in communities where water (even bad water) is relatively easily available from a traditional source, it can be extraordinarily difficult to inculcate the attitudes of

water conservation and careful use that are absolutely essential to making a PV project practical.

In defining a water program strategy with a PV focus, it is important to realize that PV occupies a niche, and that this niche is confined to a certain community size, well depth, and service level. It is far better to include PV in a mix of implementation options as one of the tools for meeting a rural need than to stipulate PV for a determined number of projects. PV can be a good option in some very difficult circumstances where other solutions are impossible, but it may not be suited to broad application within a region.

The flow chart in Figure 8.29 summarizes the key technical points for considering when PV is a likely feasible method for a water-pumping system. The selection process considers such parameters as distance to the grid, hydraulic workload, and the solar energy resource available at the site. As is often the case with water-pumping, no matter what the technology is, every project requires a somewhat customized and individualized approach.

For PV power systems, the energy needed to power the pump is provided by the Sun. Solar energy is captured and transformed into electrical energy by solar cells, which are the building blocks of a PV module. The solar energy is typically coupled directly to power a pump motor.

8.8.3 PRESSURE

A column of water enclosed in a pipe or tank exerts a force due to the weight of the water. This force is described as *water pressure,* also known as *head.* Water pressure is expressed in terms of pounds per square inch (*psi*) or in kilograms per square centimeter (*kg/cm²*). Head is a useful indicator of water pressure that refers simply to the height of the column of water. For example, a column of water 20 m high would be said to have 20 m of head. Knowing the head, one can calculate the pressure and vice versa.

8.8.4 STATIC HEAD

A system where the water is not in movement is static. Regardless of whether or not water in a pipe or tank is actually flowing, the water pressure always exists and is referred to as the static head. In a static system, the water pressure is dependent exclusively on the height of the water "column" in the system. That is, a narrow column of water will have the same static head as a wide column, provided that both are at the same height. Two tanks of water filled to the same height will have the

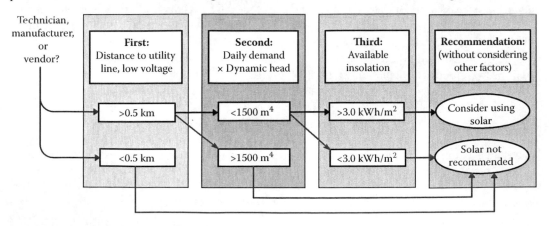

FIGURE 8.29 Basic decision-making flow chart for PV water-pumping use.

same pressure at an outlet on the bottom, even if one tank is narrower and has a smaller volume of water. The rule of thumb for calculating pressure and head is as follows:

$$1 \text{ psi} = 2.3 \text{ ft of vertical height (head)}$$

$$1 \text{ ft} = 0.43 \text{ psi}$$

or, in the metric system, as

$$1 \text{ kg/cm}^2 = 10 \text{ m}$$

$$1 \text{ m} = 0.1 \text{ kg/cm}^2$$

Example 8.1

A static column of water in a pipe indicates on a gauge a pressure of 37 psi. What is the vertical height of the column of water (i.e., head)?
Solution:

$$37 \text{ psi} * 2.3 \text{ ft/psi} = 85 \text{ ft}$$

Example 8.2

A column of water in a pipe runs to a point 25 m above its starting point. If the pipe is closed and full of water, what is the pressure at the starting point?

Solution:

$$25 \text{ m} * 0.1 \text{ kg/cm}^2/\text{m} = 2.5 \text{ kg/cm}^2$$

The column exerts a pressure of 2.5 kg/cm^2.

8.8.5 Pumping Requirements

For solar water-pumping systems, it is important to think in terms of how much water is required each day. Many water users, such as ranchers, are accustomed to pumping all of their water in a relatively short timeframe with an oversized gasoline- or diesel-powered pump. Solar pumping gradually pumps the same quantity of water during the course of the daylight hours. The pumping requirement is QH (meters to the fourth power/day), where Q is flow (cubic meters/day) and H is the dynamic head (m) (1 m^3 = 1,000 L). For surface water resources (rivers, streams, reservoirs), the water capacity needs to be determined by season or month. For wells, it is very important to determine the capacity and drawdown for different pumping rates. In both cases, the dynamic head needs to be determined correctly in order to select the right pump and the total solar energy power system required.

8.8.6 Dynamic Systems

When there is movement of water in a system, it is a *dynamic system*. The water pressure in a dynamic system is dependent not only on the water "column" height, but also on the friction from the movement of water in a pipe, as well as any drop in the static water level due to pumping. In the dynamic system it is necessary to take the following into account:

- *Length* of the pipe. The longer the pipe is, the greater is the pressure drop due to friction.
- *Diameter* of the pipe. The smaller the pipe is, the greater is the pressure drop.

- *Flow* of water. The greater the flow is, the greater is the pressure drop.
- *Roughness* of the inside of the pipe. The rougher the interior surface is, the greater is the pressure drop. PVC pipe is smoother than galvanized iron pipe.
- *Fittings and joints.* Each union or elbow has an additional associated pressure drop.
- *Change in static water level.* As water is pumped, the water level may drop.

The length of the pipe, the speed of the water, drop in static water level, etc. increase resistance and cause an increased pressure drop and higher pumping power requirements.

The *static head* (*SH*) refers to the height from the static water level in the well up to the discharge level. This is often divided into two components, as expressed in the formula that follows this paragraph. The *total dynamic head* (*TDH*) is the sum of all the components that contribute to the total pumping height, expressed in feet or meters. Dynamic head includes drawdown and all frictional and pressure losses. Frictional losses depend on size of pipe, flow (volume/time), number of elbows, etc. (Figure 8.30).

$$TDH = SH + \text{well drawdown} + \text{friction} \tag{8.1}$$

The friction factor can be handled in two ways: with pipe friction tables or with an estimated value. To use the friction table, the length of pipe used (vertical and horizontal distance), the type of pipe (PVC, GI), the flow rate, and diameter of the pipe must be known. The tables in Appendix B will provide the total friction loss. Because the friction losses are usually not a significant portion of the total, their value can also be reasonably approximated for the TDH equation. A standard default is to consider 2–5% friction loss for a well designed distribution system. If there are long pipe runs, this number may have to increase.

To use the friction tables found in Appendix B, first find the pumping flow rate (in liters per second) for the system by dividing the total daily water pumped by the number of seconds in the solar pumping cycle. A typical solar pumping cycle is around 6 h, between 9 a.m. and 3 p.m. In that period there are 21,600 s (6 h × 60 min × 60 s). If, for example, the daily water pumped is 25,000 l, then the average flow rate will be 25,000 L/21,600 s = 1.16 l/s.

This flow rate is applied to the tables along with the type of pipe material and pipe size (see Appendix B). In the case of a 2 in. PVC pipe, the friction factor for a flow of 1.16 l/s is 0.71, or more precisely, 0.71 m of friction for every 100 m of pipe distance (horizontal or vertical). If the pipe is 300 m long, the total friction factor would be about 2 m (0.71 × 3 = 2.13 m).

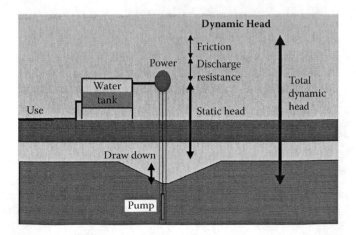

FIGURE 8.30   Total dynamic head includes all components.

Example 8.3

A small community has the following characteristics:

Static level of water in well	**31 m**
Drawdown	5 m
Height from well head to discharge at tank	16 m
Pipe run from well to tank	80 m
Daily water need (for 360 persons)	18,000 L
Pipe material and size	1.5 in. PVC
Average solar insolation	6 sun-hours/day

Determine the TDH.

Solution:

First, the height components are added (do not include the horizontal measurement):

$$31 \text{ m} + 5 \text{ m} + 16 \text{ m} = 52 \text{ m}$$

Then 2% is added for friction (using the default method):

$$52 \text{ m} * 0.02 = 1.04 \text{ m}$$

Round to the nearest meter.
The TDH is found to be

$$\text{TDH} = 31 \text{ m} + 5 \text{ m} + 16 \text{ m} + 1 \text{ m}$$

$$\text{TDH} = 53 \text{ m}$$

Alternatively, determine the friction factor from the friction tables:
Divide 18,000 l/day water need by the number of seconds for the average sun-hours: 6 sun-hours = 21,600 s

$$18,000 \text{ L}/21, 600 \text{ s} = 0.83 \text{ l/s}$$

By the table for a flow 0.83 l/s, 1.5 in. pipe, PVC material: 1.70 m of friction for every 100 m of distance. For the 80 m pipe run in the example:

$$1.70 * 0.80 = 1.36 \text{ m}$$

Round to nearest meter.
The TDH is then likewise found to be

$$\text{TDH} = 31 \text{ m} + 5 \text{ m} + 16 \text{ m} + 1 \text{ m}$$

$$\text{TDH} = 53 \text{ m}$$

8.8.7 WATER DEMAND

The average daily demand (cubic meters/day) is estimated for the month of high demand and/or the solar design month (month with lowest average solar insolation). Also, the demand must take into account any growth during the design period, which should be at least 10 years.

The water demand for livestock can be up to 90 l/day (Table 8.1). Evaporation, especially in windy and dry areas, will require even more water. Also, animals will only travel a limited distance from the water source, so the water sources need to be spaced around one source per 250 ha of rangeland. If the water supply and grassland are communal, then there is the distinct possibility that the growth in the size of the herds will result in overgrazing, especially close to the water supply.

The domestic water demand depends on number of people, usage, and type of service (Table 8.2). What is considered necessary in some countries or regions would be considered a luxury in other locations. In addition, people will consume more water during hot, dry periods. Local water consumption is the best guide; however, remember that usage per person will probably increase if water availability improves.

Village water supply includes clinics, stores, schools, and other institutions. Growth in demand will depend primarily on water availability, growth in size of herds or flocks, and growth in population for villages. Again, the growth in population should be estimated from present local trends (i.e., not from national trends).

Water demand for irrigation (low or high volume) will depend on local conditions, season, crops, and evapotranspiration. These data are generally available from regional or national government agricultural agencies.

8.8.7.1 Water Resources

For surface water resources (rivers, streams, reservoirs, etc.), the capacity needs to be determined by season or month. For wells, it is very important to determine the capacity and drawdown for different pumping rates. In both cases, the dynamic head needs to be determined. Dynamic head includes drawdown and all frictional and pressure losses. Frictional losses depend on size of pipe,

TABLE 8.1
Livestock Water Requirements

Animal	Liters/day
Cattle, beef	40–50
Cattle, dairy	60–75
Camels	40–90
Sheep and goats	8–10
Swine	10–20
Horses	40–50
Chickens (100)	8–15
Turkeys (100)	15–25

TABLE 8.2
Typical Water Consumption per Person

Service	Liters/day
Standpost	40
Yard tap	75
Home connection	100
World Health Organization recommendation	45

flow (volume/time), number of elbows, etc. If drawdown is not known and frictional losses are not calculated, these can be estimated but should be verified, especially for larger pumping projects. Smaller capacity sources may need a bigger storage tank for domestic, livestock, or village use, or even multiple wells.

Thousands of solar pumping systems are in operation throughout the world. They provide for a wide range of needs, including water for cattle and small-scale irrigation as well as for human needs, aquaculture, and industrial applications. They are reliable and low in maintenance when properly engineered and installed. Since the 1990s, the quality of solar pumps has increased significantly and the costs have dropped. Sometimes a solar pump costs no more to install than an engine-driven pump system.

A typical solar pumping system is shown in Figure 8.31. The main components consist of an array of PV modules, a controller, a motor, and a pump. The array can be mounted on a solar tracker to lengthen the daily pumping period and increase the daily water volume. The motor may be either a traditional type (with brushes) or an electronic "brushless" motor. The pump may use either a centrifugal or a positive displacement (volumetric) mechanism. Most often, water is stored in a tank instead of energy being stored in batteries. A nonbattery system is called "PV-direct" or "solar array direct." In this section, the pump, motor, and controller are briefly explained.

8.8.8 Storage of Water versus Storage of Energy in Batteries

To make water available at all times, some form of storage is required. Storing water in a tank is more economical than storing energy in batteries. Batteries are expensive and must be replaced every few years, while the useful life of a storage tank can be many decades. A battery system

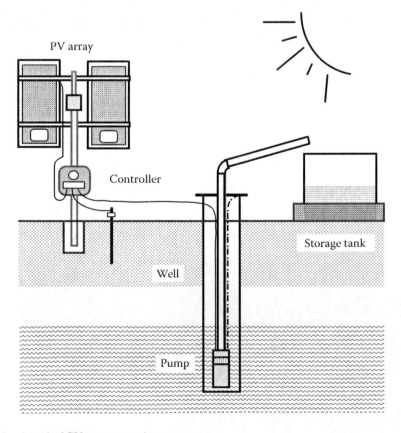

FIGURE 8.31   A typical PV water pumping system.

requires shelter from temperature extremes and controls to prevent overcharge and overdischarge of the batteries. Battery round-trip efficiencies are typically only about 70%, so they lose much of the energy that cycles through them. The introduction of batteries into a PV pumping system will reduce its reliability and increase cost and maintenance requirements. In general, it is best to size a solar pumping system to supply the required water volume without batteries, even if it necessitates installing two pumps in the same well or constructing an additional well and pump. A battery system might be used in cases where a water tank is not practical or the water must be pressurized beyond what is available from natural elevation of a tank.

8.8.9   Pumping Mechanisms Used for Solar Pumps

Conventional water well pumps are designed to run at a constant speed from a stable power source. However, the power from a solar array varies with the intensity of solar radiation and with the angle of the sunshine on the array. The speed of a solar pump varies accordingly. For this reason, some manufacturers have designed pumps for solar power. From a mechanical point of view, these pumps fall under two categories: *centrifugal* and *positive displacement* (*volumetric*).

8.8.9.1   Centrifugal Pumps

These pumps have one or more impellers that spin the water to subject it to centrifugal force. To attain high lift, a centrifugal pump may have a multitude of stages, each consisting of an impeller. Each stage adds to the pump's lift capacity. Conventional electric well pumps are built this way (Figure 8.32).

Centrifugal pumps may use over 20 stages to attain high lifts. Each stage adds pressure but also imposes friction, resulting in an efficiency loss of about 5% per stage. Centrifugal pumps with many stages can have poor energy efficiency and are not always optimum for solar pumping.

Centrifugal pumps are most efficient for flow in excess of about 40 l/m and for lifts less than 40 m. At lower flow rates and higher lifts, the efficiency is poor. At reduced speeds such as those that occur during low-sun conditions, centrifugal pumps lose efficiency in a disproportionate manner. For these reasons, positive displacement pumps are used for most systems that require high lift, especially at modest volumes.

8.8.9.2   Positive Displacement Pumps

A positive displacement pump draws water into a sealed chamber and then forces it out mechanically. A piston pump is a classic example. A solar pump may use a diaphragm, instead, or a helical rotor that traps water in cavities that progress upward as it turns. These pumps have high lift capacity and high energy efficiency. They are optimum for lower flow rates (e.g., 50 l/m), especially when the lift exceeds 15 m.

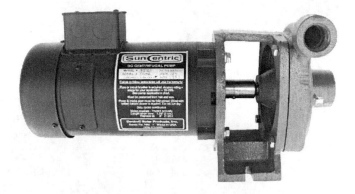

FIGURE 8.32   Surface centrifugal pump.

Positive displacement pumps are used for most solar pumps in the power range of 500 W (0.5 hp) or less. The efficiency and lift capacity of these pumps remain high even at low rotational speeds, such as those that occur in a solar-direct pump during low-light conditions. This is not true for centrifugal pumps.

8.8.9.3 Surface Pumps versus Submersible Pumps

A *surface pump* is one that cannot be submerged in water (see Figure 8.33). It can be installed above the water source, but nature imposes a strict limit on the height to which water can be drawn by suction. The pump must not be more than 3–6 vertical meters above the water source level. Otherwise, it will extract bubbles from the water and will fail to pump. A surface pump can draw from a river, irrigation ditch, pond, or water tank, but not from a deep well. It may be less expensive than a submersible pump and more efficient for high-volume pumping. However, a submersible pump is often simpler to install, better protected from the environment, and less likely to be damaged from running dry (Figure 8.35).

Some solar *submersible pumps* use the same centrifugal mechanism as a surface pump. Others use a positive displacement mechanism.

Centrifugal submersible pumps are the dominant technology for deep well pumping (see Figure 8.34). Solar pumps of this type are similar, except for the use of a specialized motor and controller.

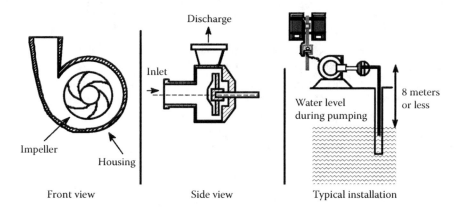

FIGURE 8.33 Diagram of a surface centrifugal pump.

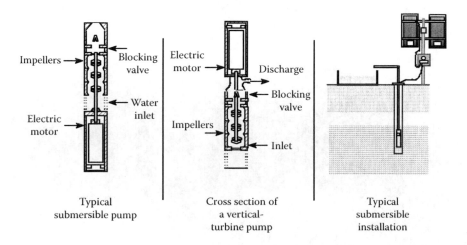

FIGURE 8.34 Diagram of a submersible centrifugal pump.

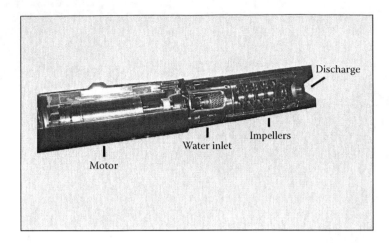

FIGURE 8.35 Grundfos submersible pump.

FIGURE 8.36 Submersible ETA helical rotor pump with controller.

The *helical rotor submersible pump* is a positive displacement pump mechanism that is mounted to a submersible motor. The motor is similar to that used for centrifugal submersibles. Like the centrifugal submersible, the helical rotor can last for many years with no regular maintenance. Many of the newer solar pumps use this type of design (Figures 8.36 and 8.37)

Diaphragm submersible pumps (Figure 8.38) displace water by means of a diaphragm made of flexible synthetic material. Diaphragms fail after about 2 or 3 years of continuous use. Manufacturers of these pumps provide diaphragm replacement kits. If the diaphragm fails in use, water floods the motor and destroys it. Therefore, preventive maintenance should be scheduled to replace the diaphragm before it fails. These pumps also use a brush-type motor that requires brush replacement at intervals of 3–5 years.

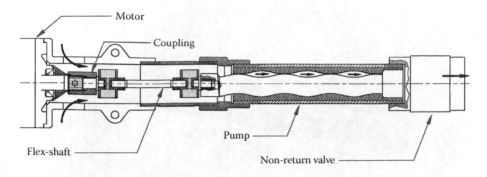

FIGURE 8.37　Diagram of a helical rotor pump.

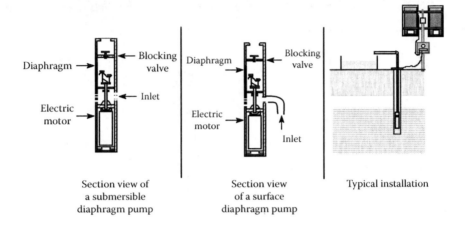

FIGURE 8.38　Submersible diaphragm pumps.

A diaphragm pump may be used when the initial cost must be minimal, when the water volume requirement is very low, and when the future cost of pump maintenance or replacement is acceptable.

8.8.10　Types of Motors Used with Solar Pumps

A PV array generates DC power at a power level that varies with the intensity of the sunshine that falls upon it. To run a pump directly from this unique source of energy requires a special kind of motor or motor/control system. There are two major types of solar pump motors: brush-type motors and brushless motors.

The *brush-type motor* is the traditional DC motor technology that has been used in battery-powered applications for many decades. The "brushes" are small blocks of electrically conductive carbon-graphite. They rub against the spinning part of the motor (commutator) and conduct current into it. This causes the current to alternate (to become AC) within the motor. This simple technology has two major disadvantages: (1) The brushes wear out and must be replaced periodically, and (2) the motor must be filled with air (not liquid) and must be 100% sealed against water leakage. These are major disadvantages for submersible pumps. Brush-type motors are often used for surface pumps where they are kept dry and access is easy.

The term *brushless DC motor* refers to a special type of AC motor driven by an electronic controller that converts DC power into variable AC power. The controller does the job of the brushes and commutator in a brush-type motor. The brushless motor has two major advantages: (1) There are no brushes to wear, and (2) the motor can be filled with oil or water. The safest solar submersible pumps use water inside as a lubricant, eliminating potential oil contamination.

8.8.11   Solar Pump Controllers

There are two types of solar pump controllers for both motor types:

Controllers (linear current boosters) for brush-type motors. A positive displacement pump requires a surge of current for start-up and must come up to speed against the constant pressure imposed by the water in the pipe. A PV array may not be sized large enough to produce the required starting surge, especially in low-light conditions, when it produces reduced current. A linear current booster (LCB) can be used to reduce the voltage from the PV array while it boosts the current. This starts the pump motor and prevents it from stalling during low-light conditions. A brush-type centrifugal pump is often supplied without an LCB because it starts easily and its current draw diminishes with speed. An LCB controller will increase its efficiency during low-sun periods, but the performance gain is relatively small.

Controllers for brushless solar pump motors. A brushless motor controller contains a special type of inverter (a device that converts DC to AC). It performs the LCB function and matches the motor speed to the available power. The three-phase AC power is optimum for starting and running the motor at high efficiency. The controller varies the motor speed by varying the frequency of the AC power. A brushless pump is normally sold with a controller that is engineered specifically for it.

8.8.11.1   Additional Features of Pump Controllers

Solar pump controllers incorporate other control functions to make solar pumping practical and efficient. A typical controller has connections for a *float switch* to prevent the storage tank from overflowing. When the tank fills, the switch signals the controller to turn the pump off. When the water level drops, the float switch resets. This prevents flooding, unnecessary pump wear, and waste of water.

Most solar pumps can be damaged if they run dry, so most pump controllers have a *dry-run prevention* system. This may use a sensor mounted above the pump's intake. If the water level drops below the probe, an electric current is opened and the controller will stop the pump. When the water level recovers, the controller will wait for the level to rise (typically a 20-min delay) and will then restart the pump. Other pumps use a thermal switch so that if the temperature begins to rise due to dry running, the pump automatically shuts off.

A controller also has *overload protection* to prevent damage if the pump is stopped by dirt, ice, crushed pipe, or a closed valve. A controller should also be installed with appropriate overload and surge protection (Figure 8.39). See Section 8.9.

The controller should also have *indicator lights* so that an observer can easily determine when the pump is running, when the tank is full, and when there is a fault in the system.

A function called *maximum power point tracking* (MPPT) is commonly used on most solar pump controllers. This is an improvement on the basic linear current booster. It helps the pump to draw the maximum power from the solar array even as solar cell characteristics vary with temperature and sun intensity.

Location of the pump controller. A brushless submersible solar pump may have its controller built into the motor (Grundfos SQFlex), mounted aboveground (ETA pump), or partly above and partly inside the motor (Sun Pumps). A submerged controller is isolated from the weather and from

FIGURE 8.39 Typical pump controller with overcurrent protection for PV water-pumping system in Chihuahua, Mexico.

human interference. However, if there is a problem with the electronics in the motor, the entire pump and pipe assembly must be removed from the well and the entire motor assembly replaced.

8.8.12 Pump Selection

The process of selecting a pump is critical to the success of a project. A solar pump must use energy efficiently because the PV array that powers it is the most expensive part of the system. Centrifugal and volumetric pumps offer different characteristics for different ranges of application. The pump-selection process can appear complicated due to the multitude of technologies available and the many models available. For help in selecting the best type of pump for a given application, refer to Figure 8.40 and Table 8.3. Manufacturers who produce both helical rotor and centrifugal submersibles (Grundfos SQFlex, ETA) have combined these into a single product line. The manufacturer's selection guide, often computerized, will indicate the best pump for a particular application.

8.8.13 Installation, Operation, and Maintenance

Good operation and maintenance practices are important to ensure the long-term reliability of a PV water-pumping system. Although a well designed and installed PV pumping system is safe, reliable, and requires little attention, there may be times when basic maintenance is required, especially for the pump. The operator should know how to run the system and perform routine maintenance and operation procedures, such as system shut-off/start-up procedures. All of this information should be included in an operation and maintenance manual from the original system provider. The operator

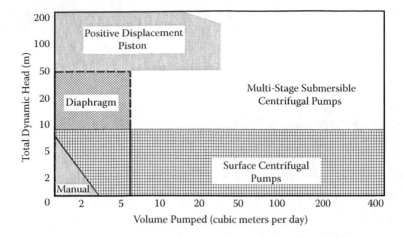

FIGURE 8.40   Approximate pump selection based on lift and volume requirements.

TABLE 8.3
Pump Characteristics

Type of solar pump	Advantages	Disadvantages
Submersible centrifugal	Simple, one moving part	Poor efficiency at low volumes
	Regular maintenance not required	(<30 L/m)
	Efficient at high flow rates	Lift capacity is greatly reduced at slow
	Good tolerance for moderate amounts of sand and silt	speeds (during low-sun conditions)
Submersible helical rotor	Simple, one moving part	
	Regular maintenance not required	
	Highly efficient at low to medium flow rates (4–50 L/m)	
	Maintains full lift capacity even at low speed	
	Good tolerance for moderate amounts of sand and silt	
Diaphragm submersible	Low initial cost	Requires regular preventive maintenance
	Efficient at very low flow rates (4–20 L/m)	Poor tolerance for sand and silt
	Maintains full lift capacity even at low speed	
Surface centrifugal	Low cost	Suction limit is about 6 m
	Efficient for low lift and very high flow rates	May be damaged by running dry if it loses prime
	Easy to inspect and maintain due to surface location	May be damaged by freezing in cold climates
	Good tolerance for moderate amounts of sand and silt	

should understand the expected system output in cubic meters per day, the flow rate on a sunny day, and the significance of indicator lights, as well as basic array, wiring, and pump features.

8.8.14   System Installation

Any water-pumping system component can fail if it is not properly installed and maintained. Because solar pumping systems are assembled in the field, qualified personnel are essential for a safe and professional installation. The installer should follow local electrical safety codes and is responsible to ensure that all materials and tools are available during installation. The pump manufacturer's installation recommendations should be followed. Additional special measures may be required, depending on location and local conditions (freezing, flooding, lightning, vandalism, theft, etc.) For a successful installation:

- verify the water source (seasonal production);
- check civil works (foundations, piping, and storage system);
- test mechanical and electrical field connections;
- run through system operational modes;
- quantify component and system performance (acceptance test);
- conduct basic system training for the system owner/operator; and
- provide an operation and maintenance manual to the system owner/operator.

Experience has shown that it is important to pay attention to detail during installation to avoid later unexpected system malfunctions that are often caused by poor initial electrical or mechanical connections. For example, thermal cycling of poor electrical connections over the years can cause a system to decrease in performance or fail. The system controller box may not be properly sealed, which allows moisture to enter and corrode circuit boards or connections eventually. These original simple problems can cause a halt in operation and later high repair costs.

The designer should correctly specify the gauge and type of conductor to be used for the current, voltage, and operation conditions of components and the system. All exposed cables should be approved for outdoor use (e.g., USE or SE wire) or installed in electrical conduit. Cables should be protected and adequately secured. In some cases, it may be necessary to bury conductors; underground cable or conductor approved for direct burial should be used (e.g., USE or SE). All connections should be made in accessible junction boxes where they can be inspected, repaired, and mechanically secured. All electronic equipment and electrical connections should be protected against water, dust, and insect intrusion. It is important to protect cables against physical abuse, especially where a pump cable enters a well. Excessively long wire runs should be minimized to avoid increased voltage losses. All connections should use strain relief. Any cable ties used should be sunlight (UV) resistant (i.e., black nylon).

Extra caution needs to be used for the installation of submersible pump cables. This cable may remain submerged in water for decades; consequently, it should be perfectly waterproof and have adequate strain relief to avoid failure. For pump cable splices, cylindrical butt connectors, sized for the wire, are normally used. If the wire gauge of the submersible cable is greater than that of the original manufacturer's pump cable, a connector sized for the submersible cable should be used and the pump cable doubled up in order to make a secure connection. Ratcheted crimping tools should be used for maximum force. Insulation of the pump splice connections should be done with epoxy and rubber-sealed thermal shrink tubing. Each splice connection should be separately insulated to avoid short circuits. The manufacturer's installation instructions should be followed carefully. The weight of the pump should never be supported on the electrical pump cable and a separate inorganic rope, corrosion proof cable, or rigid pipe should always be used to support the pump and to haul it into or out of a well.

8.8.14.1 Civil Works

Array support foundations are critical. The array support should be able to withstand a wind loading of at least 160 km/h (category 2 hurricane). Concrete must be allowed to cure adequately. For a well, the combined weights of the pump, motor, pipe, and water column must be considered for well supports.

8.8.14.2   Piping

The piping and fittings used for the system should be corrosion resistant. The piping that is used from ground level down to the well should be able to withstand the pressure caused by the column of water. The fittings should be able to withstand these forces without developing leakage over time. Leaks reduce productivity and, in the case of surface pumps, they cause loss of suction. Friction losses contribute significantly to the overall head and as a result decrease system productivity. To cut down on friction losses, long piping runs and small diameters of pipe should be avoided. The use of elbows and valves should also be minimized whenever possible. Corrosion-resistant mounting structures and fasteners should always be used.

It is advisable to protect the PV array against physical abuse from animals. A fence may be constructed around the array. Care should be taken not to shade the array (from trees, fences, buildings, etc.) between the hours of 9:00 a.m. and 4:00 p.m.

8.8.14.3   Surface-Pump Installation

Ground-level pumps should be mounted to a structure (typically concrete) placed over the surface of the water source (Figure 8.41). The structure and mounting fasteners should be sufficiently firm to withstand pump vibration and the weight of the water column in the piping that runs from ground level down to the well. Surface-mounted centrifugal pumps have a maximum suction capacity of about 7 or 8 m. Surface-mounted piston and diaphragm pumps also have suction limitations. For this reason, the vertical distance from the pump to the water level in the well should be minimized. To reduce friction losses, wide diameter pipes should be installed with valves and a discharge water flow meter.

A check valve is recommended for positive displacement pumps. Water must be present in the suction pipe in order for the pump to operate. After priming the pump, the check valve should keep the suction pipe full of water, including when the pump is off for a period of time. If a check valve is not installed, the system will require manual priming (filling the suction pipe with water) each time the pump is started. If the water-distribution line is long, it is important to install a check valve on the discharge side of the pump to avoid damage due to "ram" (water hammer). Any pump intake should be installed far enough away from the well bottom and sides to avoid pumping mud, sand, and debris, which can all cause damage to pump seals and components. If it is probable that the water level will fall below the intake, it will be necessary to install a switch (a float or electrode) to avoid pumping dry.

FIGURE 8.41   PV-powered surface jack pump in Chihuahua, Mexico, for a 170-meter deep well.

Sand is one of the main causes of pump failure because it destroys seals, fills impellers, etc. If the well is located where sand or dirt can penetrate into the pump, a sand filter should be installed. Most pump manufacturers who sell this kind of filter can recommend ways to reduce the risk of damage.

8.8.14.4   Surface Water Pumps: Preventing Cavitation and Noise

Excessive suction causes cavitation, which is the formation and collapse of bubbles. When water pressure is reduced beyond a critical point, water vapor and/or dissolved gasses are released similarly to when a carbonated beverage is opened. When a bubble reaches the pressure side of the pump, gas returns to the liquid state. Bubbles collapse in sudden implosion. This causes water to strike violently, like tiny hammer blows, against the working surfaces of the pump. Cavitation causes loud noise and excessive pump wear. It is not the fault of the pump, but rather of the installation. To prevent cavitation, follow these precautions:

- Refer to the pump's specification sheet and instructions and observe the limits of vertical suction lift.
- Water should flow easily for intake lines. Use large intake pipe (larger than the pump's intake port). This is especially critical in cases of long intake piping (see pipe sizing chart.)
- Avoid 90° elbows. Use pairs of 45° elbows to reduce friction losses.
- Carefully choose intake screens or intake filters for low friction and make sure that they will be easy to clean.
- Work carefully to minimize the possibility of air leaks.
- Avoid high spots in the intake pipe. They can trap bubbles that will restrict the flow (like in a siphon). If a high spot is unavoidable, install a pipe tee at the highest point, with a cap or a ball valve above it. When water is poured in at the high point, it will displace all of the air to prime the intake line fully.

8.8.14.5   Installation of Submersible Pumps

Good submersible pump installation requires experience. For example, submersible centrifugal pumps utilize components that must be installed within the well. Manual installation can be difficult without the use of mechanized equipment. The structure to which the equipment is connected should be robust in order to support the combined weight of the water column, the metal piping from the surface to the well, and the well point. Each manufacturer provides installation instructions.

It is important to install a safety cable on submersible pumps (Figure 8.42). It is best if the casing (but never the power cable) supports the weight of the pump and the water column. For centrifugal pumps, it is recommended that the piping from ground level to the well be sized to reduce friction losses.

8.9   GROUNDING AND LIGHTNING PROTECTION FOR SOLAR WATER PUMPS

Surges induced by lightning are one of the most common causes of electronic controller failures in solar water pumps. Damaging surges can be induced from lightning that strikes a long distance from the system. The risk of damage can be greatly reduced by taking the following steps:

- *PV array wiring.* Array wiring should use minimum lengths of wire, tucked into the metal framework and then run through metal conduit. Positive and negative wires should be of equal length and be run together when possible. This will minimize induction of excessive voltage between the conductors. Long outdoor wire runs should be buried instead of run overhead and placed in grounded metal conduit if maximum protection is required. The negative conductor should be grounded to meet electrical code specifications.
- *Location of pump controller.* In general, the input circuit of a pump controller is more sensitive than the output circuit. Therefore, in cases where a long wire run is required between

FIGURE 8.42 Installation of a PV submersible pump in Sonora, Mexico.

the PV array and the water source, it is usually best to locate the controller near the array to minimize the length of the input wires.

• *Construct a discharge path to ground.* A properly made discharge path to ground (earth) will discharge static electricity that accumulates in the aboveground structure. This helps prevent the attraction of lightning. When there is a nearby lightning strike, it is hoped that a well grounded structure will divert the surge around the power circuitry, greatly reducing the probability or the intensity of damage. Most solar pump controllers have built-in surge protectors that function only if they are effectively grounded.

8.9.1 Bond (Interconnect) All Metal Structural Components and Electrical Enclosures

The PV module (solar panel) frames, mounting rack, and ground terminals of the disconnect switch and the controller should be interconnected using wire of minimum size AWG#8 (6 mm²) and the wire run to an earth connection. When connecting dissimilar metals, connectors approved for the materials involved should be used. For example, at the aluminum framework of the solar array, connectors labeled "AL/CU" and stainless steel fasteners should be used. This will reduce the potential for corrosion.

8.9.2 Ground

One or more 8 ft (2.5 m) copper-plated ground rods should be installed, preferably in moist earth. Where the ground gets very dry (poor conductance), more than one rod, spaced at least 10 ft (3 m) apart, should be installed. One can also bury AWG#6 (16 mm²) or double AWG #8 (10 mm²) or larger bare copper wire in a trench at least 100 ft (30 m) long. One end should be connected to the array structure and controller. If a trench is to be dug for burial of water pipes, ground wire can be run along the bottom of the trench. A steel well casing near the array can be used as a ground rod. A hole should be drilled and tapped to make a strong bolted connection to the casing. Concrete footers with rebar of a ground-mounted array will not provide adequate grounding alone.

8.9.3 Float Switch Cable

A long run of control cable to a float switch can pick up damaging surges from nearby lightning strikes. The best protection is to use shielded, twisted-pair cable. Shielded cable has a metallic foil or braid surrounding the two wires. The cable shield should be grounded at the controller end rather than at the float switch.

8.9.4 Additional Lightning Protection

Lightning protection devices (surge arrestors) are intended to bypass excessive voltage such as that from a lightning strike. Most pump controllers have built-in surge protection devices metal oxide varistors (MOVs) that are useful but limited in their capacity. Additional grounding measures or surge protection devices are recommended under any of the following conditions:

- isolated location on high ground in a severe lightning area;
- dry, rocky, or otherwise poorly conductive soil; and
- long wire run (more than 100 ft/30 m) from the controller to the wellhead or to the float switch.

Solar pump wiring should be kept away from electric fence systems and the pump system should not be connected to the same ground rod as an electric fence system. A float switch cable should not be run near an electric fence.

8.10 SOLAR TRACKING FOR SOLAR WATER PUMPS

A solar tracker is a PV rack that rotates on an axis to face the sun as it crosses the sky. Two-axes tracking can increase energy yield by about 25% annually, depending on latitude. For solar pumping, tracking can improve performance while reducing overall system costs. Tracking offers more water out of smaller, less expensive PV array by increasing performance.

Some solar pumps (particularly centrifugal pumps) experience a disproportionate drop in performance when the sun is at a low angle (early morning and late afternoon). When the PV array output is less than 50%, a centrifugal pump may produce insufficient centrifugal force to achieve the required lift. By causing the pump to run at full speed through a whole sunny day, tracking can greatly increase the daily water yield (30% or more). In the case of positive displacement pumps, the gain from tracking is more closely proportional to the actual energy capture.

The tracking decision is a variable in the design process. Often a proposed system produces a little bit less than is needed, but the next larger system costs much more. A tracker is a low-cost means to increase the yield of the smaller system sufficiently to meet the demand. Tracking is least effective during shorter winter days and during cloudy weather. If the need for water is constant during the year or greatest in the winter or if the climate is substantially cloudy, then it may be more economical to design the system with more solar Watts and no tracker.

8.10.1 Passive Trackers

Passive trackers have been in regular production since 1983. The tracking process uses no moving parts and no electrical parts, but rather only a fluid/vapor flow that tips a balance. An automotive type shock absorber may need replacement every 5–10 years. Passive trackers rarely fail, even after many years. In a case of failure, the tracker can be made to hold at mid-day position and the pump will still function, or it can be tracked by hand. New Mexico's Zomeworks Corporation invented the first widely used passive trackers.

8.10.2 ACTIVE TRACKERS VERSUS PASSIVE TRACKERS

An active tracker uses one or more electric motors powered by solar electricity. This is a more precise method of tracking the sun. High accuracy is necessary for a solar device that uses optical concentration and must be aimed accurately. However, with conventional flat-plate PV modules, a tracking error of as much as 10% will have no significant effect on the power. Therefore, either type of tracker may be considered. Most active trackers have a nighttime or early morning return mechanism that will deliver power earlier in the morning than a passive tracker, which may take a half-hour to wake up. Active trackers are much more complex and generally require more maintenance than passive techniques and may need to be replaced every 4 or 5 years as motors and gears wear out.

8.11 OPERATION AND MAINTENANCE OF THE SYSTEMS

Well designed and installed PV water-pumping systems are relatively simple to operate and maintain. Typically, the system has to start and stop depending on the demand and availability of water and sunshine. With the use of switches (float or electrode), the majority of the systems can be automated at a relatively low additional cost. Manual shutoff is necessary for repair or modification of the water distribution system and the electrical system, as well as when the pump is extracted from the well for inspection, maintenance, and repair.

Personnel responsible for operation and maintenance of the PV water-pumping system should be trained by the installer. The system installer should provide an operation and maintenance manual, which establishes the operational principles of the system, a routine maintenance program, and service requirements. The manual should also include information related to safety and common problems that might surface.

The most effective means of maximizing the benefits of PV water-pumping systems is through preventive maintenance. A preventive maintenance program should be designed to maximize the useful life of the system. Clearly, each type of system has different maintenance requirements; some pumps may operate 10–20 years without any maintenance actions, while others require maintenance in the first year. Specific operational and water conditions will determine frequency.

In general, maintenance of a PV water-pumping system requires the following:

- *Routine maintenance and minor repairs.* Included is monitoring of system performance, water level, and water quality. On-site inspections can detect small problems before they become big ones. It is necessary to look for unusual noises, vibrations, corrosion, loose electrical connections, water leaks, algae, etc. The system operator (typically the owner) should be able to perform routine maintenance and minor repairs. Routine maintenance will help detect and correct the majority of small problems that crop up from time to time before they become major problems (Figures 8.43 and 8.44).
- *Preventive and corrective repairs.* This may require the replacement or repair of components such as diaphragms and impellers as well as defective parts. This type of maintenance may require special tools and knowledge beyond that possessed by the system owner. In the majority of cases, it is necessary for trained personnel to perform the repairs. Pump failures are typically the most common problem found with PV water-pumping systems; PV modules rarely fail (Richards, 1999).

FIGURE 8.43 PV pump damaged cable splice where conductors have worn through from hauling the pump up and down on the pump cable rather than the security rope (Roatan, Honduras).

FIGURE 8.44 FIRCO engineers in Mexico learning how to inspect PV water-pumping systems with NMSU and Sandia Labs.

8.12 THE PV ARRAY

One of the most important points with regard to the PV array is the prevention of shade. Nearby weeds and trees can grow up over time and cause shade over the pump, so they must be controlled. It is not necessary to clean PV modules; heavy buildup of dust will reduce efficiency only 2–4% and will wash off with the next good rainstorm. If the mounting structure permits, the array inclination can be adjusted twice a year to ensure better productivity between summer and winter pumping seasons. Field maintenance of controllers consists of assuring a good seal to avoid the infiltration of dust, water, and insects.

8.12.1 PUMPS AND MOTORS

From an operational point of view, it is very important to avoid dry pumping, which will cause a motor to overheat and fail. Water in the pump is necessary for lubrication and heat dissipation. In the case of surface-mounted centrifugal pumps, if priming is frequently required, inspection should be made to ensure that there are no leaks in the suction pipe or the check valve. The operator should never allow pumping against an obstructed discharge, which could cause the motor to overheat.

Both surface-mounted and submersible centrifugal pumps require little maintenance. The majority of problems that arise are due to excessive sand and corrosive water with high mineral content. These agents can degrade impellers and pump seals. In some cases, the pump may not fail completely, but its productivity may diminish significantly as impellers fill with mud. All that may be required is a good cleaning of the impellers to bring a pump back to 100% capacity. Some pumps can be reconstructed with new impellers and water seals. Algae and other organic material can obstruct the entrance to the pump, which can be reduced with the use of intake screens. Submersible pumps are made of corrosion-resistant stainless steel.

Positive displacement pumps use more components that are subject to wear. Under normal operating conditions, diaphragms should be replaced every 2 or 3 years (more frequently for sandy water). The seals on piston pumps typically last 3–5 years, but can be damaged sooner due to freezing. Diaphragms and seals all fail prematurely in the presence of sand, which wears the components more rapidly. Many positive displacement pumps can be rebuilt several times in the field by replacing diaphragms.

Brushless AC and DC motors do not require field maintenance and can last 10–25 years under ideal operating conditions. The brushes on brush-type motors must be replaced periodically. This is a simple task in most designs. The brushes should be replaced with components supplied by the manufacturer to guarantee good equipment performance. Small motors with brushes can last 4–8 years, depending on use.

8.12.2 WATER SUPPLY SYSTEMS

Finally, it does no good to install a PV water-pumping system to provide water if the rest of the water supply system is not well designed and maintained. Poorly made wells can collapse and destroy hardware. Community water supply systems should be designed with health in mind and there must be drainage to avoid creating a swamp (breeding insects) through which people have to walk to obtain their water.

8.13 PV WATER-PUMPING RESULTS

PV systems have proven to be an excellent option in meeting water-pumping needs when electrical grid service does not exist. Between 1994 and 2005, over 1,700 PV water-pumping systems were installed throughout Mexico, initially as part of the USAID/DOE MREP–Fideicomiso de Riesgo Compartido (FIRCO) program and later with the GEF/World Bank renewable energy for agriculture program. PV water pumping was largely unknown in Mexico prior to 1994, and MREP paved the way for widespread adoption there; the country now leads Latin America in this application.

FIRCO, NMSU, and Sandia conducted a review in 2004 on 46 of the initially installed PV-pumping systems. Typical system configurations included a PV array (~500 W_p on average), pump, controller, inverter, and overcurrent protection. Over three-fifths of the surveyed systems were

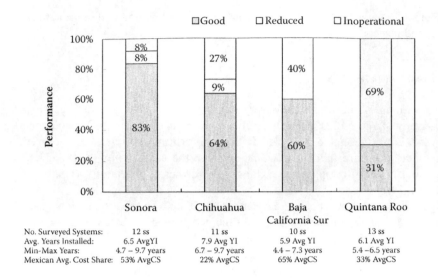

FIGURE 8.45   Performance of Mexican PV water-pumping systems.

operating appropriately after as much as 10 years. The surveys were conducted in Baja California Sur, Chihuahua, Quintana Roo, and Sonora. A total of 85% of users thought that PV systems had excellent to good reliability (Figure 8.45; Cota et al. 2004).

Fully 94% of users classified water production as excellent or good, with only 2% unsatisfied. The survey found that over four-fifths of the rural Mexican users were satisfied with the reliability and performance of their PV water-pumping systems. When system failures occurred, they were typically specific to pump technology and installer.

When problems have occurred, they have been mostly due to failure of pump controllers and inverters, well collapses, or drying out due to drought. There were no PV module failures. Investment payback for the PV water-pumping systems has averaged about 5 or 6 years, with some systems reporting paybacks in half that time (Cota et al. 2004).

REFERENCES

Arriaza, H. 2005. Diagnóstico del sector energético en el área rural de Guatemala. Organización Latinoamericana de Energia, Canadian International Development Agency and the University of Calgary.

CEPAL. 2007a. Estrategia Energetica Sustentable Centroamericana 2020, CEPAL, Mexico.

_____. 2007b. La Energia y las Metas del Milenio en Guatemala, Honduras y Nicaragua, CEPAL, Mexico.

Cota, A. D., R. E. Foster, et al. 2004. Ten-year reliability assessment of PV water pumping systems in Mexico. SOLAR 2004, ASES, paper 322A, Portland, OR, July 9–144.

Bihn, D. 2005. Japan takes the lead. *Solar Today* 19 (1): 20–23.

Foster, R. E., L. Estrada, M. Gomez, and A. Cota. 2004. Evaluación de la Confiabilidad de los Sistemas FV SOLISTO en Chihuahua AR27-02, 12th International Symposium on Solar Power and Chemical Energy Systems, SolarPACES, 28th Semana de Energía Solar—ANES, Oaxaca, Mexico, October 4–8.

Foster, R., L. Estrada, S. Stoll, M. Ross, and C. Hanley. 2001. Performance and reliability of a PV hybrid ice-making system. 2001 ISES Solar World Congress, Adelaide, Australia, November 25–30.

Hammill, Mathew. 2007. Pro-poor growth in Central America. *Serie Estudios y perspectivas*, N° 88, México, CEPAL, October 2007.

Ikki, O. 2004. PV activities in Japan. RTS Corporation, 10 (11).

Ikki, O., T. Ohigashi, I. Kaizuka, and H. Matsukawa. 2005. Current status and future prospects of PV deployment in Japan: Beyond 1 GW of PV installed capacity. EUPVSEC-20, Barcelona, Spain, June 6–10.

JAERO (Japan Atomic Energy Relations Organization), Tokyo. 2004.

Japanese Standards Association. 2004. Technical standard of electric facilities. Tokyo, Japan.

JET. 1998. Guidelines of the technical requirements for grid-interconnection. Tokyo, Japan, March 10.

_____. 2002. Test procedure for grid-connected protective equipment, etc. for photovoltaic power generation systems. Tokyo, Japan, October.

Jones, J. 2005. Japan's PV market: Growth without subsidy. *Renewable Energy World* March–April: 36–42.

Kadenko. 2004. Technical standards for electrical standards. Japanese Standards Association, Tokyo.

Ley, D., H. Martinez, E. Lara, R. Foster, L. Estrada. 2006. Nicaraguan renewable energy for rural zones program initiative, paper A185, Solar 2006, American Solar Energy Society, Denver, CO, July 2006.

MHLW (Ministry of Health, Labor and Welfare). 2002. National livelihood survey. Tokyo, Japan.

Opto-electronic Industry and Technology Development Association. 2004. Tokyo.

Palma, A., and R. Foster. 2001. Guatemalan PV project development for rural uses. *Proceedings International Solar Energy Society Solar World Congress,* Adelaide, Australia.

Richards, B., C. Hanley, R. Foster, et al. 1999. PV in Mexico: A model for increasing the use of renewable energy systems. In *Advances in solar energy: An annual review of research and development,* vol. 13. Boulder, CO: ASES. ISBN: 0-89553-256-5.

Romero-Paredes, A., R. E. Foster, C. Hanley, and M. Ross. 2003. Renewable energy for protected areas of the Yucatán Peninsula. SOLAR 2003, ASES, Austin, TX, June 26.

Sharp, T. 2005. Policy switchback. *Renewable Energy World* March–April: 92–99.

TEPCO (Tokyo Electric Power Co.). 2004. *FY2004 electric utility handbook.* Tokyo, Japan.

UNDP (United Nations Development Program)/ESMAP (Energy Sector Management Assistance Program)/ World Bank. 2005. Energy services for the millennium development goals: Achieving the millennium development goals.

9 Economics

Contributing Author Vaughn Nelson

9.1 SOLAR ENERGY IS FREE, BUT WHAT DOES IT COST?

"Solar energy is free, but it's not cheap" best sums up the major hurdle for the solar industry. There are no technical obstacles per se to developing solar energy systems, even at the utility megaWatt level (e.g., 14 MW utility scale PV system at Nellis AFB or a 64-MW CSP system in Nevada); however, at such large scales a high initial capital investment is required. Over the past three decades, a significant reduction of the cost of solar products has occurred, without including environmental benefits; yet, solar power is still considered a relatively expensive technology. For small- and medium-scale uses, in some applications, such as passive solar design for homes, the initial cost of a home designed to use solar power is essentially no more than that of a regular home, and operating costs are much less. The only difference is that the solar-energy home works with the Sun throughout the year and needs smaller mechanical systems for cooling and heating, while poorly designed homes fight the Sun and are iceboxes in the winter and ovens in the summer.

Industrial society and modern agriculture were founded on fossil fuels (coal, oil, and gas). The world will make a gradual shift throughout the twenty-first century from burning fuels to technologies that harness clean energy sources such as sun and wind. As energy demand increases as developing countries modernize and fossil fuel supply constricts, increased fuel prices will force alternatives to be introduced. The cost of technologically driven approaches for clean energy will continue to fall and become more competitive. Eventually, clean energy technologies will be the inexpensive solution. As the full effect and impact of environmental externalities such as global warming become apparent, society will demand cleaner energy technologies and policies that favor development of a clean-energy industrial base. By the end of the twenty-first century, clean-energy sources will dominate the landscape. This will not be an easy or cheap transition for society, but it is necessary and inevitable.

Already, solar energy is cost effective for many urban and rural applications. Solar hot-water systems are very competitive, with typical paybacks from 5–7 years as compared to electric hot-water heaters (depending on the local solar resource). PV systems are already cost competitive for sites that are remote from the electric grid, although they are also popular for on-grid applications as environmental "elitists" try to demonstrate that they are "green." However, one should beware of "green-washing" as people and companies install grid-tied PV systems without making efforts to install energy-efficient equipment first. Far more can be achieved through energy conservation than solar energy usage alone for reducing carbon emissions.

The decision to use a solar energy system over conventional technologies depends on the economic, energy security, and environmental benefits expected. Solar energy systems have a relatively high initial cost; however, they do not require fuel and often require little maintenance. Due to these characteristics, the long-term life cycle costs of a solar energy system should be understood to determine whether such a system is economically viable.

Historically, traditional business entities have always couched their concerns in terms of economics. They often claim that a clean environment is uneconomical or that renewable energy is too expensive. They want to continue their operations as in the past because, sometimes, they fear that if they have to install new equipment, they cannot compete in the global market and will have to reduce employment, jobs will go overseas, rates must increase, etc.

The different types of economics to consider are pecuniary, social, and physical. Pecuniary is what everybody thinks of as economics: *dollars*. Social economics are those borne by everybody and many businesses want the general public to pay for their environmental costs. If environmental problems affect human health today or in the future, who pays? Physical economics is the energy cost and the efficiency of the process. There are fundamental limitations in nature due to physical laws. In the end, the environment and future generations always suffer the corollary of paying now or probably paying more in the future.

An economical analysis should be looking at life cycle costs, rather than at just the ordinary way of doing business and low initial costs. Life cycle costs refer to all costs over the lifetime of the system. Also, incentives and penalties for the energy entities should be accounted for. What each entity wants is to earn subsidies for itself and penalties for its competitors. Penalties come in the form of taxes and fines; incentives may come in the form of tax breaks, unaccounted social and environmental costs, and also what the government (society) could pay for research and development.

9.2   ECONOMIC FEASIBILITY

The most critical factors in determining the value of energy generated by renewable energy systems are the (1) initial cost of the hardware and installation, and (2) amount of energy produced annually. In determining economic feasibility, renewable energy must compete with the unit worth of energy available from competing technologies. If the system produces electrical energy for the grid, the price for which the electrical energy can be sold is also critical. For renewable energy to have widespread use, the return from the energy generated must exceed all costs in a reasonable time. Systems and applications of renewable energy vary from the Watts for a light and radio to mega-Watts for large-scale solar farms and solar electric systems producing electric energy for the grid. Economics is intertwined with incentives and penalties, so actual life cycle costs are hard to determine, especially when externalities of environmental impact and government support for research and development are not included.

For faster investment payback on residential or small systems connected to the grid, most of the energy should be used on site. That energy is worth the retail rate while selling to the utility is generally valued less because most utilities do not voluntarily want to purchase energy at the retail level from their customers. However, net energy billing (also called net metering) allows for larger size systems because the system can be sized for producing all the energy needed on site. Net metering typically needs to be mandated by government to be adopted by often uncooperative utilities.

First, though, passive solar and energy efficiency can be implemented before even considering active solar energy systems. A solar home has to be an energy-efficient home first. Conservation and energy efficiency measures are the cheapest to install and generally have paybacks within 2–4 years. Every home is a solar home, either working with the sun or fighting against it. Designing and orienting homes and buildings with the sun in mind is the first and foremost solar application.

9.2.1   PV Costs

For many applications, especially remote-site and small-power applications, PV power is the most cost-effective option available. Generating clean electric power on site without using fossil fuel is an added benefit. Capital costs are high for PV, but fuel costs are nonexistent. PV module costs have dropped by an order of magnitude over the past two decades. New PV modules generally cost about $3 per Watt, depending on quantities purchased. Off-grid PV systems with battery storage typically run from about $12 to $15 per peak Watt installed, depending on system size and location. Grid-tie PV systems are averaging $6–$8 per Watt installed, also depending on system size and location.

Larger PV water pumping systems with all balance-of-system components, including the pump, can be installed for under $10 per Watt.

A well designed PV system will operate unattended and requires minimal maintenance, which can result in significant labor and travel savings. PV modules on the market today are guaranteed for as long as 25 years and quality crystalline PV modules should last over 50 years. It is important when designing PV systems to be realistic and flexible and not to overdesign the system or overestimate energy requirements. PV conversion efficiencies and manufacturing processes will continue to improve, causing prices gradually to decrease. It takes many years to bring PV cells from the laboratory into commercial production, so overnight breakthroughs in the marketplace should not be expected.

9.3 ECONOMIC FACTORS

The following factors should be considered when purchasing a renewable energy system:

1. *load* (power) and energy, calculated by month or day for small systems;
2. *cost of energy* from competing energy sources to meet need;
3. *initial installed cost:*
   a. purchase price;
   b. shipping costs;
   c. installation costs (foundation, utility inter-tie, labor, etc.); and
   d. cost of land (if needed);
4. *production of energy:*
   a. type and size of system:
      i. system warranty; and
      ii. company (reputation, past history, years in business, future prospects);
   b. solar resource:
      i. variations within a year and from year to year;
   c. reliability, availability;
5. *selling price of energy* produced and/or unit worth of energy and anticipated energy cost changes (escalation) of competing sources;
6. *operation and maintenance costs:*
   a. general operation, ease of service;
   b. emergency services and repairs;
   c. insurance; and
   d. infrastructure (are service personnel available locally);
7. *time value of money* (interest rate, fixed or variable);
8. *inflation* (estimated for future years and how conventional energy source costs will increase)
9. *legal fees* (negotiation of contracts, titles, easements, permits);
10. *depreciation* if system is a business expense; and
11. any national or state *incentives.*

9.4 ECONOMIC ANALYSIS

Economic analysis is both simple and complicated. Simple calculations should be made first. Commonly calculated quantities are: (1) simple payback (2) cost of energy (COE), and (3) cash flow. More complicated analysis factoring in time value of money, discount rates, etc., can be conducted later.

A renewable energy system is economically feasible only if its overall earnings exceed its overall costs within a time period up to the lifetime of the system. The time at which earnings equals cost is called the payback time. The relatively large initial cost means that this period is often a number of years, and in some cases earnings will never exceed the costs. Of course, a short payback is preferred and a payback of 5–7 years is often acceptable. Longer paybacks should be viewed with extreme caution.

How does one calculate the overall earnings or value of energy? If no source of energy for lights and a radio is available, a cost of $0.50–$1/kWh may be acceptable for the benefits received. If a solar hot-water system will be bought, it is necessary to compare the costs of that system against a conventional gas or electric hot-water system. Many people are willing to pay more for renewable energy because they know it produces less pollution. Finally, a few people want to be completely independent from the utility grid, with little regard to cost.

9.4.1 Simple Payback

A simple payback calculation can provide a preliminary judgment of economic feasibility. The difference between borrowing money for a system and lost interest if there is enough money to pay for the system is usually around 5–7%. The easiest calculation is the cost of the system divided by cost displaced per year, assuming that operation and maintenance are minimal and will be done by the owner:

$$SP = \frac{IC}{AKWH \times \$/_{kWh}} \tag{9.1}$$

where

    SP = the simple payback in years
    IC = initial cost of installation ($)
    $AKWH$ = energy produced annually (kWh/year)
    $\$/kWh$ = price of energy displaced

Example 9.1

You purchased a solar hot-water heater to replace an electric hot-water heater (70 gal/day for a family of four). Installed cost = $3,000, and displacing electricity is 6,000 kWh/year at $0.10/kWh. You are assuming that the price of electricity will stay the same over the lifetime for this simple analysis:

$$SP = \frac{\$3000}{6000 \ kWh/_{year} \times \$0.1/_{kWh}} = 5 \ \text{years}$$

If your hot-water heater needs replacement anyway, you have an initial cost, $400, and then you pay for the electricity, $50/month for approximately 500 kWh/month. Reducing the IC cost to $2,600 means that now the simple payback on the solar hot-water system would be less:

$$SP = \frac{\$2600}{\$600/year} = 4.3 \ \text{years}$$

Example 9.2

The installed cost for a solar hot-water system is $3,000. Go to a store that sells electric hot-water heaters. Information for electric hot-water heater: costs are $500/year:

$$SP = \frac{\$3000}{\$500/year} = 6 \ \text{years}$$

The next calculation would include the value of money, borrowed or lost interest, and annual operation and maintenance costs:

$$SP = \frac{IC}{AKWH \times \$/_{kWh} - IC \times FCR - AOM}$$

(9.2)

where

SP = the simple payback in years
IC = initial cost of installation ($)
$AKWH$ = energy produced annually (kWh/year)
$\$/kWh$ = price of energy displaced or price obtained for energy generated
FCR = fixed charge rate per year
AOM = annual operation and maintenance cost ($/year)

Example 9.3

The scenario is the same as that in Example 9.1, except that you are losing interest at 5% on the installed cost:

$$SP = \frac{3000}{500 - 3000 \times 0.05} = 8.5 \text{ years}$$

Notice that if you had to borrow the money at 12% interest, the payback would be longer. However, if electric costs increase in the future, then payback would be shorter. The FCR could be the interest paid or the value of interest received if you displaced money from savings. An average value for a number of years (five) will have to be assumed for dollars per kiloWatt-hour.

Example 9.4

IC = $5,000
FCR = 0.10 = 10%
AOM = 1% of IC = 0.01 * 5,000 = $50/year
AKWH = 12,000 kWh/year
$/kWh = $0.10/kWh

Value of energy displaced per year = 12,000 * 0.10 = $1,200

$$SP = \frac{5000}{1200 - 500 - 50} = 7.7 \text{ years}$$

Equation 9.2 involves several assumptions: the same number of kiloWatt-hours are produced each year, the value of the electricity is constant, and no inflation occurs. More sophisticated analysis would include details such as escalating fuel costs of conventional electricity and depreciation. These factors might reduce the payback to around 5 years.

9.4.2 Cost of Energy

The cost of energy (COE) is primarily driven by the installed cost and the annual energy production. For PV systems, that cost is determined primarily by the cost of the modules. For on-grid systems, PV costs are from about $6–$8/Wp. After losses, each Watt produces 2–6 Wh/day, depending on solar resource; this translates to about $0.22–$0.35/kWh.

The cost of remote stand-alone PV systems with batteries will be from 1.5–2 times more than grid-connected systems. High-quality industrial batteries last 7–9 years; others last 3–5 years.

Automobile batteries, which are not designed for deep cycling, last only 1–1.5 years. Battery life depends greatly on how much batteries are cycled.

The COE (value of the energy produced by the renewable energy system) provides a levelized value over the life of the system (assumed to be 20–30 years):

$$COE = \frac{IC \times FCR + AOM}{AKWH} \qquad (9.3)$$

The COE is one measure of economic feasibility, and when it is compared to the price of energy from other sources (primarily the utility company) or to the price for which that energy can be sold, it gives an indication of feasibility. If the COE is within 30% above these prices, further analysis is justified.

The annual energy production for a PV system can be estimated as follows:

$$AKWH = EF \times AKWH = EF \times Wp \times \overline{PSH} \times 365 \qquad (9.4)$$

where

EF = system efficiency factor—typically about 50% off grid and 75% grid tie
Wp = array rating (peak kiloWatts)
$\overline{PSH}$ = average daily solar insolation (sun-hours) (kWh/m²/day)

See the NREL Web site for average sun-hours for a particular location.

Example 9.5

Find the COE for a 2-kWp grid-tie PV system for a home in El Paso, Texas, with an average of 6 kWh/m² per day, displacing electricity at an average of $0.12/kWh over 25 years.

Solution:

$AKWH$ = 75% * 2 kWp 6 sun-hours/day * 365 days/year = 3,285 kWh/year
IC = $12,000
FCR = 0.08
AOM= $100/yr
COE = (12,000*0.08 + 100)/3,285 = 1,060/3,285 = $0.32/kWh

9.5 LIFE CYCLE COST

In order to gain a true perspective as to the economic value of solar energy systems, it is necessary to compare solar technologies to conventional energy technologies on a life cycle cost (LCC) basis. This method permits the calculation of total system cost during a determined period of time, considering not only initial investment but also costs incurred during the useful life of a system. The LCC is the "present value" life cycle cost of the initial investment cost, as well as long-term costs directly related to repair, operation, maintenance, transportation to the site, and fuel used to run the system. Present value is understood as the calculation of expenses that will be realized in the future but applied in the present.

An LCC analysis gives the total cost of the system, including all expenses incurred over the life of the system. There are two reasons to do an LCC analysis: (1) to compare different power technology options, and (2) to determine the most cost-effective system designs. For some renewable energy applications, there are not any options to small renewable energy systems because they produce power where there is no power. For these applications, the initial cost of the system, the infrastructure to operate and maintain the system, and the price people pay for the energy are the main concerns. However, even if small renewable systems are the only option, a life cycle cost analysis can be helpful for comparing costs of different designs and/or determining whether a hybrid system would be a cost-effective option.

An LCC analysis allows the designer to study the effect of using different components with different reliabilities and lifetimes. For instance, a less expensive battery might be expected to last 4 years, while a more expensive battery might last 7 years. Which battery is the best to buy? This type of question can be answered with an LCC analysis:

$$LCC = C + M_{pw} + E_{pw} + R_{pw} - S_{pw}$$

(9.5)

where

LCC = life cycle cost.

C = initial cost of installation—the present value of the capital that will be used to pay for the equipment, system design, engineering, and installation. This is the initial cost incurred by the user.

M_{pw} = sum of all yearly O&M (operation and maintenance) costs—the present value of expenses due to operation and maintenance programs. The cost of O&M includes the salary of the operator, site access, guarantees, and maintenance.

E_{pw} = energy cost, sum of all yearly fuel costs—an expense that is the cost of fuel consumed by the conventional pumping equipment (e.g., diesel or gasoline fuel). This should also count the cost of transporting fuel to remote sites.

R_{pw} = sum of all yearly replacement costs—the present value of the cost of replacement parts anticipated over the life of the system.

S_{pw} = salvage value—net worth at end of final year, typically 10–20% for mechanical equipment

Future costs must be discounted because of the time value of money, so the present worth is calculated for costs for each year. Life span for PV is assumed to be 20–25 years. Present worth factors are given in a table in Appendix C or can be calculated.

Life cycle costing is the best way of making purchasing decisions. On this basis, many renewable energy systems are economical. The financial evaluation can be done on a yearly basis to obtain cash flow, breakeven point, and payback time. A cash flow analysis will be different in each situation. Cash flow for a business will be different from that for a residential application because of depreciation and tax implications.

The LCC of various alternatives can be compared directly. The payback time is easily seen, if the data are graphed. The option with the lowest LCC is the most economic over the long term. Note that social, environmental, and reliability factors are not included here but could be added if they are deemed important. These factors are difficult to quantify in conventional economic terms, but they should be considered when important to the user (e.g., risk of fuel spill in a delicate natural protected area).

Example 9.6

A residential PV application (done when there was a tax credit) resulted in the following:

installed cost = $20,000
down payment = $6,600
loan = 7 years at 19%
maintenance = 2.5% * IC = $500/year
energy production = 50,000 kWh/year, 75% consumed directly, displacing $0.08/kWh electricity and 25% sold to the utility at $0.04/kWh with utility escalation at 5%/year

In this analysis, the breakeven point is at the end of year 5 and the payback time is in year 8. There are a number of assumptions about the future in such an analysis. A more detailed analysis would include inflation and increases on costs for operation and maintenance as the equipment becomes older.

A cash flow analysis for a business with a \$0.015/kWh tax credit on electric production and depreciation of the installed costs would give a different answer. Also, all operating expenses are a business expense. The economic utilization factor is calculated from the ratio of the costs of electricity used at the site and electricity sold to the utility.

The government of Canada has developed a useful solar analysis tool called RETScreen, which includes economic comparisons. The tools consists of a standardized and integrated renewable energy project analysis software that can be used to evaluate the energy production, life cycle costs, and greenhouse gas emission reductions for following renewable energy technologies: wind, small hydro, PV, passive solar heating, solar air heating, solar water heating, biomass heating, and ground-source heat pumps (see http://retscreen.gc.ca/ang/menu.html).

9.6 PRESENT VALUE AND LEVELIZED COSTS

Money value increases or decreases with time, depending on interest rates for borrowing or saving and inflation. Many people assume energy costs in the future will increase faster than inflation. The same mechanism of determining future value of a given amount of money can be used to move money backward in time. If each cost and benefit over the lifetime of the system were brought back to the present and then summed, the present worth could be determined.

The discount rate determines how the money increases or decreases with time. Therefore, the proper discount rate for any life cycle cost calculation must be chosen with care. Sometimes the cost of capital (interest paid to the bank or, alternately, lost opportunity cost) is appropriate. Possibly the rate of return on a given investment perceived as desirable by an individual may be used as the discount rate. Adoption of unrealistically high discount rates can lead to unrealistic life cycle costs.

If the total dollars are spread uniformly over the lifetime of the system, this operation is called levelizing.

Present value (PV) is the adjusted cost, at present, of future expenses using the real discount rate (defined later). The future payment can represent a single payment of an annual payment. The present value of a single payment made in the future is

$$PV = FV * (1 + i_r)^{-n} \qquad (9.6)$$

where

PV is the present value
FV is the future value amount to be paid in the future
i_r is the real discount rate
n is the number of years between now and the year of the payment

For a given discount rate and number of years, the present value factor for a future payment, given by $(1 + i_r)^{-n} = FVP$ can be calculated or simply read from an FVP factor table such as Table C.1 in Appendix C.

The present value of a fixed annual payment is

$$PV = AV \times [(1 - 1/(1 + I_r)^n)/I_r] \qquad (9.7)$$

where

> PV is the present value
> AV is the value amount paid annually
> i_r is the real discount rate
> n is the time period, in years, in which the annual payment is incurred

For a given interest rate and time period, the present value factor for annual payments, given by Equation 9.3, can be calculated or simply read from an PVFA factor table, such as Table C.2 in Appendix C.

To find the PV and the PVFA in the tables in Appendix C, simply locate the column that corresponds to the real discount rate and the row with the number of years. The PV and PVFA values are found in the cell where the column and the row meet.

Example 9.7

A PV water-pumping system uses a submersible centrifugal pump. According to the manufacturer, the pump has a useful life of 10 years. It is anticipated that the pump will be replaced every 10 years. The current cost of the pump is $400. The real discount rate, for purposes of this calculation, will be 7%. According to Table C.1, the PVF value for a discount rate of 7% for a period of 10 years is 0.5083. We multiply $400 by this factor to obtain the present value of the investment that will be made in 10 years:

$$PV = \$400.00 \times 0.5083 = \$203.00.$$

Real discount rate (i_r):

$$i_r = \text{interest rate} - \text{inflation rate} \tag{9.8}$$

The interest rate is the rate at which capital increases if it is invested. The inflation rate is the rate of price increases. Sometimes, especially of late with rising oil prices, the annual inflation rate of fuel is significantly different from the general inflation rate. Given that the annual fuel expense represents a sizable portion of the LCC of internal-combustion systems, a real discount rate i_r for fuel should be used in the present value calculation:

$$i_r = \text{interest rate} - \text{inflation rate for fuel} \tag{9.9}$$

Once the real discount rate and the associated time period are known, the present value of each future expense can be found as well as the LCC of the option under consideration.

Example 9.8

- The interest rate is 20% annually, the inflation rate is 10% annually, and the inflation rate for fuel is 13% annually.
- The real discount rate (I_r) is 20% – 10% = 10% = 0.10. This is the rate that we should use to determine the present value of expenses incurred in the future.
- The interest rate is 20% annually. The real discount rate (i_r) of fuel is 20% – 13% = 7% = 0.07. For this example, this is the rate that should be used to determine the present value of fuel expenses.

9.6.1 STEPS TO DETERMINE THE LCC

- Determine the period of analysis and the interest rate. To make an LCC comparison for PV equipment, 20–25 years is generally the time period used for analysis because this is

considered to be the useful lifetime of such a system and most PV modules are still under
warranty.

- Determine the initial cost of the installed system. The previous section shows how to estimate the initial cost of a solar energy system. The initial cost of an internal-combustion
system varies depending on the type of system.

9.7 ANNUALIZED COST OF ENERGY

One further step has been utilized in assessing renewable energy systems versus other sources of
energy such as electricity. This is the calculation of the annualized cost of energy from each alternative. The annualized cost calculated is divided by the net annual energy production (AEP) of that
alternative source:

$$COE = \frac{\text{Annualized Cost}}{AEP}$$

(9.10)

It is important that annualized costs of energy calculated for renewable energy systems be compared to annualized costs of energy from the other sources. Direct comparison of annualized cost of
energy to current cost of energy is not rational. Costs of energy calculated in the preceding manner
provide a better basis for the selection of the sources of energy.

9.8 EXTERNALITIES

Externalities are now playing a role in integrated resource planning (IRP) as future costs for pollution, carbon dioxide, etc., are added to the life cycle costs. Values for externalities range from zero
(past and present value assigned by many utilities) to as high as \$0.10/kWh for steam plants fired
with dirty coal. Again, values are being assigned by legislation and regulation (public utility com-
missions). It is possible to assign a societal value for using clean PV technology and include this as
part of a life cycle cost analysis. In order to understand the societal value offered by clean-energy
technologies such as PV, it is necessary to understand the environmental and political consequences
of the modern energy infrastructure. The extraction, production, distribution, and consumption of
fossil fuels significantly deteriorate the quality of the natural environment, while exacerbating geopolitical competition for scarce fuel resources. These problems affect our air and water quality,
ecosystems, land and material resources, human health, and global stability, as well as the aesthetic,
cultural, and recreational values of affected regions.

Energy production and usage, particularly through fossil fuels, has become a dominant force
related to environmental destruction and climate change. Anthropogenic emissions of carbon dioxide (CO_2), methane, and nitrogen oxides are the principal contributors to global climate change.
Energy usage is the largest contributor to emissions and includes all aspects of power production
and utilization. About three-quarters of all anthropogenic emissions related to global warming can
be directly attributed to the energy sector and to the widespread use of fossil fuels.

Anthropogenic CO_2 emissions currently generated are about 5% of total global CO_2 emissions,
with the rest coming from natural sources (Easterbrook 1995). However, this does not excuse anthropogenic emissions because the natural carbon cycle is in an approximate equilibrium state. Even small
and continuous additions to a system in equilibrium can cause large, long-term consequences. CO_2 is
the greenhouse gas responsible for 64% of human-induced changes in the climate (Dunn 1998).

Externality is a side effect that exists whenever economically productive actions (production
or consumption) of an economic agent directly affect the opportunities of some other agent,
other than through price. Externality supposedly addresses market failure because prices may not
always truly reflect the effect of all activities of an economic agent. External effects defined as

such can be either positive or negative. Externalities represent a shortcoming in classic economic theory because actions that affect environmental well being are counted as being only external. Without taking into account the externalities and environmental costs of different power generation technologies, it is difficult for PV technologies to compete fairly with other conventional energy technologies.

Positive externalities or side effects usually affect economic agents not directly involved in the production or consumption process in a positive manner by expanding their economic activity or by reducing costs. For PV applications, positive externalities exist in the form of no pollution emissions (CO_2, SO_4, NO_x, etc.), no risk of fuel spills and contamination, and no noise pollution, as well as no dependence on imported energy sources. Negative externalities are commonly associated with production or technological externalities. A firm's production processes can produce pollution or other unwanted by-products that affect the welfare of other persons (e.g., a polluted water supply).

A societal approach for determining the most efficient resource allocation for any society is needed to take into account externalities that conventional markets have failed to recognize. Social costs and social benefits need to be accounted for, rather than just the private costs and private benefits of any energy resource.

9.8.1 EXTERNALITY EVALUATION METHODS

Two basic approaches can be used to evaluate the costs and benefits of externalities. These methods can be based on market prices, which try to find a proxy measure of some sort, such as land value, to derive the value placed by society on avoiding pollution damage. The other and more popular method is based on nonmarket valuation methods, which try to estimate what the market clearing price would be if a good or service were traded in the market.

Common nonmarket techniques for evaluation of externalities are the hedonic pricing, travel cost, and contingent valuation methods. These mostly rely on survey techniques, which try to identify information from users of a resource as to how they value a certain level of good and what they are willing to pay for it.

Conventional economic theory holds that the value of all environmental assets can be measured by individual preferences for the conservation of environmental "commodities." The contingent valuation method is used to provide "true" valuation of environmental welfare measures. A consumer's "willingness to pay" represents a compensated variation about how much a consumer would be willing to pay for a welfare gain due to changes in provisions of nonmarket environmental commodities.

9.8.2 SOCIETAL PERSPECTIVES ON SOLAR ENERGY UTILIZATION

The issue of externality stems directly from concerns related to sustainability and how society views such a concept. Sustainability issues cut across a number of areas, including ecological, economic, political, and cultural concerns for all societies. Societal sustainability and benefits can be thought of as meeting the needs of the current generation within their own sociopolitical framework and resource base in a manner that enhances the quality of life and respects cultural tradition. Sustainability issues must address equity, empowerment, and local resources (people and capital). The overall societal benefit of using PV power can be described as

$$SB = CB + UB + PRB + EB \tag{9.11}$$

where

SB = society's benefits
CB = consumer's benefits

UB = utility's benefits
PRB = producer's/retailer's benefits
EB = environment's benefits

For example, one way that this benefit has been previously calculated for PV power has been to determine the average societal benefit of solar energy utilization. Another example is a situation in which externality costs have been quantified for environmental emission calculations. There is already a robust emissions trading system for sulfur oxides (SO_x) and nitrogen oxides (NO_x). In addition, some emission brokerage firms have quantified a cost for carbon dioxide emissions trading, and there is a growing international carbon trading market. Basic environmental emission fair market pricing for various pollutants can be established to value the damage done by emissions to the environment by conventional power generation technologies. This is compared to relative average emissions for electric power plants.

9.9   SOLAR IRRIGATION CASE STUDY

Small solar pumps (<2 hp) are very competitive in relation to small diesel or gasoline engines. Medium solar pumps (>2 hp), in relation to large diesel engines, are competitive for remote sites. The largest off-the-shelf commercial solar water pumping systems in use today are about 10 hp (e.g., Sunpumps). Note that 1 kW produces approximately 0.75 true KW (~1 hp) due to system inefficiencies.

Today's market for small solar pumps (less than 2 hp) is far greater than the market for larger ones. Therefore, solar pump manufacturers concentrate on products up to the 2-hp range. However, given the dramatic increases in gasoline and diesel fuel prices recently, solar water pumping systems have now been developed as large as 10 hp. Relatively few large PV pumping systems over 10 hp are in existence.

9.9.1   ESTIMATING SYSTEM COSTS

The best way to estimate the cost of a solar water pumping system is to obtain a quote from one or more local system providers or contractors. However, the cost can be estimated with the help of data related to recently installed systems. One can take into account the total cost of an installed system as follows:

- cost of materials, including all applicable taxes;
- installation cost, guarantees, and maintenance agreement; and
- company profit margin.

The cost of installation, guarantee, and O&M vary considerably according to the system provider and project-site access. However, it is rare that these costs exceed 30% of the total cost of the system.

9.9.2   TABLE OF APPROXIMATE COSTS

A cost estimate can be obtained knowing the water demand, total dynamic head, and solar resource at the site. Table D.1, Approximate Costs of PV Pumping Systems, is found in the appendix. The table shows approximate costs of materials and installation costs. It does not show the cost of system guarantees as well as applicable taxes. The table is used as follows:

- Select the column that corresponds to the amount of insolation (in peak solar hours) in the critical design month.

- Move down the column and select the water volume required (in cubic meters/day).
- Move across the row to the right and select the system cost that corresponds to the total dynamic head of the pumping project (in meters).

Example 9.9

El Jeromín Ranch in Chihuahua, Mexico, requires a system capable of pumping 12.5 m^3 of water in the summer (6 peak solar hours). The total dynamic head of the system is 40 m. The approximate cost obtained from Table D.1 is US$11,600.

Another factor affecting the cost of the system is the type of equipment used. For example, systems using a DC pump generally have a lower initial cost because they tend to be more efficient and do not require an inverter. Efficient components can reduce the required PV array size and, consequently, the initial system cost. However, AC pumps tend to last longer than DC pumps and may be cheaper over the system's lifetime. Systems that use tracking devices can also prove to be more economical because they can be used with a smaller array to accomplish but still provide the power needed by the load.

9.9.3 Comparison of Pumping Alternatives

Because of their high capital cost, PV systems generally are not competitive in locations that have access to conventional electricity. When access to the electrical grid is not available, solar and internal-combustion systems are clearly the most viable alternatives. If the solar resource is good at the project site (at least 3 peak hours per day) and a hydraulic workload of less than 1,500 m^4/day is required, solar energy systems may be more economical in the long run than internal-combustion systems. Even though internal-combustion systems generally are less costly initially, their long-term cost is considerable if the long-term costs of fuel, maintenance, and repairs are taken into account (Table 9.1). A diesel system may cost $0.40/kWh or more to operate, depending on how remote the site is.

- Estimate the annual cost of operation and maintenance. For internal-combustion systems, the cost of parts (lubricants, filters, tuning, etc.) and labor must be included. The operator's pay must also be taken into account. If the system requires frequent visits for operation and maintenance, the cost of fuel used for transportation to the site can be significant and should be considered under operation and maintenance. The pump is the only solar energy system component that is subjected to mechanical wear. Under normal operating conditions, centrifugal pumps do not need maintenance. The majority of small diaphragm pumps require replacement of the diaphragms and brushes every 3–5 years of continuous operation.
- Estimate the useful life and the replacement cost of the principal components of the system (pump, motor, generator, etc.) during the period of analysis. The useful life depends on the quality of the components and the operating conditions. The useful life of principal components and the maintenance that they require are estimated based on previous experience

TABLE 9.1
Approximate Costs of Internal-Combustion Systems

Type of system	Cost (installed)
Pump-generator set (at least 3 hp)	More than $200/hp
Diesel generator (at least 4 kW), submersible pump	More than $600/kW

TABLE 9.2
Useful Lives of Internal-Combustion and PV System Equipment

Component	Useful life (years)	Maintenance
PV array and structure	25+	None
PV power controller	10+	None
Submergible centrifugal pump/motor AC	7–25	None, or clean impellers
Surface centrifugal pump DC	7–10	None
Submergible diaphragm pump DC	3–5	Replace diaphragms every 5 years
Diesel generator (10 kW)	5–7	Oil, filters, annual tuning
Motors (3–5 hp)	3–4	Oil, filters, annual tuning
Motors (6–10 hp)	4–6	Oil, filters, annual tuning

or information found in owners' manuals or other literature provided by the manufacturers. If this information is not available, the approximate values in Table 9.2 may be used.

- Estimate the annual cost of the fuel used by the system. The annual fuel expense of an internal-combustion system depends on the characteristics of the motor used and the hours of operation needed to pump water. The minimum size of pump-generator sets commonly used is 3 hp. The annual hours of operation can be estimated using the Formula:

Annual hours of operation = 1.33 × hydraulic workload (m³/day × m)/pump efficiency × motor power (hp)   (9.12)

Note that the pump efficiency depends on total dynamic head. Field experience indicates that pump-generator sets in the range of 3–15 hp consume approximately 0.25 l of fuel per hour per unit of horsepower. Consequently, the annual fuel consumption (in liters) can be estimated by Formula 9.2:

Annual fuel consumption (liters) = 0.25 L/h/hp × motor power (hp) × annual hours of operation   (9.13)

For systems using a generator and submersible pump, the same formula is used to estimate the annual hours of operation, keeping in mind that motor power (hp) refers to the power of the electric motor that drives the pump. These systems consume more fuel because the internal-combustion motor of the generator is larger than the electric motor powering the pump. As an approximation, the annual consumption of fuel (in liters) is given by Formula:

Annual fuel consumption (liters) = 1 L/h/hp × motor power (hp) × annual hours of operation   (9.14)

9.10   WATER PUMPING EXAMPLE

The real-world example in Table 9.3 compares the LCC of a PV water pumping system with those of internal-combustion systems. The example is taken from a system installed in San Jeromín (Aldama), Chihuahua, Mexico, in 1997 that has been operational ever since for over one dozen years now (see Figures 9.1 and 9.2). Note that diesel fuel is subsidized and below international norms. It is assumed that the compared systems pump the required volume of water. In addition, the following assumptions are made as shown in Table 9.4.

TABLE 9.3
LCC Analysis Assumptions

Study period	20 years
Average interest rate for the study period	20%/year
Average inflation rate for the study period	10%/year
Average fuel inflation rate for the study period	13%/year
Operation and maintenance (PV system)	2–3% of initial cost/year
Operation and maintenance (internal-combustion system)	$200/year
Labor cost	$1/h
Cost of fuel used at the site	$0.6/L
Minimum size of the pump motor	3 hp
Minimum size of the diesel generator output	4 kW
Annual inspection visits (PV system)	12 visits/year
Annual maintenance visits (internal-combustion system)	52 visits/year
Transportation cost for each visit	$6/visit
Efficiency of the conventional pumping system (pump, generator, friction, etc.)	15%

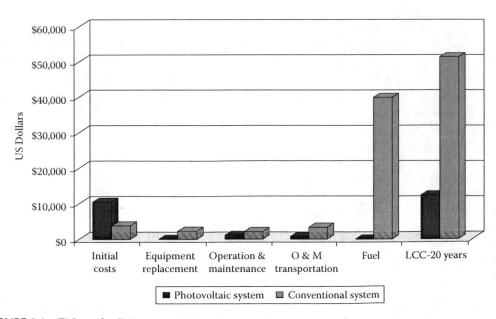

FIGURE 9.1   El Jeromín, Chihuahua, Mexico. Present value cost comparison.

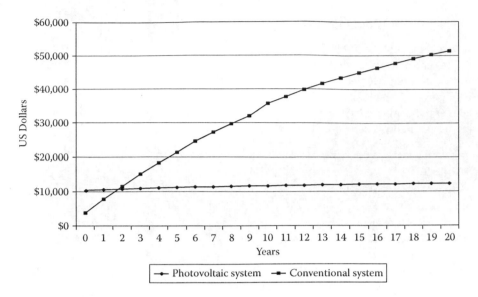

FIGURE 9.2   Investment payback period was less than 3 years even with Mexican subsidized diesel fuel rates, in El Jeromín, Chihuahua, Mexico.

9.11   SUMMARY

The environment and global warming have now become significant issues for power generation. Whether one believes in global warming or not, policies are being enacted around the globe to promote clean power. Green pricing is now available from hundreds of utilities. The old economics no longer apply as the rules of the game are changing.

The general uncertainty regarding future energy costs, dependence on imported oil, oil availability, and climate change have provided the driving forces for development of renewable energy sources. The prediction of escalation of energy costs is a hazardous endeavor because the cost of energy is driven primarily by the cost of oil. In the late 1990s, predictions were for a gradual increase to $30/barrel by 2020; however, oil prices had already reached nearly $150/barrel by 2008, and then dropped down to $40/barrel before rising again. Price increases for oil and natural gas have not been and will not be uniform in terms of time or geography. Society reached the point at which demand exceeds production, so there has been a sharp increase in the price of oil due to demand, rather than increased production costs. World oil production is at its peak now, and costs will only increase in the future. The most important factors are the estimated total reserves and the recoverable amount. As price increases, it becomes economic to recover more from existing reservoirs. It also paves the way for new and alternative energy sources, such as solar power.

Another major driving force for renewable energy is economic development and jobs at the local or state level. That is because renewable energy is local and it does not have to be shipped from another state or country. Solar hot water will become a major market.

PV production will continue to increase and electricity for grid-tie applications will be the major market due to government incentives. PV rooftop systems will gradually be integrated into building design.

Trading in carbon dioxide is growing, much as there is now trading in NO_x and SO_x. The value of renewable energy would increase by about $0.08/kWh if the avoided CO_2 is worth $10 per ton.

Whether by economics, mandates (legislation or regulation), and/or on a voluntary basis, there will be more use of renewable energy. Traditional energy sources have an advantage in that fuel costs are not taxed; for renewable energy, the fuel costs are free. The problem is the high initial costs for renewable energy and that most people would rather "pay as they go" for the fuel. Every effort

TABLE 9.4
San Jeromin LCC Assumptions

Technical Specifications:

Total Dynamic Head: 40 m

Water Pumping Capacity: 15.0 m^3

Hydraulic Cycle: 600 m^4

PV system (848 Wp):

Grundfos SP3A-10 Pump

Internal Combustion System (15 kW):

Annual Hours of Operation: 397

Annual Consumption of Fuel: 7,980 L

Total Annual Cost for Fuel: $3,770

PV system					Internal-combustion system				
	Year	**Amount**	*FVP* **or** *FVPA*	*PV*		**Year**	**Amount**	*FVP* **or** *FVPA*	*PV*
Initial cost	0	$10,491	1.0000	$10,491	Initial cost	0	$3,785	1.000	$3,785
Replacements					Replacements				
					AC pump	6	$575	0.6663	$383
					Generator	10	$2,910	0.5083	$1,479
					AC pump	12	$575	0.4440	$255
					AC pump	18	$575	0.2959	$170
O&M	Each year	$105	10.594	$1,111	O&M	Each year	$200	10.594	$2,119
Transportation for maintenance visit	Each year	$72	10.594	$763	Transportation for maintenance visit	Each year	$312	10.594	$3,305
Fuel for operating pump	Each year	$0	10.594	$0	Fuel for operating pump	Each year	$3,770	10.594	$39,939
ULCC (20 years)				$12,365	ULCC (20 years)				$51,436

should be made to take advantage of all incentives, especially federal and state ones. The cost of land is a real cost, even to those using their own land. This cost is often obscure because it occurs as unidentified lost income.

The costs of routine operation and maintenance for individuals represent the time and parts costs. Until system reliability and durability are better known for long time periods, the costs of repairs will be difficult to estimate. It is important that the owner has a clear understanding of the manufacturer's warranty and that the manufacturer has a good reputation. Estimates should be made on costs of repairing the most probable failures. Insurance costs may be complicated by companies that are uncertain about the risks involved for a comparatively new technology. However, the risks are less than those of operating a car.

Inflation will have its principal impact on expenses incurred over the lifetime of the product. The costs of operation and maintenance—and especially the unanticipated repairs—fall into this category. On the other hand, cheaper dollars would be used to repay borrowed money (for fixed-rate loans).

REFERENCES

Dunn, S. 1998. Carbon emissions resume rise. *Vital Signs*. Worldwatch Institute, Washington, D.C.
Easterbrook, G. 1995. *A moment on the Earth: The coming age of environmental optimism*. Viking Penguin Books, New York.

PROBLEMS

Information needed for problems on solar hot water: 1 kWh = 3,410 Btu, average input water temperature (ground temp below freeze line). The Florida Solar Energy Center has a site for estimating the gallons of hot water for different size households and the British thermal units needed. Of course, if one has teenagers, a lot more hot water is needed for showers and washing dirty clothes! See http://www.fsec.ucf.edu/SOLAR/APPS/SDHW/SizeProc.htm#limitations

9.1   Calculate the simple payback for a solar hot-water heater for your home, replacing an electric hot-water heater.

9.2   Calculate the simple payback for a solar still over purchasing bottled water for your location. Assume an initial capital cost of US$500 for a 1-m² single-basin still, with a water production of 0.8 L/ m² per sun-hour.

9.3   Calculate the cost of energy (use Equation 9.3) for a PV system for your home that would supply 60% of your electric needs. The PV system is connected to the grid. www.southwestPV.com has information on complete systems.

9.4   What are the two most important factors in the cost of energy formula (the factors that influence COE the most)?

9.5   Calculate the cost of energy for a PV stand-alone system with batteries that would supply 100 kWh/month. Use Equation 9.4 and assume batteries would be replaced every 7 years (use a 25-year lifetime).

9.6   Estimate the years to payback. IC = $100,000, r = 8.5%, and AKWH = 100,000 at $0.08/kWh. Assume a fuel escalation rate of 4%. This problem has to be done numerically; assume an L, calculate, and then modify l in terms of your answer and do the calculation again.

9.7   Explain life cycle costs for a renewable energy system. Make a comparison to nuclear power plants. What is the COE for the nuclear power plants installed most recently in the United States? (Do not calculate; find an estimate from any source.)

9.8   What are today's values for fuel inflation, discount rate, and interest rate? What is your estimate of average per year between now and the year 2020?

9.9   What is the price of oil (dollars per barrel) today? Estimate the price for oil for the year 2020. Estimate the price for oil when the costs for the military to keep the oil flowing from the Mideast are added. Place results in a table.

9.10   Estimate a life cycle cost using a spreadsheet for a solar hot-water heater. Assume the need is 50 gal/day and the solar hot-water heater will supply 75% of that on average. You are going to borrow the money for the installed cost and pay it back over 5 years (no salvage value).

10 Institutional Issues

10.1 INTRODUCTION

The key to any successful solar energy project is to lay the institutional framework for long-term success. Solar energy systems are commercially available today to meet a wide range of urban and rural applications economically, from small to large scale. However, without proper institutional and market frameworks to operate and maintain renewable energy systems long term, they eventually fail. For larger utility-scale projects, future operation and maintenance are usually planned out ahead of time. In contrast, for smaller scale individual applications, future follow-up and maintenance are often overlooked. In many ways, this is the most important chapter of this book in that it discusses the lessons learned related to key institutional issues that ultimately define long-term solar energy projects' successes or failures.

The greatest institutional challenges typically are for the smaller, rural-scale, solar energy applications rather than the larger, utility-scale applications. The smaller systems are often the most remote and easily forgotten, and the solar option was normally chosen because it was prohibitively expensive to extend the existing electric grid to start with. Frequently, these kinds of projects are planned by office bureaucrats who do not have much field or solar application experience. Smaller rural solar energy systems are the simplest, most cost-effective, and most appropriate applications for solar energy technologies; however, consideration of institutional issues is critical to long-term success.

The critical link for any renewable energy project is not only the technology used, but also the implementing and follow-up agencies or companies and the infrastructure required to support it. Technical aspects are important to ensure quality and successful implementation of renewable energy projects, but this is not enough to guarantee the future success of a project. Technically acceptable designs and installations often fail due to the lack of focus on institutional issues and follow-up. This is especially true in development programs that introduce new and little understood technologies such as solar energy into rural settings. However, as with any conventional mechanical and electrical system, the implementing agency and user must be prepared to conduct future maintenance on the system to ensure its long-term operation. A viable renewable energy project must take into account not only the maintenance issues, but also policy and social issues such as capacity building, technical assistance, education and training, and local infrastructure development to ensure long-term sustainability.

10.2 SUSTAINABILITY

Sustainable development, hereafter referred to as sustainability, is the achievement of continued economic and social development without long-term detriment to the environment, culture, and natural resources. For instance, in the use of solar energy technologies for water pumping in rural areas, sustainability provides users (consumers) local access to qualified suppliers, high-quality equipment, and maintenance capabilities at a cost and payment schedule that are reasonable. Due to the higher initial capital cost of solar energy systems compared to conventional technologies, access to reasonable financing is often an important factor in the sustainability for rural renewable energy technologies. Long-term sustainability is a natural consequence of local market growth. When demand for a product or service is high enough to allow for profit generation and competition, market forces eventually establish the infrastructure required for the generation of a local market.

The goal of renewable energy development programs should be to provide needed services, such as water supply, refrigeration, or communication, while contributing to local market growth and sustainability. Development programs are often carried out in economically depressed regions where consumer ability to pay is low, and the supply infrastructure is inadequate or nonexistent. Rural program implementation often takes place in the context of social programs that include various forms of subsidy by governments or other organizations. Subsidized programs are not inherently sustainable in themselves; however, they are justifiable and can make significant social contributions and can be used as a catalyst to develop carefully the local markets for renewable energy technologies, as well as to export or promote local products. Most of the global growth for on-grid photovoltaics has been fueled by large government incentive programs (e.g., California, Germany, Spain). The more sustainable and cost-effective off-grid PV markets have been largely ignored by many decision makers because they are less visible and typically benefit remote communities with little political clout.

10.3 INSTITUTIONAL CONSIDERATIONS

A number of institutional issues must be addressed to achieve sustainability for solar energy projects. The following sections discuss some of the key areas to consider for institutional development, especially for large renewable energy development programs (e.g., World Bank, USAID, GTZ).

10.3.1 POLICY ISSUES

The implementation of renewable energy projects is most successful when favorable international, national, state, and local policies are in place. Recognition of the social, environmental, and health benefits of solar energy systems can lead to sound policies regarding importation requirements, taxes, fossil fuel subsidies, and other government barriers that can artificially increase the cost of installed renewable energy systems. Established government programs that are already working in related areas such as farming, cattle ranching, and potable water can justify the direct involvement of government agencies in the implementation of renewable energy programs. A majority of U.S. states now have renewable energy portfolio standards requiring that a percentage of electrical generation be from clean-energy sources; Texas and California are leading the way. Germany has feed-in tariffs to encourage investment in PV generation. Such programs are valuable vehicles in promoting solar technologies and educating potential end-users. Favorable policies encourage entrepreneurs and widespread market growth.

Solid partnerships should be nurtured. Strong partnerships among government, industry, and development agencies should be nurtured for solar energy programs to address the diverse cultural, technical, social, and institutional issues that are faced in working to meet program goals. A solar energy program depends for its success on working with in-country organizations and with industry. In addition, members of the program team, which is composed of individuals from different organizations, must function well together. It is important to choose partners carefully and maintain honest and open communications.

10.3.2 CAPACITY BUILDING

Significant attention is required to assist partners in building the local capacity necessary to develop and independently evaluate solar projects successfully. Capacity building includes technical assistance, formal training workshops, focused field activities, and in-depth reviews of supplier quotes and designs for proposed systems.

Local support and training is crucial for successful solar energy programs. In-depth training is critical in developing the interest and knowledge required for understanding and successfully applying renewable energy technologies. A structure is essential to assist partners in building the capacity necessary to ensure long-term operation and maintenance of solar energy systems. Technical assistance and training are continual processes best served up in an incremental fashion over time. It is important not only to train project developers, but also to upgrade local industry (supply side). System suppliers also need to have the opportunity to return to old installations occasionally and rectify any problems if needed. This helps them to see what works and what does not over the long term. Success depends largely on the technical capacity of local technicians and administrators who continue to operate a solar energy system long after the day of the system inauguration. Greater technical capacity of local suppliers leads to greater consumer and development agency confidence in terms of assuring quality projects.

10.3.3 EDUCATION AND TRAINING

A successful renewable energy program absolutely requires development of local technical capabilities and knowledgeable consumers. Providing training to vendors, project developers, and government personnel is one of the many components that ensure a quality installation. In addition, training plays an important role to ensure that the technology is being used appropriately. Both vendors and end-users must recognize the importance of the locations and the applications in which solar energy makes sense, as well as recognize those that are impracticable and not economically feasible.

End-users (consumers) should receive training on the basic operation and maintenance of renewable energy systems. This training is a key component that ensures longer system lifetimes. To enhance the effectiveness of the solar energy system, end-users should learn and practice conservation and resource management. Education plays a key role in this area. The resources invested in training are justified by the better economics of more reliable and longer lasting solar energy systems.

10.3.4 TECHNICAL ASSISTANCE

Technical assistance can take a variety of forms, from working with local partners and project developers to providing technical assistance to local system suppliers. It is imperative to work with local partners (project developers) to develop practical technical specifications for solar energy systems that take into account local norms and hardware limitations. This allows for a basic understanding of what is required for a quality and safe-system installation that will provide years of useful life. It is also important to work with local suppliers to assure that they understand what is specifically required to meet the technical specifications.

The importance of including industry in all aspects of a renewable energy program cannot be overstressed. On a local level, sustainability and growth of markets can only be assured if a strong supply-chain infrastructure exists and installed systems function reliably over time. Project developers must work closely with local suppliers to help strengthen their ability to deliver high-quality systems at reasonable costs. Suppliers should be encouraged to attend training courses, learn appropriate electric codes, conduct pilot system installations, and develop their own in-house capacity-building programs.

Renewable energy resource maps of project regions are useful for determining where best to target particular technologies. These maps are valuable tools for partner organizations and system suppliers as they work to determine the most feasible regions for renewable energy technologies. These and other forms of technical assistance are part of the capacity-building process, and they assist program partners in making informed decisions about the appropriate use of solar and other renewable energy technologies.

10.3.5 LOCAL INFRASTRUCTURE DEVELOPMENT

The establishment of local infrastructure is critical for sustainability. An adequate infrastructure provides access to systems, components, and qualified technical services. In rural areas, most renewable energy vendors rely on outside suppliers for equipment and system design. However, costs are lowered when local vendors can handle design, installation, maintenance, and repairs. Healthier business relations between local vendors and their suppliers generally lower the overall costs for end-users. In a good business environment, suppliers are more likely to support local vendors with technical assistance and discount pricing.

10.3.6 Involving the Community: Sustainability and Inclusion

The message needs to be clear: *Solar technology requires a more significant social component to be successful compared with conventional technologies*. It is not just a question of technical maintenance or system administration; rather, the operational viability of the system depends on wider community engagement and commitment. The absolute need for a closer community engagement with solar projects requires policies of inclusion: Women, the elderly, children, and men all must be engaged and buy in to solar energy project development and responsibilities. If not, children may carelessly throw rocks that break panels and nobody will take ownership to repair them if it was a complete government giveaway program. Specific and focused efforts are needed to engage each stakeholder group on its own turf and instill a sense of ownership and responsibility. Ownership sentiment can be generated by asking beneficiaries to pay a reasonable portion of system costs within their means or to provide sweat equity to assist with system installation. Sometimes people are not willing to provide any in-kind contributions; it is best in these cases to walk away and go further down the road because one will surely find different attitudes in different villages.

Solar technology needs are well suited to broadening project involvement to include children and the elderly in understanding the issues and, with that knowledge, providing additional vigilance for the system. The situation parallels home lighting projects for PV where it has been long understood that children were good agents for ensuring that lights got turned off—something critical in these PV systems. Water projects of all types are especially appropriate places to push hard for more involvement of women, given that they are the most affected by water availability issues—as users and, traditionally, as procurers.

10.4 STAKEHOLDERS

For renewable energy systems, any project must satisfy the energy demand while being reasonably affordable. A participatory stakeholder approach to project development is the best way to integrate the priorities of the different groups (Table 10.1). Participation means widespread consultation and open discussion. Users need a vested interested in the project for sustainability during the design period. If the project does not meet their needs, there is no point in implementing it.

10.4.1 Panels versus Fuel or Electric Bills

Solar projects present the community with a counterintuitive problem for long-term sustainability: The absence of fuel or electric bills permits the community to become lax in the collection of tariffs (e.g., to maintain a solar water pumping system). Communities where no water is pumped if a fuel or electric bills are not paid are accustomed to stay on top of tariff collection. This points to the need to find creative ways to ensure that communities with PV systems have some "meaningful but manageable" financial responsibility on an ongoing basis.

The "panel as you go" approach to system expansion mentioned earlier is one possibility. Requiring communities to take responsibility for service agreements with vendors, with monthly payments, is another. Experimenting with a payment approximating what the fuel cost would have been (before solar power) is a possibility as well. Other methods include monthly payments by various communities

TABLE 10.1
Key Stakeholders and Their Roles in Solar Projects

Stakeholder	Role
Community, distributed users, village	Energy demand and quality of service
	Ability and willingness to pay for service
	May contribute cash and/or in-kind labor
Businesses, utilities	Design, installation
	Operation and maintenance services
Government, financial institutions	Incentives, smart subsidies
	Investment capital
State or national agencies (agriculture, energy, community development, water)	Cofinancing through grants or subsidized loans
	Incentives and/or mandates
	Approval process
Project developer	Identify project opportunities
	Bring together all stakeholders
	Ensure that priorities and expectations are known
	Orchestrate compromises when stakeholders' priorities and expectations are different
System designer	Design cost-effective technical solution
	Install and commission
	Provide training on system operation
	Provide service and maintenance support
System management	Oversight
	Tariffs and collection
	Operation and maintenance
	Review of procedures
Third-party finance, national, international NGOs	Ensure long-term sustainability (operation and money flow)

into a common capital replacement fund or monthly payments to the donating or implementing agency for a defined portion of the capital equipment (e.g., new pump). Who will hold on to the tariffs generated is often a critical point. The issue is not so much about the receipt of the payment as it is about helping the community develop the capacity to collect and manage regular tariffs.

10.4.2 COMMUNITY REDUCTION OF THEFT RISKS

Small solar energy systems are vulnerable to theft and engineering-oriented antitheft measures may not stop a thief. Use of antitheft nuts or rivets can lead thieves to cut support members under the panels. Alarms could be a possible solution, but they will add to system cost and may be ineffective if the array is isolated. Some communities (e.g., Honduras, Mexico, and Dominican Republic) have spontaneously and independently come up with a social solution to the problem. The community members take turns patrolling the solar array. This is not an ideal solution, especially for developed countries with a high value placed on labor. In a rural village, it may be a reasonable solution if outside theft is a threat. Usually, local residents will not steal from their own communal system. It is the isolated system near a well traversed road on a ranch with nobody around that faces the greatest threat of theft.

10.4.3 PV and the "Virtuous Circle"

Building energy conservation into projects is a positive concept on its own merits, as is pricing of energy service that more nearly reflects overall value. The good solar engineer is first and foremost a good energy conservation engineer. PV has the potential to be a catalyst for multiple intertwined benefits that would not likely occur in isolation—a virtuous circle where PV is the driver for positive elements that in turn make PV more viable and the overall project more valuable in terms of change. This can only happen, however, if there is a good understanding of why these elements are important to renewable energy projects as well as a willingness to depart somewhat from convention in design and implementation.

10.5 PROGRAM IMPLEMENTATION

The implementation of a renewable energy program can be successfully carried out by government, nongovernment organizations (NGOs), or private industry. Each implementing organization will have different goals and objectives; however, combinations of these agencies working collaboratively are often the most successful.

Governmental agencies have the ability to set an agenda for deployment and enforce requirements for procurement and quality control. In addition, they usually have significant human resources and infrastructure at their disposal to cover a wide geographic area. They are also in a position to promote the use of renewable energy as an alternative to conventional energy systems when they are a more practical alternative for isolated rural areas. Program developers need to work renewable energy into existing development programs as part of the solution to meeting program objectives (rather than the focus only on renewable energy). Keep in mind that government personnel often lack the technical expertise and experience needed to develop a renewable energy program on their own and often harbor unrealistic expectations. They should seek outside experienced professionals to help with realistic planning and goals.

Experience has shown that NGOs that focus their efforts on renewable energy can be quite efficient in the implementation of renewable energy programs. In recent years, some NGOs have been successful in obtaining funding to carry out development projects in rural areas. The key for an NGO or government agency to apply renewable energy successfully is to avoid the trap of becoming the system installer, but rather to work with local system installers and provide an oversight role. Unfortunately, sometimes NGOs have received funding for renewable energy programs, but have little real knowledge or commitment. In turn, they have inefficiently applied the resources and installed substandard systems that give the industry a poor image that may take years to overcome locally. This approach has retarded renewable energy development in some regions. The greatest pitfall for an NGO is to implement a system and not provide any long-term project maintenance and support.

Key steps required for successful implementation of renewable energy programs are discussed in the following sections.

10.5.1 Conduct Strategic Planning

Strategic planning with collaborative partners helps to create realistic goals to include renewable resources as part of instituted programs. Early planning must be realistic and within the bounds of available resources; in other words, it is better to do one thing well than to do many things poorly. Planning should include sufficient promotional activities to accelerate acceptance of the technology, including training. The development of a comprehensive program from the project identification stage to acceptance testing and operation is a key theme that local developers must learn to dominate, yet keep program development as simple and straightforward as possible.

In general, many more options for partnering and tapping into opportunities exist than resources can support; therefore, it is a good idea to focus, limit, and succeed in a few locations, rather than expand. Government-funded programs generally impose a 1-year cycle on which to base planning and budgeting. Renewable energy development programs greatly benefit from multiyear funding, mainly because significant results tend to be realized only after several years of diligent effort. Short-term, one-of-a-kind programs are frequently not successful in the long run.

10.5.2 Pilot Project Implementation

Pilot projects can provide an important foundation for growing, sustainable renewable energy markets. Local suppliers have the opportunity to gain a better technical understanding of the integration of renewable energy systems and have learned that, with adequate planning and design, little cost is required to maintain installed systems for the long term. As a result of pilot projects and gradually increasing demand, prices to end-users usually decline in areas where pilot project programs have been implemented well.

10.5.3 Create Sustainable Markets

Investments in cost sharing of pilot projects greatly facilitate renewable technology introduction and acceptance while fostering a sense of local ownership. As project volume increases, system costs are reduced due to increased competition. Renewable energy must be cost accessible to users through cost sharing or financing. End-user financing at an affordable level similar to that for conventional energy expenditures lowers out-of-pocket initial capital expenditures and expands the renewable energy market. Pilot projects should be used as a tool, not as an end; their goal should be beyond the initial installations, eventually helping establish growing and sustainable long-term markets. Their primary value is as a tool for training and building the capacity of implementing organizations, business, and the community (end-users). Pilot projects should never be used strictly as research projects on real people who have real hopes and needs.

10.5.4 Grassroots Development Approach

An integrated and grassroots development approach is needed for solar energy system development. A local and capable champion greatly facilitates local renewable energy development. If a rural solar energy system is going to succeed and have any lasting impacts, the system has to be installed from a development perspective first. System ownership and responsibilities need to be established early on before installation.

10.5.5 Install Appropriate Hardware

Many renewable energy programs and systems have suffered poor reputations related to the installation of substandard components and designs. Some development programs, especially when dealing with poor rural populaces, offer less than quality solutions to meet their needs. Even the poorest rural people deserve quality and safe components and designs to receive only the best service possible from renewable energy technologies. Substandard systems only create an attitude that solar energy systems are limited, do not function well, and are prone to failure. Good installations require quality components and designs that are safe, reliable, and for the long haul. For any solar energy project, the first order of business for good system design is to use energy-efficient equipment. Systems should never be installed that are poorly thought out and executed.

10.5.6 Monitoring

One characteristic of successful renewable energy development programs that differentiates them from less than stellar programs is a genuine commitment to project follow-up and monitoring. Monitoring activities should be designed into any program at its inception, and they should focus on several issues, including the technical, social, economic, and environmental impacts of the appropriate use of the technologies and applications. Monitoring data can come from a variety of sources, including interviews with partner agencies, suppliers, and end-users; site visits; and performance monitoring of installed systems. Long-term impacts cannot be evaluated without monitoring activities. It is much more useful to receive photos of and data from operational systems in the field after several years—rather than a pretty inauguration-day photo of dignitaries with a new system that may be doomed to fail due to lack of a maintenance infrastructure.

Monitoring activities should strive to develop a bed of a variety of projects and technologies for long-term evaluation. It is valuable to maintain a database of applicable project and program information collected from field personnel. Maintaining a database allows program personnel to conduct analyses and make necessary adjustments along the way during program implementation. As any program continues its transition from direct implementation of pilot projects to further replication and institutionalization of partner organizations, these monitoring efforts continually grow in importance.

10.6 INSTITUTIONAL MODELS FOR SOLAR ENERGY DISSEMINATION

Project replication, or growing sustainable markets, is a program's ultimate measure of success or failure, and it can occur in a number of ways. As partner institutions and end-users gain familiarity with the use of solar energy technologies, they begin to implement new projects on their own. This generally occurs within a specific region first and then spreads to new regions. Through such activities, other related institutions become familiar with the merits of renewable energy technologies and initiate projects as well.

The potential for this type of replication can be enormous, given that budgets for development organizations to do development work can be in the many millions of dollars, whereas relatively few funds are earmarked specifically for renewable energy. Private-sector spin-off replication occurs as a result of successful pilot projects. For replication to be substantial, several factors must be adequately addressed: The local population must know the technology and what it can provide, quality products and services must be available locally, and the ability to pay for the technology must exist. For the latter reason, access to applicable financing mechanisms is extremely helpful.

Solar energy can be prohibitive in initial cost for many potential users, especially in less developed regions, despite the fact that the levelized life cycle costs of renewable energy are often quite competitive compared to conventional fossil fuel costs, especially in rural areas. Sometimes development funds are available to buy down the system cost to make system cost accessible.

Table 10.2 provides a summary of project development models typically employed for renewable energy development. How effective a model is will depend on where the project is located, local cultural norms, degree and type of political organization, and other such factors. Program implementation by private enterprise is relatively rare in the area of rural renewable development, but some initiatives have been quite successful, especially in the area of financing. Programs headed by private interests have the advantage that sustainability is in the best economic interest of the implementing agency or consumer.

Four basic approaches used to encourage the purchase of renewable energy systems in the private sector include:

TABLE 10.2
Solar Project Development Models

Entity	Project development	Investment	Management
Business	Business	Private	Private
Government or NGO	Private or nonprofit System Designer	Government, NGO, Local	Local government, Community, hired
Villages	Cooperative	Government, NGO, Local	Government agency, Private company
Distributed systems	Private or nonprofit System Designer		
Concessions	Private	Holder will make investment, however there may be government subsidy or guarantee	Holder of concession
Private extension	Business	Business	Private

- cash sales;
- financed sales;
- leasing (energy service); and
- direct subsidies.

Of these, market-based financing and leasing approaches for renewable energy have the greatest potential for expanding the access of rural households to this technology. Solar energy also offers the potential to generate new and important business activity in economically depressed rural areas by creating jobs through local retail sales and services and even manufacturing (e.g., solar cookstoves).

In most developed as well as less developed countries, renewable energy technologies have yet to be recognized as a consumer good that can be financed like a car or refrigerator. However, some exceptions in a few countries are establishing creative opportunities for renewable technology dissemination, such as Soluz has done in the Dominican Republic and Honduras.

Sales of renewable energy technology, especially PV in less developed countries, can be classified at four different levels, as exemplified by a classic sales approach pyramid shown in Figure 10.1. At the top of the pyramid are the few direct cash sales to relatively wealthy households that can afford the high initial capital costs of a renewable energy system. Following this are many more consumers who can afford to purchase a renewable energy system if reasonable credit terms are provided. The concept also shows that still more people could afford simply to pay a service fee for energy by leasing a renewable system. Finally, the poorest households, often traditional tribal groups largely living outside any cash economy, live a subsistence lifestyle and would probably not choose to participate in any form of renewable electrification program unless it was subsidized directly by development agencies; however, it is appropriate to look for in-kind sweat equity from even these tribal groups to help generate a feeling of project ownership. The exact percentage of persons that fall within any of these particular categories varies greatly from country to country.

10.6.1 CASH SALES

Most solar energy systems are sold directly through cash sales worldwide. This is typically the only form of sale available in many countries where no credit terms are available. Many local solar energy distributors (systems houses) are smaller, family-owned entrepreneurial companies that cannot afford to offer end-user financing and only have access to supplier credit terms, thus

FIGURE 10.1   Nicaraguan National Electrical PV Code training by New Mexico State University for government and industry engineers under the World Bank rural electrification program (PERZA).

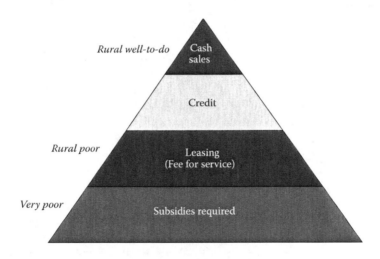

FIGURE 10.2   Institutional renewable energy sales approach pyramid.

only allowing them to make cash sales. Obviously, cash sales are restricted to only the wealthiest rural customers who can afford to purchase a renewable energy system outright. Most systems sold in industrialized countries (e.g., Japan, Germany, United States) are also done on a cash-sales basis (e.g., 50% down payment and 50% upon system commissioning).

10.6.2   Consumer Financing

One of the most important advances of the past century was the development of consumer credit. Consumer financing is a common way of increasing the sale of consumer goods all over the world. This has allowed citizens of developed nations to have widespread ownership of homes, automobiles, and appliances that the average person could not afford to purchase outright. Unfortunately, commercial banks and vendors rarely finance the purchase of consumer goods to people living in rural areas of developing countries and, if so, only at very high interest rates. Many more renewable energy systems could be installed if financing was readily available to consumers. This would allow for increased economic development of rural areas. Unfortunately, virtually no financing mechanisms for renewable energy systems can be found in most countries.

Financing should be developed at competitive interest rates and avoid a mismatch of loan and subloan maturities in order to make financing a viable business. Procedures should be as simple as possible and allow for quick disbursement when dealing with rural people who are unaccustomed to financing concepts. It is important to have parallel compliance monitoring in place, which allows for end-user audits, performance audits, product and installation standards, after-service sales, warranties, and customer satisfaction surveys. This way, financing program progress can be tracked in real time and adjustments made as needed before a program gets into trouble.

10.6.2.1 Revolving Credit Fund

A revolving credit financing fund is started with seed capital that allows families to purchase solar energy systems. As payments are made, the families replenish the fund with monthly payments that include interest. As the fund grows, additional families can be included to expand the number of systems financed. A program established for this type of renewable energy dissemination should attempt to use an integrated development approach, providing a complete institutional support system including service enterprises, technician training, and financing mechanisms.

10.6.2.2 Local Bank Credit

Another financing model that has been implemented for renewable systems is through conventional commercial banks, typically rural ones. The difficulty in getting commercial banks to finance PV systems is that the technology is relatively unknown and represents a new concept for most banks. Commercial bank financing can be successful for renewable systems implementation if the following steps can be taken:

- Bank staff becomes familiar with renewable energy system capabilities.
- Renewable systems become eligible for bank financing.
- Borrowers have convenient access to the bank.
- Loan application procedures are straightforward.
- Numerous smaller projects can be bundled together into one larger project.
- Collateral requirements are reasonable (e.g., use hardware as collateral).
- Repayment schedules are flexible and complement borrower's income flow.

10.6.3 Leasing

Another approach that has been implemented for solar energy systems in rural regions is the leased systems model. The idea behind leasing is to make solar-powered home systems even more affordable for rural people by eliminating the need for a down payment, lowering monthly charges, and reducing the customer's financial commitment to a simple month-to-month leasing arrangement for energy service. This approach has been tried in places like the Dominican Republic and Honduras with mixed success. It is a difficult model to implement for small solar energy home systems because administrative costs are high.

10.6.3.1 Dealer Credit

The dealers that sell solar energy systems are sometimes able to offer their customers credit. When a dealer provides consumer financing, the dealership is provided a second income stream based on interest payments. The difficulty for most dealers is that they are typically small, family-owned enterprises with limited access to credit that they can pass on to their customers.

10.6.4 SUBSIDIES

Subsidies are often poorly applied and designed by planners. Subsidies for renewable energy technologies that do not create any local infrastructure for maintaining systems or creating an infrastructure base forego sustainable market development. When subsidies are going to be provided, they should be done with a vision toward establishing a sustainable future (i.e., "smart subsidies"). Subsidies must be able to sustain cost-reduction pressures in the technology; however, they should not stifle competition by providing subsidies to only a single entity. Subsidies should be technology and supplier neutral and leveraged by a reasonable cost share from the users to create a sense of ownership. For example, Japan successfully used subsidies in the late 1990s to mid-2000s to help bring down the cost of PV; these were then eliminated by the late 2000s as PV became economically competitive with national electric prices on a life cycle basis (electric rates of about US\$0.25/kWh). Japan PV installation growth leveled off as a result, but a sustainable steady-state industry was achieved.

Subsidies are better placed if they finance results and not investment costs. Capital cost subsidies provide incentives to install systems, but not to utilize them over the long term. For instance, there is no reason that a subsidy for renewable energy water pumping could not be implemented in such a way as to allow for a fee-for-service approach. This would help assure working systems over the long run while establishing a viable local supply and service base.

Subsidies should also be used to assure that they are meeting the needs of communities, as prioritized by those communities. Participating households should also be appropriately selected and have a genuine interest in the service provided, whether it be water, electricity, ice, etc.

10.7   MANAGEMENT AND OWNERSHIP

For private entities, financing is from individuals, business, farms, ranches, and local lending institutions without subsidies. Because of the relative high initial costs and low income levels of remote areas, these kinds of projects rely more on grants and subsidies from government (local to national), international government funding, and national and international NGOs. More local funding from users, the community, the project developer, and/or local business results in better long-term viability of the project. A revolving credit fund for small systems is also possible (e.g., Honduras World Bank PIR [Proyecto de Infraestructura Rural] program).

10.7.1 AUTHORIZATION ARRANGEMENT

The community or distributed system is managed by community or cooperative committee, which includes user representatives. The committee usually has the following roles: articles of association, procedures and regulations for the community water system, who pays and how much, hiring of an operator, and distribution of revenues between operation/maintenance and replacement repairs.

10.7.2 CONTRACTS

The system owners contract the operation to an individual or a business. The contract primarily stipulates the service and the cost of that service. The contractor takes full responsibility for the system, may or may not collect a tariff, and does ordinary operation and maintenance. Remaining funds or a percentage of the tariff is used to pay for the contract.

10.7.3 LEASES

A lease has more legal requirements than a contract because the tangible assets are leased to an operator or business. The lessee can be an individual or a business that then takes on a greater degree of responsibility for long-term maintenance. A detailed lease arrangement for the distribution of revenues for long-term maintenance is mandatory. The system owners will usually set up basic guidelines; however, operator and end-users will normally determine the service.

10.7.4 OWNERSHIP TRANSFER (FLIP MODEL)

The system developer and/or governmental agencies that provided funding transfer the system to a private individual or business for a fee after a period of time. Then system development becomes a private sector enterprise. The transfer agreement needs to include transfer fee or requirements, minimum levels of service, and guidelines on tariffs (probably with maximum values for different time periods). This is similar to the privatization of government agencies such as telephone, electricity, etc. The wind industry has begun using this model for large wind farms.

10.7.5 ASSOCIATIONS AND COOPERATIVES

Rural water associations or cooperatives can both manage and operate the system. They should be involved in all aspects of the project, from system development to assistance in raising capital.

10.8 TARIFFS AND PAYMENT

The economic viability of a solar energy system depends on long-term payment for management, operation, and maintenance and replacement repairs. It can be entirely appropriate to establish service tariffs with users for village power or community water pumping systems. The implementation of a use-based water tariff in a community where people are accustomed to paying a minimal flat fee for water can provide a capital fund for future maintenance actions. This helps create an attitude of ownership. Tariffs can be classified into the following: free, nominal charge (subsidized), and full charge.

10.8.1 FREE

Often rural and indigenous people in less developed regions largely live outside a cash economy and do not have enough income to pay for operating costs, as well as part of the capital costs. Therefore, they consider such systems welfare projects and that the energy should be free. Such systems often fail fairly soon because no funds are available for system operation and maintenance. Also, they generally consider that there is no limit on the amount of energy consumption, especially for growth. For example, in Sudan, water wells were placed in arid regions and then the herds stripped all vegetation in the region surrounding the wells. In Mexico where free hybrid solar/wind village power systems were installed in 1990s, loads doubled within a year, and the hybrid systems often failed within a couple of years.

10.8.2 NOMINAL (SUBSIDIZED)

Rather than a complete subsidy, only a partial nominal subsidy is provided by the government. For communal systems, questions must be answered related to the amount to be charged for management, operation and maintenance, replacement repairs, and tariff per family, businesses, etc. There are two possibilities: (1) subsidy for energy consumption and capital cost, which is not viable because money from government is needed on a continual basis; or (2) a tariff designed to generate enough revenue to cover management, operation and maintenance and replacement repair costs (e.g., a centralized battery-charging station where a nominal payment is made each time a battery is charged).

In general, the subsidy is for the capital cost; however, a tariff may or may not cover part of the capital cost or low-interest loan for capital cost. If there are taps for residences, institutions, and businesses, is the energy consumption metered and what is a fair charge?

10.8.3 FEE FOR SERVICE

This tariff is designed to generate enough revenue to cover management, operation and maintenance, and replacement repair costs and to pay all or part of capital costs (probably low-interest loans). A fee for energy consumption has the additional problem in that meters add to the cost of the system. Therefore, the fee could be based on size and type of use: residence, institution, business, number of livestock, etc.

10.8.4 PAYMENT

Payment by month is common for urban areas with meters; however, this is problematic in rural areas, especially those where sale of agriculture products and livestock is seasonal. When a tariff is communal, payment could be once or twice per year. Now the question is when to pay and whether payment should be before or after the energy is consumed. Are in-kind payments acceptable? This places an additional burden on management and operators, who may have little expertise in the livestock or agricultural products that would normally be traded.

10.9 OTHER CRITICAL ISSUES

Other critical issues include legal concerns, permitting, training, technical assistance, and warranty and after-sales service by vendors and manufacturers. Training may include factory training and there must be local training for operation and maintenance. Warranty should include at least 1 year of technical assistance and product replacement for failure of components. Permitting can become a real obstacle in industrialized countries. Operation and maintenance should include provision of spare parts and, for remote locations, delivery times for additional spare parts in the future.

10.10 SUMMARY

For solar energy systems to be a viable and sustainable energy solution for both urban and remote village applications, an adequate and manageable institutional structure must accompany the technology deployment. Key lessons learned from successful renewable energy experiences are as follows:

- Local support infrastructure and training is crucial.
- Long-term planning is required for all renewable energy development projects.
- System ownership and responsibilities need to be established clearly early on.
- Permitting and approvals should be simple and straightforward.
- Maintenance is critical for long-term system survival.
- Implementing agencies should strive to work with industry to conduct project installations, thus strengthening local industry while developing a local infrastructure for system maintenance.

Preventative maintenance steps should be included in project planning from the start. Maintenance activities can often be funded with revenues generated by local end-users. However, the lack of attention to institutional issues often leads to inadequate system maintenance and eventual system degradation to the point of failure.

To minimize failure, renewable energy systems must be realistically sized and include proper institutional controls from the onset. Planning must allow for anticipated growth, a realistic tariff structure, and a means to meet future maintenance requirements. Only then can renewable energy systems provide long-term, reliable service to users. Solar energy systems represent a relatively simple yet elegant technology that can meet a wide range of applications; with proper attention to institutional details, these systems can provide decades of reliable service.

PROBLEMS

10.1 The World Bank and the government of Nicaragua Ministry of Energy and Mines have agreed to establish a rural PV electrification program (PERZA) for the Miskito Indians in the RAAN region along the Atlantic coast. This is a traditional tribal group that conducts subsistence farming and largely lives outside the cash economy. Discuss what sustainable development approaches might work best for supplying basic solar electricity services and what social impacts you can envision that electrifying traditional nonelectrified communities could have.

10.2 Discuss what the energy utilities could do in your community to promote the use of solar energy for water heating and electricity.

10.3 What kind of government incentives, if any, do you think are justifiable for promoting renewable energy development? Why or why not?

10.4 Determine the local electrical code and permitting requirements for installing a PV system in your city or county.

11 Energy Storage

11.1 INTRODUCTION

Solar energy is a nondispatchable energy technology that only captures energy during daylight hours. Some type of energy storage is thus required to make the energy available during nonsunny periods. Energy storage can take a number of forms, most commonly electrochemical energy storage through batteries. But energy can also be stored in the form of compressed air, pumped hydrostorage, hydrogen, or thermal mass. Many types of batteries and charge controllers are used in stand-alone PV systems to provide energy when the sun is not shining. The focus of this chapter is to look at the most common type of electrochemical storage systems used for off-grid PV systems.

11.2 BATTERIES IN PV SYSTEMS

A storage battery is an electrochemical device. It stores chemical energy that can be released as electrical energy. When the battery is connected to an external load, the chemical energy is converted into electrical energy and current flows through the circuit (Harrington 1992; Lasnier and Gan Ang 1990).

The three main functions of a PV system battery are to:

- store power produced by the PV system;
- supply the power required to operate the loads (e.g., lighting, pumping) for the end-use application; and
- act as a voltage stabilizer in the electrical system. The battery smoothes out or reduces temporarily high voltages (transient voltages) that may occur in the PV electrical system. High transient voltages can be generated in the electrical system (this could occur in making or breaking a circuit). The battery partially absorbs and greatly reduces these peak voltages and protects solid-state components from being damaged by these excessively high voltages.

PV systems do not charge batteries in the same manner as that to which battery manufacturers are accustomed. Designers can have difficulties choosing and optimizing batteries for PV applications. Improved understanding of the PV environment can help improve the lifetimes of batteries. PV designers may not use batteries exactly as they were intended, but they can use them more effectively through better design.

Many types of secondary, or rechargeable, batteries are used in stand-alone PV systems. Of the lead-acid battery types, the flooded (wet) traction or motive power battery is the most suitable for PV applications due to its deep-cycle capabilities and long life. Starting, lighting, and ignition (SLI) batteries, commonly used in automobiles, are not recommended for PV applications due to their limited deep-cycle capabilities. Valve-regulated lead-acid (VRLA) sealed batteries are popular but have some requirements that are hard to meet in a PV system. To be successful, these batteries will need more attention regarding how they are treated.

The life of a lead-acid battery is proportional to the average state of charge (SOC) of the battery if the battery is not overcharged, overdischarged, or operated at temperatures exceeding manufacturers' recommended specifications. A typical flooded, deep-cycle, lead-acid battery that is

FIGURE 11.1 Example of quality battery bank in Chiapas, Mexico, for a remote ecotourist lodge. Note the spark arrestors on top of each 2 V Exide battery cell.

maintained above 90% SOC can provide two to three times more full charge/discharge cycles than a battery allowed to reach 50% SOC before recharging (Figure 11.1). Similar and more dramatic results are found with sealed VRLA and lead-calcium alloyed grid batteries.

11.2.1 LEAD-ANTIMONY BATTERIES

For *flooded lead-antimony, open vent* batteries, capacities range from 80 Ampere-hours (Ah) to over 1,000 Ah. Typically, these are the most widely available and appropriate type of battery for PV applications because of their deep discharge capability and ability to take abuse. These batteries require the addition of water to maintain electrolyte levels. Loss of electrolyte occurs from evaporation and gassing. Gassing rates are determined by the charging algorithm and set points. Water loss can be significantly reduced by the addition of catalytic recombiner caps (CRCs).

These batteries have the best tolerance to charging algorithms and misadjusted set points. They are physically rugged and tolerate temperature extremes, although temperature compensation is recommended when determining charge controller set points. The electrolyte is easily adjusted for batteries in continuous temperature extremes. Health of the battery can be checked by reading the specific gravity of the electrolyte. This reading is not an accurate measure of capacity by itself but does indicate the health of each individual cell and the level of sulfation or electrolyte stratification.

Advantages:
  Antimony adds strength to lead.
  Can accept large charge currents and deliver large discharge currents.
  Can be repeatedly discharged to 50–80% of capacity.
  Less shedding of active material.

Disadvantages:
  High self-discharge rate.
  Battery bubbles (gasses) early during recharge due to high gassing rate because of the antimony in the plates.
  Require high maintenance; water must be added periodically.

11.2.2 Lead-Calcium Batteries

Flooded lead-calcium, open-vent batteries, commonly called stationary batteries, are typically nominal 2 V cells at 1,000 Ah or greater. These are not adapted for deep cycling but have the advantage of low self-discharge rates and lower water losses. If treated properly, they will last more than 10 years in continued standby use. They typically experience shortened lifetime due to sulfation and stratification of the electrolyte. This hazard can be reduced by using charging techniques appropriate for PV charging regimes.

Flooded lead-calcium, sealed-vent batteries were initially adapted from the automotive industry for the PV industry. These batteries are called *maintenance free*. They are not adapted for deep cycling, but they have the advantage of low self-discharge rates and lower water losses. They do not tolerate more than 20% depth of discharge very well, which drastically reduces the lifetime. Typically, these batteries are 12 V and have between 80 and 100 Ah of storage. They do not do well in hot climates or when overcharged. These batteries need controlled charging and minimal gassing. Because they are flooded, they need to be gassed to mix the electrolyte. If gassed too much, the electrolyte is lost through the vents forever, reducing battery life. Because of the sensitivity to overcharging, these batteries usually suffer from undercharging more often than overcharging.

Lead-antimony/calcium hybrid batteries are typically a flooded battery with large Ampere-hour ratings—300 Ah and up. These batteries are designed for lead-calcium tubular positive electrodes and pasted plate negative electrodes, which combine the advantages of both lead alloys. Because they are designed as a hybrid, the expectation is reduced electrolyte loss and extended depth of discharge and lifetime in cyclic applications.

Advantages:
Calcium adds strength to lead.
Calcium reduces gassing and loss of water (higher electrolysis threshold).
Low maintenance required.
Low self-discharge rate.

Disadvantages:
Poor charge acceptance after deep depth of discharge.
Battery life is shortened greatly if deep-discharged repeatedly (>15–20% depth of discharge [DOD]).

11.2.3 Captive Electrolyte Batteries

Captive electrolyte batteries are manufactured in sealed configurations. Lead-calcium grids are typically employed, but some grids use lead-antimony/calcium hybrids. The sulfuric electrolyte solution is immobilized. Captive electrolyte batteries have sophisticated valve (pressure)-regulated mechanisms for cell vents, often referred to as VRLA batteries. These batteries are spill proof and there is no need to add water.

Sealed lead-acid, gelled electrolyte (gel) batteries were initially designed for use in electronic instruments and controlled environments. Gel batteries are typically lead-calcium pasted plates. Battery technology is very sensitive to charging methods, set points, and temperature extremes. Charging voltage regulation (VR) set points vary from manufacturer to manufacturer for the same technology. The recommended charging algorithm is constant voltage, with temperature compensation required. The battery's chemistry has a definite upper temperature limit. When exceeded, this produces irreversible damage. Some batteries function better than flooded batteries in cold environments because the electrolyte is suspended in a silicon gel rather than water, thus reducing the susceptibility to freezing.

New technology has been applied to reduce the detrimental effect of deep discharging of these batteries by adding phosphoric acid to the electrolyte. As a battery is discharged, the concentration of sulfuric acid is decreased, thus increasing the resistance of the battery and reducing its ability to recharge properly. Phosphoric acid is used to minimize oxidation between the grid and paste, which occurs during low states of charge.

Sealed lead-acid, absorbed glass matte (AGM) technology is sensitive to charging methods, set points, and temperature extremes. Charging VR set points vary from manufacturer to manufacturer for the same technology. The recommended charging algorithm is constant voltage, with temperature compensation required. The battery's chemistry has a definite upper temperature limit of approximately 135°F; exceeding this limit produces catastrophic, irreversible damage. The reduced maximum recharge rate of gel type electrolyte batteries also applies here.

11.2.4 Nickel-Cadmium Batteries

Nickel-cadmium (NiCad) batteries are adapted for deep cycling. They are manufactured as either sealed types with a pressure relief valve built into the cell (some valves may not reclose) or vented types with resealable vents that open and close under small pressure changes. Voltage output is about 1.2 V per cell and remains relatively stable until the battery is nearly discharged. NiCads can accept a relatively high charge rate (C/1) and are capable of operation under continuous overcharge, provided that the charging current does not exceed the C/15 rate. A general comparison of nickel-cadmium and lead-acid batteries is given in Table 11.1.

Advantages:
  Long life.
  Reduced maintenance.
  Can be deep-discharged without damage.
  Performance much less affected by temperature.
  Voltage regulation not as important.
  Excellent charge retention and high capacity at low temperatures.

Disadvantages:
  Cost per Ampere-hour is high.
  Display a "memory" of battery discharge history.

11.3 LEAD-ACID BATTERY CONSTRUCTION

11.3.1 Plate Grids

The plate grids are the supporting structure for the active materials of the plates. They are made of pure lead or of an alloy of lead and another material (e.g., antimony) added to strengthen and stiffen the soft lead. Batteries with grids containing calcium for strength have reduced gassing and thus have less water usage and low self-discharge rates (Lasnier and Gan Ang 1990; Vinal 1951).

11.3.1.1 Positive and Negative Plates

A paste mixture of lead oxide, sulfuric acid, and water is pasted onto a grid to form the positive and negative plates. Fibers may be added to help bind the active material together. The paste applied to the negative plate uses an expander material to prevent the negative material from contracting back to a dense inactive state during service.

TABLE 11.1

Comparative Features of Lead-Acid and Nickel-Cadmium Storage Batteries

	Lead-acid	Nickel-cadmium
Type	Medium rate, deep discharge, lead-calcium grid	Medium rate, cycle service, vented pocket plate
Rated capacity at 77°F, 8 h	100–900 Ah per cell	10–400 Ah per cell
Nominal discharge cut-off voltage	1.75 V per cell	1.0 V per cell
Nominal voltage	2.45 V per cell, varies with state of charge	1.25 V per cell, fairly constant with state of charge
Available capacity against temperature (% of rated capacity)	70% at 32°F 20% at –20°F	90% at 32°F 65% at –20°F
Nominal energy efficiency	70–80%	60–70%
Nominal cycle life for 80% discharge cycle	1,000–1,500	1,500–2,000
Nominal calendar life without cycling	10–20 years	24 years
Energy density	6–13 Wh lb^{-1}	9–10 Wh lb^{-1}
Internal resistance	0.6–3.0 mΩ/100 Ah	0.2–1.5 mΩ/100 Ah
Charge control	Sensitive to long overcharging	Can accept 5–10% overcharge
Required maintenance	Water replacement; charge equalization; protection against freezing and temperature extremes	Water replacement; occasional full discharge

Sources: Kiehne, H. A. 1989. *Battery Technology Handbook.* New York: Marcel Dekker, Inc.; Lasnier, F., and T. Gan Ang. 1990. *Photovoltaic Engineering Handbook.* New York: Adam Hilger.

A forming charge is then applied by immersing the plates in a dilute sulfuric acid solution. When the forming charge is applied in a particular direction, the lead oxide on the positive plate is electrochemically converted to lead dioxide (PbO_2), while the lead oxide of the negative plate is converted to a gray sponge lead. The PbO_2 and the sponge lead are highly porous and allow the electrolyte to penetrate the plates easily.

11.3.1.2 Separators

Separators are placed between the positive and negative plates to prevent a short circuit from discharging all stored energy in the plates. A separator is a thin sheet of finely porous, electrically insulating material that allows the passage of charged ions of the electrolyte between the positive and negative plates. The high porosity of the separators ensures low resistance to current passing between the plates. The separators must provide good insulation to prevent metallic conduction between plates of opposite polarity. Suitable separator materials include microporous rubber, plastic, and glass-wool mats.

11.3.1.3 Elements

Each cell of a lead-acid battery contains an element assembled by placing one group of positive plates with one group of negative plates together with separators between the positive and negative plates (Figure 11.2). An element can contain any size or number of plates, although the open-circuit voltage of the cell will be about 2.1 V regardless of the number or size of the plates. However, an increased total plate surface area per element results in an increased current output during discharge at high rates. The cells are connected together in series by welding the post straps of one cell together with the post straps of the adjacent cell (positive group to negative group). The voltage of the battery equals the sum of the voltages of the individual cells. The battery is not active until the electrolyte mixture of sulfuric acid and water is added.

11.3.1.4 Cell Connectors

Connectors are intercell straps comprising lead-plated copper that are burned or bolted to positive and negative posts of adjoining cells or elements. They are low resistance and typically carry currents equal to about 5C (C is the nominal Ampere-hour capacity).

11.3.1.5 Containers

Battery containers are individual boxes for a single element or multiple compartment cases housing two to six elements; each cell comprises a separate electrical unit insulated from the adjacent cells. A 12 V battery consists of six elements housed in a six-compartment container. The plates rest on ribs or grids molded into the base, while also providing a bottom area for the accumulation of sediment. The sediment acts as an electrical conductor and, to avoid a short circuit, should not be allowed to come in contact with the plates.

11.3.1.6 Vent Plugs

Vent plugs provide access to the battery interior for adding electrolyte or water and measuring specific gravity. The plugs are fitted into the holes of the battery cover and are baffled to allow gases to escape freely while returning acid spray into the cell.

11.4 LEAD-ACID BATTERY OPERATION

When two unlike metals such as the positive and the negative plates are immersed in sulfuric acid (the electrolyte), the battery is created and a voltage is developed that is dependent on the types of metals and the electrolyte used. It is approximately 2.1 V per cell in a typical lead-acid battery. Electrical energy is produced by the chemical action between the metals and the electrolyte. The chemical actions start and electrical energy flows from the battery as soon as there is a circuit between the positive and negative terminals (whenever a load such as the head lamps is connected to the battery). The electrical current flows as electrons through the outside circuit and as charged portions of acid (ions) between the plates inside the battery (Lasnier and Gan Ang 1990; Vinal 1951).

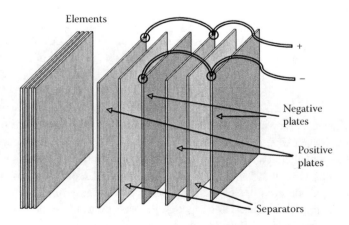

FIGURE 11.2   Lead-acid battery plate construction configuration.

The action of the lead-acid storage battery is characterized by the following equation:

$$PbO_2 + Pb + 2H_2SO_4 \leftrightarrows 2PbSO_4 + 2H_2O \tag{11.1}$$

where

lead dioxide (PbO_2) = the material on the positive plate
sponge lead (Pb) = the material on the negative plate
dilute sulfuric acid (H_2SO_4) = the electrolyte

11.4.1 DISCHARGE CYCLE

When a battery is connected to an external load, current flows. The lead dioxide (PbO_2) in the positive plate is a compound of lead (Pb) and oxygen (O_2). Sulfuric acid (the electrolyte) is a compound of hydrogen (H_2) and the sulfate radical (SO_4). As the battery discharges, Pb in the active material of the positive plate combines with the SO_4 of the sulfuric acid, forming lead sulfate ($PbSO_4$) in the positive plate. Oxygen (O) in the active material of the positive plate combines with H_2 from the sulfuric acid to form water (H_2O); the concentration of acid is reduced due to SO_4 removed from solution (into the plate $PbSO_4$). A similar reaction is occurring at the negative plate at the same time. Lead of the negative active material combines with SO_4 from the sulfuric acid to form $PbSO_4$ in the negative plate (See Figure 11.3).

As the discharge progresses, the sulfuric acid in the electrolyte is diluted; thus, its specific gravity becomes lower. The specific gravity can be measured with a hydrometer or refractometer, giving an accurate and convenient method for determining the SOC of a battery. During the discharge, the active material of both plates is changing to $PbSO_4$. The plates are becoming more alike and the acid is becoming weaker. Therefore, the voltage is becoming lower because it depends on the difference between the two plate materials and the concentration of the acid. Eventually, the battery can no longer deliver electricity at a useful voltage and is said to be discharged.

A battery discharges quickly when it is subjected to a high discharge rate. Under high discharge rates, the acid circulation into the pores of the plates and the diffusion of electrolyte from the pores of the plates are too slow to sustain the discharge. Only a small percentage of the electrolyte and plate active materials near the plate surface in the cell take part in the chemical reaction during the relatively short duration of a high discharge rate.

The acid circulation diffusion has less of an effect on battery performance at lower discharge rates. At slow discharge rates, practically all of the acid may be consumed, and the material near the centers of the plates has more of an opportunity to take part in the chemical reaction.

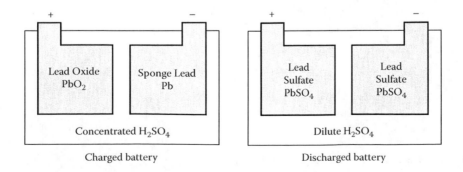

FIGURE 11.3 Lead-acid battery operation for charging and discharging modes.

The lead-acid storage battery is chemically reversible. A discharged storage battery can be charged (pass electrical current through it in the direction opposite to the direction of discharge) and its active chemicals will be restored to the charged state. The battery is again ready to deliver power. This discharge/charge cycle can be repeated multiple times until the plate's or separator's deterioration or another factor causes the battery to fail.

11.4.2 Charge Cycle

The chemical actions that take place within a battery during charge are basically the reverse of those that occur during discharge (Figure 11.4). The sulfate ($PbSO_4$) in both plates is split into its original form of Pb and SO_4. The water is split into H and O. As the sulfate leaves the plates, it combines with the hydrogen and is restored to sulfuric acid (H_2SO_4). At the same time, the oxygen combines chemically with the lead of the positive plate to form lead dioxide (PbO_2). The specific gravity of the electrolyte increases during charge because sulfuric acid is being formed and is increasing concentration.

A battery will generate gas when it is being charged. Hydrogen is given off at the negative plate and oxygen at the positive. These gases result from the decomposition of H_2O. As a battery gases, it uses water. Generally, a battery will gas near the end of a charge because the charge rate is too high for the almost charged battery. A charger that automatically reduces the charge rate as the battery approaches the fully charged state eliminates most of this gassing. It is extremely important not to charge low-water-loss batteries for long periods of time at rates that cause them to gas. No battery should be overcharged for a long period of time.

11.4.3 Electrolyte and Specific Gravity

The electrolyte in a lead-acid storage battery is a dilute sulfuric acid solution. Specific gravity is a unit for determining the sulfuric acid content of the electrolyte. A battery with a fully charged specific gravity of 1.265 corrected to 80°F (26.7°C) contains an electrolyte with approximately 36% sulfuric acid by weight or 25% by volume. The remainder of the electrolyte is water. Pure (concentrated) sulfuric acid has a specific gravity of 1.835. The sulfuric acid in the electrolyte is one of the

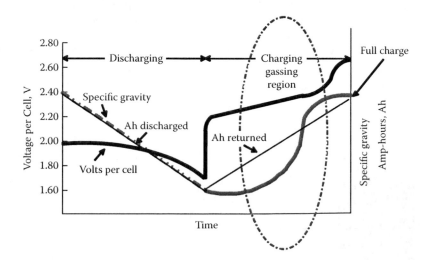

FIGURE 11.4 Lead-acid battery charging and discharging

necessary ingredients in the chemical actions taking place inside the battery. It supplies the SO_4 that combines with the active material of the plates. It is also the carrier for the electric current as it passes from plate to plate. When the battery terminals are connected to an external load, the sulfate combines with the active material of the positive and negative plates forming $PbSO_4$ and releasing electrical energy.

The recommended fully charged specific gravity of most 12 V batteries is 1.265 corrected to 80°F (26.7°C). Water has been assigned a value of 1.000. Therefore, when an electrolyte has a specific gravity of 1.265, it is 1.265 times heavier than pure water.

If it is necessary to dilute concentrated sulfuric acid to a lower specific gravity, the acid must always be poured into the water—slowly—and water should never be poured into acid. A dangerous "spattering" of the liquid, caused by extreme heat generated whenever strong acid is mixed with water, would result. The liquid should be stirred continually while acid is being added.

11.4.4 WATER

The most satisfactory water to use when preparing electrolyte is distilled water. This is also true for routine water additions to the battery. Water of a known high mineral content or rainwater collected from metallic troughs (e.g., tin roof) should not be used. Use of metallic containers (except lead or lead-lined containers) should be avoided. Metal impurities in the water will lower the performance of the battery. Many liquids, such as salt water, vinegar, antifreeze, and alcohol or harmful acids such as nitric, hydrochloric, or acetic, will ruin a battery.

11.4.5 BATTERY ROUNDTRIP EFFICIENCY

Due to the laws of thermodynamics, batteries cannot possibly deliver all of the energy that was stored in them. System losses occur in the form of bubbling electrolyte and internal resistance that produces heat that reduces the overall efficiency. For instance, new lead-acid batteries typically have roundtrip efficiencies of about 70–75%. Internal resistance increases with age as plates' sulfate and efficiencies may drop to 60%. Faster discharging time also decreases roundtrip efficiency as internal resistance grows (Figure 11.5).

11.5 LEAD-ACID BATTERY CHARACTERISTICS

11.5.1 AMPERE-HOUR STORAGE CAPACITY

The Ampere-hour capacity is the quantity of discharge current available for a specified length of time at a specific temperature and discharge rate. For example, a 12 V battery rated at 100 Ah over 20 h can deliver 5 Ah for 20 h, which is equivalent to 1.2 kWh of power (12 V × 100 Ah). This same battery might provide only 84 Ah at a 10 h discharge rate, 70 Ah at a 5 h discharge rate, and only 44 Ah at a 1 h discharge rate. Battery size, construction, temperature, concentration of electrolyte, plate history, and discharge rate all affect battery capacity (Lasnier and Gan Ang 1990; Vinal 1951).

A battery has a larger Ampere-hour capacity at longer discharge rates because more time is available for the acid in the electrolyte to penetrate more deeply into the battery plates. At high discharge rates, only a small amount of the electrolyte and active plate materials are used because the acid penetration into the plates and diffusion of electrolyte from the plates are too slow to sustain the high-rate discharge.

FIGURE 11.5   Battery testing at Sandia National Laboratories' PV systems lab.

Decreased temperatures result in less available battery capacity due to slower chemical reactions. A lead-acid battery's storage capacity decreases roughly about 1% for every 1°C drop in temperature. At slow discharge rates, temperature effects on capacity are somewhat reduced.

11.5.2   Battery Cycle Life

Battery life is expressed in cycles. The cycle life of a battery is the number of lifetime cycles expected from a battery at a specified temperature, discharge rate, and depth of discharge. Typically, the end of battery life is when the battery capacity falls 20% below its rated capacity. Battery lifetime is increased with a shallower depth of discharge; it is decreased with deeper depth of discharge due to greater internal stresses resulting from more complete utilization of active materials. Cycle life also decreases with increasing battery temperature. Longer discharge rates will increase available battery capacity, but will also shorten battery life due to deeper penetration of the acid into the plates—for example, 1,500 cycles at 40% DOD at 25°C for a 20 h discharge rate (C/20). Figures 11.6 and 11.7 demonstrate a battery life cycle curve.

Decreased battery life can be caused by several factors: External corrosion increases the interconnect resistance, internal (grid) corrosion reduces the physical size of the current-carrying grid wires (stratification causes this to occur faster in the plate bottoms), or excessive gassing causes loss of electrolyte and plate damage. Battery capacity may be lost if the electrolyte level falls below the top of the plates, thus preventing active materials from reacting there. Violent gassing can physically damage the plates by scrubbing off active materials.

Good system designers do not allow for greater than 10–15% DOD for flooded lead-acid batteries.

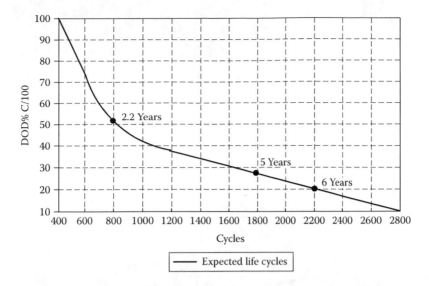

FIGURE 11.6   Cycle life characteristics at a low discharge rate (C/100) for a thick-plate Trojan T105 battery. (Graph courtesy Sandia National Labs.)

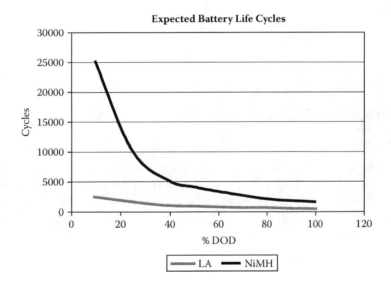

FIGURE 11.7   Cycle life characteristics comparing lead-acid and nickel metal hydride batteries.

11.5.3   BATTERY CONNECTIONS

Just like PV modules, batteries can be connected in series and parallel connections to vary the voltage and current delivered, respectively. Capacity is determined by number of batteries (or cells) in parallel. Good installation techniques should prevail, especially because thousands of Amperes can be delivered in short circuit instantaneously, causing fire and explosion hazards. Dangerous high voltages are possible (normally, battery banks > 250 V should not be designed due to increased instability. Figures 11.8 and 11.9 show series and parallel battery connections, which increase battery bank voltage and current, respectively).

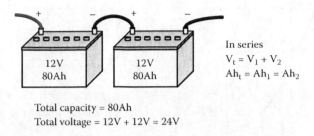

Total capacity = 80Ah
Total voltage = 12V + 12V = 24V

FIGURE 11.8   Battery series connections increase system voltage.

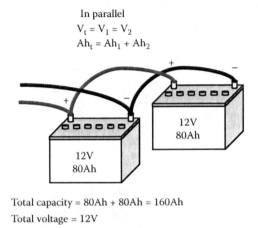

Total capacity = 80Ah + 80Ah = 160Ah
Total voltage = 12V

FIGURE 11.9   Battery parallel connections increase system current.

11.6   BATTERY PROBLEM AREAS

11.6.1   OVERCHARGING

Overcharging a battery produces corrosion of the positive grids and excessive gassing, which can loosen the active plate material. This loosened material is deposited as fine brown sediment at the bottom of the cell between the separators and plates. Overcharging can also increase battery temperature to the point where damage to the plates and separators occurs. Frequent replacement of lost water is necessary due to excessive gassing.

11.6.2   UNDERCHARGING

Consistent undercharging of a battery results in a gradual running down of the cells and progressively lower specific gravity and lighter color plates (Figure 11.10). With prolonged undercharging, fine white powder sediment of lead sulfate is deposited on the bottom of the cells. Undercharging is also one of the most common causes of plate buckling due to the plate strain caused by the lead sulfate, which occupies more space than the original active plate material.

11.6.3   SHORT CIRCUITS

Battery short circuits can be caused by a breakdown of one or more separators, by excess sediment accumulation at the bottom of cells, or by the formation of tree-like structures of lead from the

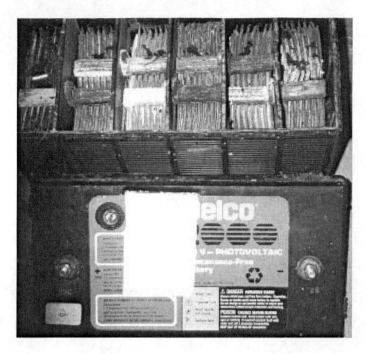

FIGURE 11.10 Excessive plate sulfation and buckling as a result of consistent undercharging of a battery in Ceara, Brazil, for a small residential PV home system.

negative to the positive plates ("treeing"). Treeing can be caused by the presence of certain grid materials (e.g., cadmium), which causes the tree-like growth at the element side or bottom (treeing is counteracted by the presence of antimony), or by a process called "mossing," in which sediment brought to the surface of the electrolyte by gassing settles on top of the plates and bridges over the separator tops. Battery short circuits are indicated by continued low specific gravity, even though the battery has received normal charging; rapid loss of capacity after full charge; and low battery open-circuit voltage.

11.6.4 SULFATION

Fine sulfate crystals are formed during discharging. Sulfation occurs when large lead sulfate crystals grow on the plates instead of the fine crystals normally present. The larger crystals increase the internal resistance of the cell, resulting in lower discharge and high charge voltages, thus lowering the voltage efficiency. A heavily sulfated battery is difficult to recharge and can become permanently damaged by plate fracture due to crystal growth.

Sulfation occurs when a completely or partially discharged battery remains unused for long periods of time, when a battery operates at partial state of charge for several days without a finishing or equalizing charge, or when battery temperature variations occur. Sulfation is partially caused by the increase in the solubility of the lead sulfate at higher electrolyte temperatures. The small lead sulfate crystals are dissolved during the high-temperature periods and are slowly recrystallized into large crystals when the temperature is reduced. Cycling of electrolyte temperature is caused by ambient temperature changes or by heat generated during battery charge or discharge.

11.6.5 WATER LOSS

Water loss occurs during electrolysis, when water is converted into hydrogen and oxygen gas. When a battery reaches full charge, the lead sulfate on the plates and the lead sulfate ions in the electrolyte are exhausted, and the rise in the plate potential beyond a certain cut-off voltage causes gassing. Gassing begins when the terminal voltage of the battery reaches about 2.3 V per cell, and the quantity of gas generated depends on the portion of the energy not absorbed by the battery.

To minimize water loss, several approaches are possible. A two-step charging process can be applied or a fast charge rate followed by a tapering or finishing charge. The charge rate should be limited to less than a C/50 rate. Finally, catalytic recombiner caps (e.g., Hydrocaps®) can be used to reduce water loss (Figure 11.11).

11.6.6 SELF-DISCHARGE

In a car parked for weeks on end, the battery will eventually completely discharge because all batteries will naturally self-discharge. The rate of self-discharge varies with battery type and age. Some approximate rules of thumbs for self-discharge are as follows (Figure 11.12):

lead-antimony
    new: ~1% per day at 25°C
    old: ~5% per day at 25°C
lead-calcium
    new: <0.5% per day at 25°C
    old: <0.5% per day at 25°C

11.7 BATTERY MAINTENANCE

Add distilled water to flooded lead-acid batteries (not to sealed batteries). The level of electrolyte in flooded lead-acid batteries should be checked about once a month, unless catalytic recombiner caps (e.g., Hydrocaps), which reduce the need for battery watering, are used. If there is no level indicator, water should be added only up to 0.5 in. (13 mm) above the separators. Water should be added when the battery is fully charged. If the battery is not fully charged, less water should be added (but equally to all cells) and the water level topped when the battery is fully charged. As a discharged battery charges, the electrolyte level rises. If too much water is added to a discharged battery, acid will bubble out of the top of the battery once it becomes fully charged (Lasnier and Gan Ang 1990).

Battery charge condition can be determined by measuring electrolyte specific gravity with a hydrometer or refractometer. For sealed batteries, the state of charge can be determined by measuring

FIGURE 11.11 Catalytic recombiner caps used on deep-cycle LA batteries for a PV system. Note the rubber container to contain any possible acid spills and appropriate battery cables.

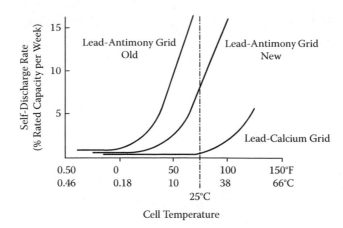

FIGURE 11.12   Lead-acid battery self-discharge rates vary with temperature.

battery open-circuit voltage. Battery voltage should ideally be tested at least an hour (24 h is best) after the PV module has been disconnected (or after sundown). All loads should be disconnected. The battery state of charge can be determined from Table 11.2.

Charge equalization should be carried out periodically for flooded lead-acid batteries (about once a month, depending on the battery) to bring all batteries up to 100% state of charge. An equalizing charge is a prolonged charge at the finishing rate or less (terminated when specific gravity or voltage readings are constant for about 3 h). This overcharging of the battery creates a gassing condition that prevents stratification of the electrolyte and subsequent sulfation.

A thin film of electrolyte can accumulate on the tops of the battery and nearby surfaces. This material can cause flesh burns; it is conductive and can cause leakage currents to discharge the battery or even a shock hazard in high-voltage battery banks. The escaped sulfuric acid should be washed away periodically with an appropriate neutralizing solution. For lead-acid batteries, a dilute solution of baking soda (sodium bicarbonate) and water works well. A mild vinegar solution works well on nickel-cadmium batteries. Anticorrosion sprays and greases that reduce the need to service the battery bank are available from automotive and battery supply stores.

TABLE 11.2
Lead-Acid Battery Approximate State of Charge

Open Circuit Voltage and Specific Gravity Values for 1.265 Specific Gravity Initial Full Charge

Charge level	Specific gravity	Voltage (12)	Voltage (6)
100%	1.265	12.68	6.3
75%	1.225	12.45	6.2
50%	1.190	12.24	6.1
25%	1.155	12.06	6.0
Discharged	1.120	11.89	6.0

Source: Battery Council International. 1987. *Battery Service Manual.* Chicago, IL.

11.7.1 HYDROMETER DESCRIPTION AND USE

The state of charge of a lead-acid battery can be determined by the specific gravity of the electrolyte (its weight compared to water). The specific gravity can be measured directly with a hydrometer or determined by the stabilized voltage.

A hydrometer is a bulb-type syringe that will extract electrolyte from the cell. A glass float in the hydrometer barrel is calibrated to read in terms of specific gravity. A common range of specific gravity used on these floats is 1.160–1.325. The lower the float sinks in the electrolyte, the lower its specific gravity is. The barrel must be held vertically so that the float is not rubbing against the side of it. An amount of acid is drawn into the barrel so that, with the bulb fully expanded, the float will be lifted free, not touching the side, top, or bottom stopper of the barrel. One's eye should be on a level with the surface of the liquid in the hydrometer barrel. One can disregard the curvature of the liquid where the surface rises against the float stem and the barrel due to surface tension. Table 11.2 illustrates typical specific gravity values for a lead-acid cell in various stages of charge. A fully charged specific gravity of 1.265 corrected to 80°F (26.7°C) is assumed (Battery Council International 1987).

A fully charged battery has all of the sulfate in acid. As the battery discharges, some of the sulfate begins to appear on the plates. The acid becomes more dilute and its specific gravity drops as water replaces more of the sulfuric acid. A fully discharged battery has more sulfate in the plates than in the electrolyte. Please note that the hydrometer float sinks lower in the electrolyte as the specific gravity becomes lower. A hydrometer reading should never be taken immediately after water is added to the cell. The water must be thoroughly mixed with the underlying electrolyte.

11.7.2 TEMPERATURE CORRECTION

Hydrometer floats are calibrated to give a true reading at one temperature only. A correction factor must be applied for any specific gravity reading made when the electrolyte temperature is not 80°F (26.7°C). Some hydrometers use a reference temperature of 60°F (15.5°C). A temperature correction must be used because the electrolyte will expand and become less dense when heated. The float will sink lower in the less dense solutions and give a lower specific gravity reading. The opposite occurs if the electrolyte is cooled.

Regardless of the reference temperature used as a standard, a correction factor of 0.004 specific gravity (sometimes referred to as four "points of gravity") is used for each 10°F (5.5°C) change in temperature. Four points of gravity are added to the indicated reading each 10°F (5.5°C) increment above 80°F (26.7°C) and four points are subtracted for each 10°F (5.5°C) increment below 80°F (26.7°C). This correction is important at temperature extremes. The thermometer should be of the mercury-in-glass type with a scale reading as high as 125°F (52°C). The smaller the bulb immersion is, the better; it should not exceed 1 in. (25 mm) (Battery Council International 1982).

11.7.3 TROPICAL CLIMATES

Most batteries used in temperate climates have a fully charged specific gravity in the 1.250–1.280 range. A fully charged electrolyte specific gravity of 1.210–1.230 is used in tropical climates. A tropical climate is defined as one in which water never freezes. This milder strength electrolyte does not deteriorate the separators and grids as much as the higher strength electrolyte. This increases the service life of the battery. The lower specific gravity decreases the electrical capacity of the battery. However, these losses are offset by the fact that the battery is operating at warm temperatures where it is more efficient.

Table 11.3 shows the approximate specific gravity values of lead-acid batteries at various states of charge. One column shows values for batteries whose electrolyte specific gravity has been prepared

TABLE 11.3
Typical Specific Gravity Values at Various Charge Levels for Temperate and Tropical Climates

State of charge	Specific gravity used in cold and temperate climates	Specific gravity used in tropical climates
Fully charged	1.265	1.225
75% charged	1.225	1.185
50% charged	1.190	1.150
25% charged	1.155	1.115
Discharged	1.120	1.080

Source: Lasnier, F., and T. Gan Ang. 1990. *Photovoltaic Engineering Handbook.* New York: Adam Hilger.

for use in a temperate climate; the other column is for batteries prepared for use in a tropical climate. The table illustrates that batteries may be fully charged and yet have different values of specific gravity. The values shown are for a cell in various states of charge at 80°F (26.7°C). The specific gravity values shown will vary depending on the ratio of electrolyte volume to active material and the battery construction. Table 11.4 gives typical specific gravity values for different types of lead-acid batteries in both temperate and tropical climates. Table 11.5 shows properties of sulfuric acid based on an electrolyte temperature of 59°F (15°C). A compensation for other temperatures can be made using the equation:

$$SG_T = SG_{59} + C(59\text{-}T) \qquad (11.2)$$

where

SG_T = the specific gravity at the desired temperature
SG_{59} = the specific gravity at 59°F (15°C)
C (°F^{-1}) = the temperature coefficient
T (°F) = the temperature of the electrolyte

11.8   BATTERY SAFETY PRECAUTIONS

Batteries should be handled with caution because most contain toxic substances such as lead and sulfuric acid. They may also contain explosive mixtures of hydrogen and oxygen gases. In

TABLE 11.4
Specific Gravity Range for Various Types of Lead-Acid Batteries

Battery type	Temperate climate		Tropical climate	
	Fully charged	Discharged	Fully charged	Discharged
SLI	1.260–1.280	1.120	1.210–1.240	1.080
Motive power	1.260–1.280	—	1.210–1.240	—
Stationary	1.200-1.225	—	1.200–1.225	—

Source: Lasnier, F., and T. Gan Ang. 1990. *Photovoltaic Engineering Handbook.* New York: Adam Hilger.

TABLE 11.5
Properties of Sulfuric Acid Solutions

Specific gravity at 59°F (15°C)	Temperature coefficient		H_2SO_4 concentration		Freezing point	
	Per °F	Per °C (10^{-5})	(wt.%)	(vol.%)	(°F)	(°C)
1.000	—	—	0.0	0.0	32	0
1.010	10	18	1.4	0.8		
1.020	12	22	2.9	1.6		
1.030	14	26	4.4	2.5		
1.040	16	29	5.9	3.3		
1.050	18	33	7.3	4.2	26	−3.3
1.060	20	36	8.6	5.0		
1.070	22	40	10.1	5.9		
1.080	24	43	11.5	6.7		
1.090	26	46	12.9	7.6		
1.100	27	48	14.3	8.5	18	−7.8
1.110	28	51	15.7	9.5		
1.120	29	53	17.0	10.3		
1.130	31	55	18.3	11.2		
1.140	32	58	19.6	12.1		
1.150	33	60	20.9	13.0	5	−15
1.160	34	62	22.1	13.9		
1.170	35	63	23.4	14.9		
1.180	36	65	24.7	15.8		
1.190	37	66	25.9	16.7		
1.200	38	68	27.2	17.1	−17	−27
1.210	38	69	28.4	18.7		
1.220	39	70	29.6	19.6		
1.230	39	71	30.8	20.6		
1.240	40	72	32.0	21.6		
1.250	40	72	33.4	22.6	−61	−52
1.260	40	73	34.4	23.6		
1.270	41	73	35.6	24.6		
1.280	41	74	36.8	25.6		
1.290	41	74	38.0	26.6		
1.300	42	75	39.1	27.6	−95	−71

Source: Vinal, G. W. 1951. *Storage Batteries,* 4th ed. New York: John Wiley & Sons.

general, the National Electrical Code (NEC) Articles 480 and 690-71, -72, and -73 should be followed for PV installations using storage batteries. Battery storage poses several safety hazards (Battery Council International 1987; Sandia National Laboratories, Design Assistance Center 1990; Vinal 1951):

- hydrogen gas generation from charging batteries;
- high short-circuit currents;
- acid or caustic electrolyte; and
- electric shock potential.

11.8.1 BATTERY ACID

When working with acid, such as filling batteries, one should use a face shield, gloves, and protective clothing. Extreme care should be taken to avoid spilling or splashing electrolyte (which is dilute sulfuric acid) because it can destroy clothing and burn the skin. When a plastic cased battery is handled, excessive pressure placed on the end walls could cause electrolyte to spew through the vents. Therefore, a battery carrier should be used to lift these batteries or they can be lifted with hands placed at opposite corners. If electrolyte is spilled or splashed on clothing or the body, it should be neutralized immediately and then rinsed with clean water. A solution of baking soda and water may be used as a neutralizer.

Electrolyte splashed into the eyes is extremely dangerous. If this should happen, the eye should be forced open and flooded with clean water for approximately 15 min. A doctor should be called immediately when the accident occurs and immediate medical attention given if possible. Eye drops or other medication should not be added unless one is acting on a doctor's advice. If acid (electrolyte) is taken internally, large quantities of water or milk should be drunk, followed with milk of magnesia, beaten egg, or vegetable oil.

If it becomes necessary to prepare electrolyte of a desired specific gravity, the concentrated acid should always be poured slowly into the water; water should not be poured into acid. Heat is generated when acid is mixed with water. Small amounts of acid should be added slowly while stirring. The mixture should be allowed to cool if noticeable heat develops. Except for lead or lead-lined containers, nonmetallic receptacles and/or funnels should be used. Acid should not be stored in excessively warm locations or in direct sunlight.

11.8.2 HYDROGEN GAS

Hydrogen and oxygen gases are produced during normal battery operation. When flooded, non-sealed, lead-acid batteries are charged at high rates or when the terminal voltage reaches about 2.4 V per cell, batteries produce hydrogen gas. These gases escape through the battery vents and may form an explosive atmosphere around the battery if ventilation is poor. Explosive gases may continue to be present in and around the battery for several hours after it has been charged. Even sealed batteries may vent hydrogen gas under certain conditions. If it is confined and not properly vented, this gas poses an explosive hazard. The amount of gas generated is a function of the battery temperature, the voltage, the charging current, and the battery bank size. Small battery banks (i.e., one to eight 220 Ah, 6 V batteries) placed in a large room or a well-ventilated area do not pose a significant hazard. Venting manifolds may be attached to each cell and routed to an exterior location.

The use of vent caps having a flame barrier feature has increased; although such vent caps are designed to inhibit ignition of gases within the battery by external ignition sources, it is advisable to keep sparks, flames, or other ignition sources well away from the battery. Anyone in the vicinity of the battery when it explodes could receive injuries, including eye injury from flying pieces of the case or cover or acid thrown from the battery.

A catalytic recombiner cap (e.g., Hydrocap) may be attached to each cell to recombine some of the hydrogen with oxygen in the air to produce water. If these combiner caps are used, they will still require occasional maintenance. If hydrogen gas remains a concern, the batteries may be installed in a battery box with outside venting. It is rarely necessary to use forced ventilation.

Certain charge controllers are designed to minimize the generation of hydrogen gas by keeping the battery voltage from climbing into the vigorous gassing region where the high volume of gas causes electrolyte to bubble out of the cells. However, lead-acid batteries need periodic overcharging to equalize the cells. This produces gassing that should be dissipated.

In no case should charge controllers, switches, relays, or other devices capable of producing an electric spark be mounted in a battery enclosure or directly over a battery bank. Care must be exercised when routing conduit from a sealed battery box to a disconnect. Hydrogen gas may travel in the conduit to the arcing contacts of the switch.

Because batteries expel explosive gases, sparks, metallic objects, flames, burning cigarettes, or other ignition sources should be kept away at all times. Safety goggles and a face shield should always be worn when one is working with batteries.

11.8.3 BATTERY ENCLOSURES

It is recommended that storage batteries for small residential PV systems be placed in battery boxes, especially if children are present. Batteries are capable of generating thousands of Amperes of current when shorted. A short circuit in a conductor not protected by overcurrent devices can melt tools, battery terminals, and cables. Exposed battery terminals and cable connections must be protected. Batteries should be accessible by only qualified persons. A battery or acid should not be placed within the reach of children. A locked room, battery box, or other container will minimize hazards from short circuits and electric shock. The NEC generally requires about 3 ft of space around battery enclosures and boxes for servicing (Article 110-16). Battery voltages must be less than 50 V in residences unless specific protective criteria are met (Article 690-71). It is recommended that live parts of any battery bank be guarded (Article 690-71b(2)).

11.9 DETERMINATION OF BATTERY FAILURE

After a battery has been tested and found to be defective, a determination can be made to find the cause of any electrical system problems. Important factors in determining the cause of battery failure are battery application, installation, service history, condition, and age (Figures 11.13 and 11.14). Factors in the following sections will aid in making an accurate determination as to battery failure (Battery Council International 1987).

11.9.1 BATTERY APPLICATIONS AND INSTALLATION

- Is the battery being used in the application for which it was designed (i.e., PV, automotive, golf cart, electric)? For example, an automotive battery used for deep-cycle service in a PV system will greatly shorten the lifetime of the automotive battery.
- Is the battery sized properly for the application?
- Does the PV system have excessive electrical requirements for which it was not originally designed? If so, additional battery capacity may be required.
- Do the battery cables fit the battery terminals properly and are they properly adjusted and cleaned? Is there proper clearance for terminals from metallic parts?

11.9.2 BATTERY SERVICE HISTORY

Obtaining the service history of the battery and any history of problems for a system may help in determining the cause of failure.

- Has the battery been used in applications other than the present one? Other applications may have adversely affected battery life.
- Has the PV electrical system been repaired or altered recently and is it in proper operating condition? Charging system operation has a significant effect on battery life.

FIGURE 11.13 Improper bolts used for battery terminal connections that melted. Larger and appropriate battery cables or copper bars are required for proper and safe connections.

FIGURE 11.14 Exploded battery at Chiapas residence from ungrounded PV system hit by lightning.

- Has the PV system been supplying ample charging to the battery? Batteries self-discharge with time; extended periods of undercharge may have a detrimental effect on battery life.
- Has the battery required frequent water additions in one or more cells? Excessive water loss in one cell may indicate a short. Excessive water loss in all cells may indicate overcharging, a worn out battery, or both.

11.9.3   Visual Inspection

Visual inspection of the battery may reveal signs of the abuse that may have caused failure.

- Do the terminals show signs of having been hammered, twisted, or driven down into the cover? Even minor abuse can cause internal damage.
- Does the container or cover show signs of stress, breakage, high temperature, or vibration damage, which might have caused leakage or internal damage?
- Are the vents installed properly? Are they plugged with foreign material? Improperly installed, missing, or plugged vents can be a cause of explosions, leakage, or contamination.
- Is there excessive buildup of acid or foreign material on the cover? A buildup of foreign material mixed with acid around or between the posts can cause high self-discharge rates or inadequate recharge.
- Are electrolyte levels below the tops of the plates in any cell? This could indicate over-charging, lack of maintenance, or internal shorts.
- Is the electrolyte cloudy, discolored, or contaminated with foreign material? Cloudy electrolyte can indicate active material shedding due to overcharge or vibration. Electrolyte contamination can cause high self-discharge rates and poor performance.
- Are the separators cracked or broken below the vent openings? Misuse of hydrometers or other tools could cause cell shorts.
- Are alternate plates dark and light colored? In a charged cell, the positive plates should be dark in color and the negative plates light. If all plates are very light, severe undercharging could be indicated.

11.9.4   Battery Age

A battery's age can be an important factor in determining the cause of failure. The length of time in service determines whether the battery failed prematurely or simply wore out. All battery manufacturers date their product by stamping a data code into the cover or container. This code can be used to determine the age of the battery. Often the month is coded chronologically A through M (excluding I), and the numerals zero through nine indicate the last digit of the year. The individual manufacturer should be consulted regarding its specific date codes. More important is the date the battery was sold. This date determines the time the battery has been in service. The date of purchase is usually indicated on a label with the month and year.

11.9.5   Overcharging and Undercharging

The PV charging system can have a profound effect upon the life of a battery. A high-voltage setting can cause excessive gassing and water loss, thermal runaway, and eventual damage to plates and separators. If the voltage setting is too low, the battery will be in a constant state of discharge. If this happens over a long period of time, the sulfate that deposits on the plates can become hard and crystalline. The plates may not accept a charge under normal conditions and may even cause short circuits through the separators due to a buildup of lead sulfate through the pores, which is converted to lead shorts during recharge.

Voltage settings vary among charge controller manufacturers and may not be adjustable. They should be checked with the individual manufacturer. Different battery types using different grid alloys and manufacturing processes may require different charge settings.

11.9.6 INTERNAL EXAMINATION

Internal examinations of batteries should be conducted with extreme caution and only by knowledgeable personnel. An internal examination of a battery should not be attempted without proper protective clothing and tools. First, an attempt is made to charge the battery fully. The specific gravity of the electrolyte in each cell is recorded. The battery open-circuit voltage is recorded. All cell voltages should be recorded if possible. The battery is allowed to stand for 3 days and the specific gravity readings are recorded a second time. An excessive specific gravity drop (35 points) in one or more cells is an indication that shorts exist in those cells.

The plates and separators of a battery that has been in service for a long period of time may have a generally worn out appearance. On the other hand, the cause of failure may not be evident. The low performance could have been caused by a processing error, unsatisfactory expander, high-resistance separators, etc., which cannot be detected visually. These more subtle causes of failure can be detected only by examining the battery with sophisticated electrical or chemical tests.

11.9.7 CONTAINER

If container abrasion is noted, the battery may have been subjected to severe vibration (in transport), the plates will wear notches in the element rests, and/or the rests will have worn deep notches in the bottom of the separators. Another problem area concerns a battery that has required repeated recharging, but the charging system is fine; although rare, it is possible that a cracked partition is causing the discharge. Two adjacent cells with gravity readings considerably lower than the others is a good indication that electrical leakage exists between the two cells and is discharging them.

A crack in the container and/or cover could be due to abuse (an impact blow), freezing, or an explosion. An exploded battery will often have a piece or pieces of the container or cover missing or torn outward. External electrolyte leakage can generally be detected easily. If there is doubt concerning whether a leak actually exists, the battery should be washed, dried, and set on a clean, dry piece of paper overnight. If there is a leak in the wall of the container, the wet spot will reappear. If there is a leak in the bottom of the container, it will produce a wet spot on the paper.

Another bad sign is if the battery container is distorted; this could be due to a top hold-down that was too tight. High temperatures may permit the container to bulge. The container materials become softer when heated to high temperatures and may distort under the steady pressure of the weight of liquid in the cells.

11.9.8 ELECTROLYTE

The specific gravity of the electrolyte should be recorded to identify the failing cell or cells. These cells will have the lowest specific gravity readings. The cell with the lowest reading will be examined first to attempt to locate the cause of failure. If the electrolyte had a "muddy" appearance when the specific gravity readings were taken, the battery probably failed due to the shedding of the active material or vibration damage. An internal examination of the battery would reveal which condition caused the failure. If electrolyte is not clear, but rather has a color other than the muddy appearance or emits an odor, it may contain an impurity that has caused the battery to fail.

11.10 BATTERY SELECTION CRITERIA

In selecting a battery, the following factors should be considered:

- mode of operation;
- charging characteristics and specific needs;
- required days of storage (autonomy);

- amount and variability of load;
- maximum allowable depth of discharge;
- daily depth of discharge requirements;
- accessibility of location;
- ambient temperature and environmental conditions;
- cyclic life and/or calendar life;
- maintenance requirements;
- sealed or unsealed;
- self-discharge rate;
- maximum cell capacity;
- energy storage density;
- number of battery cells/modules in series;
- size and weight;
- gassing characteristics;
- susceptibility to freezing;
- electrolyte concentration;
- availability of auxiliary hardware;
- battery subsystem disconnect arrangement;
- terminal configuration;
- overcurrent protection;
- toxicity and recyclability;
- reputation of manufacturer; and
- cost and warranty.

11.10.1 Battery Procurement Considerations

When batteries are purchased for PV systems, the following may be considered for specifying batteries:

- type of battery (e.g., flooded or sealed lead-acid);
- useful Ampere-hour capacity of battery at a specified current;
- operating temperature (e.g., –15 to 65°C);
- maximum allowable depth of discharge (e.g., 20% DOD);
- average daily depth of discharge (e.g., 5% DOD);
- nominal charging current (e.g., 20 A);
- nominal battery subsystem bus voltage (e.g., 12 V);
- maximum number of strings in parallel;
- terminal and interconnect wiring specification (e.g., stud T872);
- battery cap requirements (e.g., Hydrocaps);
- shipping requirements (e.g., dry shipping); and
- recyclability.

11.10.1.1 Additional Battery Manufacturer Specifications

The following additional information may be provided by the manufacturer in response to a battery specification:

- expected cycle life;
- battery cell/module dimensions;
- battery cell/module weight (unpacked and packed);
- battery subsystem area and volume requirements;
- maximum battery cell charging voltage;

- battery subsystem voltage window;
- equalization charge requirements;
- average energy efficiency per discharge–charge cycle; and
- shipping requirements.

11.10.2 ADDITIONAL BATTERY SYSTEM CONSIDERATIONS

The following additional information may be used for specifying additional battery system components for PV systems:

- state-of-charge and instrumentation provisions;
- structural requirements;
- voltage regulation requirements (see charge controller section); and
- auxiliary equipment and hardware specification.

11.10.2.1 Small-System Considerations

In a small system, the battery is greatly affected by the design of the balance of systems (BOS). Any mismatch that the designer has not accounted for typically reduces battery lifetime significantly. Excessive voltage drops between the controller and the battery can have a detrimental effect on the original function or performance of the controller. Wire size and poor-quality fuses are the most common problems. A charge controller for a small stand-alone system may be identical in functionality to controllers in larger systems. Small problems can result in early battery failures.

11.10.2.2 Large-System Considerations

Typically, large systems do not show BOS mismatches very quickly. The effect of an inadequate charging regime is not felt until 2 years into the life of a 10-year battery. By then it is too late to correct the problem; sulfation from undercharging is difficult or impossible to reverse. There is concern over the minimum charging current required for charging large batteries. When extended days of autonomy are desired, designers must increase the battery Ampere-hour storage capacity and use larger size batteries. It is easy to discharge a battery with thicker plates, but a minimum current is required to recharge it fully.

The charge controllers for larger systems need to control larger currents. This means the control (switching) element needs special attention, along with the algorithm. This translates to different switching techniques; a linear or shunt controller becomes unmanageable. The on/off technique is most predominant. A more feasible regulation method for large currents is the series-interrupting subarray switching technique described earlier.

11.11 CHARGE CONTROLLER TERMINOLOGY

Specific control methods and algorithms vary among charge controllers, yet all have basic parameters and characteristics. Manufacturers' data provide the limits of controller application for PV and load currents, operating temperatures, set points, and set point hysteresis values. Set points may be dependent upon the temperature of the battery and/or controller, as well as the magnitude of the battery current. The four basic charge controller set points are as follows (Harrington 1992; Foster 1994):

Regulation set point (VR). VR is the maximum voltage that a controller allows the battery to reach. At this point a controller will either discontinue battery charging or begin to regulate the amount of current delivered to the battery. Proper selection of this set point depends on the specific battery type, chemistry, and operating temperature. The voltage drop between the battery and the charge controller during peak charging periods may change the actual voltage at which the battery is charged. Temperature compensation of the VR set point is often incorporated in controller design (internal or external probe) and is particularly desirable if battery temperature ranges exceed ±5°C at ambient temperatures (25°C). For flooded lead-acid batteries, a widely accepted temperature compensation coefficient is –5 mV/°C/cell[4] and –6 mV/°C/cell[5] for sealed gel cell batteries. If electrolyte concentration has been adjusted for local ambient temperature (increase in specific gravity for cold environments, decrease in specific gravity for warm environments) and temperature variation of the batteries is minimal, compensation may not be needed.

Regulation hysteresis (VRH). VRH is the voltage span or difference between the VR set point and the voltage at which the full array current is reapplied. The greater the voltage span is, the longer the array current is interrupted from charging the battery. If the VRH is too small, then the control element will oscillate, inducing noise and possibly harming the switching element or any loads attached to the system. The VRH has been shown to be an important factor in determining the charging effectiveness of a controller.

Low voltage disconnect (LVD). LVD represents the nominal voltage at which the charge controller disconnects the load from the battery bank to prevent overdischarge—generally the lowest voltage experienced by the battery bank in the system if all loads operate through the LVD. LVD defines the actual allowable maximum depth of discharge and available capacity of the battery. The available capacity must be carefully estimated in the PV system design and sizing process. LVD does not need temperature compensation unless the batteries operate below 0°C on a frequent basis. The proper LVD set point will maintain good battery health while providing the optimum available battery capacity to the system. The LVD is dependent on the designed maximum allowable depth of discharge and the rate of discharge.

Low voltage disconnect hysteresis (LVDH). LVDH is the voltage span or difference between the LVD set point and the voltage at which the load is reconnected to the battery. If LVDH is too small, the load may cycle on and off rapidly at low battery SOC, possibly damaging the load and/or controller. If LVDH is too large, the load may remain off for extended periods until the array fully recharges the battery. With a large LVDH, battery health may be improved due to reduced battery cycling, but with a reduction in load availability. The proper LVDH selection for a given system will depend on the battery chemistry and size, PV and load currents, and load availability requirements.

11.12   CHARGE CONTROLLER ALGORITHMS

Two basic methods exist for controlling or regulating the charging of a battery from a PV module or array: series and shunt regulation. Although both of these methods can be effectively used, each may incorporate a number of variations that alter basic performance and applicability. Following are descriptions of the two basic methods and variations of these methods (Harrington 1992; Foster 1994).

11.12.1   Shunt Controller

A shunt controller regulates the charging of a battery by interrupting the PV current by short-circuiting the array. A blocking diode is required in series between the battery and the switching element to keep the battery from being shorted when the array is shunted. This controller typically requires a large heat sink to dissipate power. Shunt type controllers are usually designed for applications with PV currents of 20 A or less, due to high-current switching limitations.

The *shunt-interrupting* algorithm terminates battery charging when the VR set point is reached by short-circuiting the PV array. This algorithm has been referred to as "pulse charging" due to the pulsing effect when reaching the finishing charge state. This should not be confused with pulse width modulation (PWM).

The *shunt-linear* algorithm maintains the battery at a fixed voltage by using a control element in parallel with the battery. This control element turns on or closes when the VR set point is reached, shunting power away from the battery in a linear method (not on/off) and maintaining a constant voltage at the battery. A relatively simple controller design utilizes a Zener power diode, which is the limiting element.

11.12.2 SERIES CONTROLLER

There are several variations to this type of controller, all of which use some type of control element in series between the array and the battery:

Series interrupting. This algorithm terminates battery charging at the VR set point by open-circuiting the PV array. A blocking diode may or may not be required, depending on the switching element design and nighttime control. Some series controllers may divert the array power to a secondary load.

Series interrupting, two step, constant current. This is similar to the series-interrupting algorithm; however, when the VR set point is reached, instead of totally interrupting the array current, a limited constant current remains applied to the battery. The designer should know this constant current rate.

Series interrupting, two step, dual set point. This is similar to the series-interrupting algorithm; however, there are two VR set points. A higher set point is only used during the initial charge each morning. The controller then regulates at a lower VR set point for the rest of the day. This allows a daily equalization of the battery.

Series interrupting, pulse width modulated (PWM). The algorithm uses a series element that is switched on/off at a variable frequency with a variable duty cycle to maintain the battery at the VR set point. This can also be accomplished by decreasing the VRH but there needs to be a limit to the frequency of switching. This is similar to the series-linear, constant-voltage algorithm when the integrated current applied to the battery is considered. Power dissipation is significantly reduced with the series-interrupting PWM algorithm.

Series interrupting, subarray switching. These are typically used in systems with more than six PV modules or current greater than 20 A. The array is subdivided into sections and switched individually (three to five subarrays). As the battery becomes charged, the subarrays are switched off in a sequence of voltage steps to reduce current and maintain battery voltage and not overcharge the system. One subarray is directly connected to the batteries, determining the finishing charge for the system. This minimizes the need to use high-current switching gear and reduces problems associated with high current and voltage drops in the system. Modularity provides simple maintenance. As a by-product, when subarrays are switched off charging, the power can be used for a secondary, noncritical load such as water heating.

Series linear, constant voltage. This algorithm maintains the battery voltage at the regulation set point (VR). The series control element acts like a variable resistor in series, with the PV array used to maintain the battery at the VR set point. The current is controlled by the series element and the variable voltage drop across it. Problems with voltage drops between the battery and charge controller, reducing the actual voltage at which the battery is being charged, are minimized with this algorithm. As the battery becomes charged, the current tapers off, reducing the voltage drop between the battery and controller and allowing the battery voltage to increase because of the reduction in voltage drop between the battery and controller. This is the recommended charge algorithm for sealed, valve-regulated batteries.

11.13   CHARGE CONTROLLER SELECTION CRITERIA

The following list is included to provide some basis for determining what specifications need to be addressed when selecting a charge controller (Harrington 1992; Foster 1994). Selection criteria and procurement specifications for charge controllers may include the following:

- long-term reliability;
- type of regulator; number of charging steps;
- maximum array current;
- adjustability of set points and hysteresis;
- optional relays for alarms, backup system start-up, etc.;
- parasitic power consumption (during operation);
- mounting provisions and considerations;
- instrumentation, LEDs, metering, remote display;
- availability of parts;
- other options and accessories;
- reputation of manufacturer and availability for help; and
- cost and warranty.

11.13.1   CHARGE CONTROLLER PROCUREMENT SPECIFICATIONS

The following may be used for specifying charge controllers for PV systems:

- type of regulator; number of charging steps (e.g., constant voltage or on/off);
- type of battery to be charged;
- operating temperature (e.g., 0–70°C);
- nominal charging requirements (e.g., twice the array current);
- operating voltage (e.g., 11–26 V);
- temperature compensation, internal or external sensor;
- voltage regulation (VR) set point (e.g., 14.1 V);
- low-voltage disconnect (LVD) set point (e.g., 11.8 V);
- adjust ability of set points and hysteresis;
- optional relays for alarms, backup system start-up, etc.;
- cycle life (e.g., 4,000 cycles);
- type of switching elements (solid state or relay);
- reverse polarity protection;
- load management features (e.g., priority load shedding);
- overcurrent protection; lightning protection (e.g., 10 A fuse); and
- shipping requirements (e.g., shipping/storage temperature).

11.13.1.2   Additional Charge Controller Manufacturer Specifications

The following additional information may be provided by the manufacturer in response to a charge controller specification:

- voltage regulation hysteresis (VRH);
- low voltage disconnect hysteresis (LVDH);
- parasitic power consumption (during operation);
- input and output terminals, size and type;
- materials, corrosion resistance, NEMA rating;
- dimensions and weight; and
- shipping requirements.

REFERENCES

Battery Council International. 1982. Battery technical manual. Chicago, IL.

Battery Council International. 1987. *Battery service manual.* Chicago, IL.

Foster, R., O. Carrillo, S. Harrington, and S. Durand, 1994. *Battery and charge controllers for photovoltaic systems.* Sandia National Laboratories. Guatemala City: NMSU.

Harrington, S. R. 1992. Balance of system (BOS) workshop: Charge controller technology. SOLTECH '92, Sandia National Laboratories, Design Assistance Center, Albuquerque, NM, February 17, 1992.

Kiehne, H. A. 1989. *Battery technology handbook.* New York: Marcel Dekker, Inc.

Lasnier, F., and T. Gan Ang. 1990. *Photovoltaic engineering handbook.* New York: Adam Hilger.

Sandia National Laboratories, Design Assistance Center. 1990. Working Safely with Photovoltaic Systems. Albuquerque, NM.

Vinal, G. W. 1951. *Storage batteries,* 4th ed. New York: John Wiley & Sons.

PROBLEMS

11.1   Please explain the difference between shallow-cycle and deep-cycle batteries and provide examples of typical applications for each.

11.2   Describe the internal chemistry of the electrolyte and plates of a flooded lead acid battery when it goes from a fully charged state to a completely discharged state.

11.3   What is meant by the term depth of discharge (DOD) and how does it impact battery cycle life?

11.4   A battery bank is comprised of eight 12 V, 100 Ah batteries connected 2 in series and 4 in parallel (2s x 4p). What are the battery bank voltage and the amp-hour energy capacity?

11.5   Describe the problems associated with excessive undercharging and overcharging of a flooded lead-acid battery.

11.6   Discuss how the pulse-width modulation algorithm is used for battery charging.

Solar Energy Glossary

BATTERIES

Active Material (Battery): Lead dioxide in the positive plates and metallic sponge lead in the negative plates that reacts with sulfuric acid during charging and discharging of a lead-acid battery.

Capacity (Battery): The ability of a fully charged battery to deliver a specified quantity of electricity (ampere-hour, Ah) at a given rate (ampere, A) over a definite period of time (hour). The capacity of a battery depends upon a number of factors, such as active material weight, density, adhesion to grid, number, design and dimensions of plates, plate spacing, design of separators, specific gravity and quantity of available electrolyte, grid alloys, final limiting voltage, discharge rate, temperature, internal and external resistance, age, and life history of the battery.

Cell (Battery): The basic electrochemical current-producing unit in a battery, consisting of a set of positive plates, negative plates, electrolyte, separators, and casing. There are six cells in a 12 V lead-acid battery.

Corrosion (Battery): The action of liquid electrolyte on a corrodible material (e.g., dilute sulfuric acid on steel), producing corrosion products, such as rust. Battery terminals are sometimes subjected to corrosion.

Cycle (Battery): In a battery, one discharge plus one recharge equals one cycle.

Days of Autonomy: The maximum length of time that the PV system can provide power to the load in the absence of solar power provided from the PV array. Directly related to the "usable" battery capacity and the average daily AC energy required by the load. Depending on local climatic conditions, 3–5 days is commonly considered adequate for a residential stand-alone PV system.

Discharging: When a battery is delivering current, it is discharging.

Efficiency: The ratio of how much energy derived out of a system compared to how much energy was put in (efficiency = power out/power in).

Electrolyte: In a lead-acid battery, the electrolyte is sulfuric acid diluted with water. It is a conductor and a supplier of water and sulfate for the electrochemical reaction
$$PbO_2 + Pb + 2H_2SO_4 \leftrightarrows 2PbSO_4 + 2H_2O.$$

Element: In a battery, a set of positive and negative plates assembled with separators.

Forming: During manufacturing, this is the process of charging the battery for the first time. Electrochemically, forming (also known as formation) changes the lead oxide paste on the plate grids into lead dioxide in the positive plates and to metallic sponge lead in the negative plates.

Grid (Battery): A lead alloy framework that supports the active material of a battery plate and conducts current.

High-Voltage PV Array Disconnect (HVD): For voltage set point driven controllers, the HVD is the nominal voltage at which the charge controller disconnects the PV array from the battery bank to prevent overcharge. This is generally the highest voltage experienced by the battery bank in the system and is also referred to as the voltage regulation (VR) set point. The HVD is selected based on battery type, chemistry, and battery temperature.

Hydrometer: A float type instrument used to determine the state of charge of a battery by measuring the specific gravity of the electrolyte (i.e., the concentration of sulfuric acid in the electrolyte).

Load Reconnect Voltage (LRV): The LRV is the nominal voltage at which the charge controller reconnects the load to the battery bank. The LVD subtracted from the LRV will give the load voltage disconnect hysteresis (LVDH).

Load Tester: An instrument that draws current (discharges) from a battery using an electrical load while measuring voltage. It determines the battery's ability to perform under actual loading conditions.

Load Voltage (V): The nominal voltage at which the system and charge controller operate—generally, the nominal battery bank voltage for the PV systems. Systems with higher load power demands generally dictate a higher nominal voltage to reduce voltage drop and losses in the system and allow use of smaller wire sizes.

Load Voltage Regulation: Regulating the load voltage may be needed for loads that are voltage input sensitive, such as in radios or consumer electronics; the supply to the load may need to be controlled. The battery voltage (VR) can sometimes range from 14 to >15 V when equalizing to less than 11.5 (V_{oc} when a battery fuse blows on poorly designed systems). The load may blow fuses or experience nuisance automatic turn-offs (typically transmitters).

Low-Voltage Disconnect (LVD): LVD represents the nominal voltage at which the charge controller disconnects the load from the battery bank to prevent overdischarge, generally the lowest voltage experienced by the battery bank in the system if all loads operate through the LVD. LVD defines the actual allowable maximum depth of discharge and available capacity of the battery. The available capacity must be carefully estimated in the PV system design and sizing process. LVD does not need temperature compensation unless the batteries operate below 0°C on a frequent basis. The proper LVD set point will maintain good battery health while providing the optimum available battery capacity to the system. The LVD is dependent on the designed maximum allowable depth of discharge and the rate of discharge.

Low-Voltage Disconnect Hysteresis (LVDH): LVDH is the voltage span or difference between the LVD set point and the voltage at which the load is reconnected to the battery. If LVDH is too small, the load may cycle on and off rapidly at low battery state of charge (SOC), possibly damaging the load and/or controller. If LVDH is too large, the load may remain off for extended periods until the array fully recharges the battery. With a large LVDH, battery health may be improved due to reduced battery cycling, but with a reduction in load availability. The proper LVDH selection for a given system will depend on the battery chemistry and size, PV and load currents, and load availability requirements.

Open Circuit Voltage (Battery): This is the voltage of a battery when it is not delivering or receiving power. This voltage is 2.11 V for a typical fully charged lead-acid battery cell.

Peak Load Current (A): The maximum load current at which the charge controller can operate for a short period, typically less than a few seconds. The load switching elements must be able to handle surge currents to devices such as pumps and compressor motors.

Primary Battery: This type of battery can store and deliver electrical energy, but cannot be recharged.

Rate of Charge (or Discharge): The rate of charge or discharge of a battery is expressed in Amperes as the battery's rated capacity divided by a time factor.

Regulation Hysteresis (VRH): VRH is the voltage difference between the VR set point and the voltage at which the full array current is reapplied. The greater the voltage difference, the longer the array current is interrupted from charging the battery. If the VRH is too small, then the control element will oscillate, inducing noise and possibly harming the switching element or any loads attached to the system. The VRH has been shown to be an important factor in determining the charging effectiveness of a controller.

Regulation Set Point (VR): VR is the maximum voltage that a controller allows the battery to reach. At this point, a controller will either discontinue battery charging or begin to regulate the amount of current delivered to the battery. Proper selection of this set point depends on the specific battery type, chemistry, and operating temperature. The voltage drop between the battery and the charge controller during peak charging periods may change the actual voltage to which the battery is charged. Temperature compensation of the VR set point is often incorporated in controller design (internal or external probe) and is particularly desirable if battery temperature ranges exceed ±5°C at ambient temperatures (25°C). For flooded lead-acid batteries, a widely accepted temperature compensation coefficient is –5 mV/°C/cell and –6 mV/°C/cell for sealed gel cell batteries. If electrolyte concentration has been adjusted for local ambient temperature (increase in specific gravity for cold environments, decrease in specific gravity for warm environments) and temperature variation of the batteries is minimal, compensation may not be needed.

Secondary Battery: A battery that can store and deliver electrical energy and be recharged by passing direct current through it in a direction opposite to that of discharge.

Separator: A divider between the positive and negative plates that allows the flow of current to pass through it.

Set Point Adjustment: Adjustments range from simple VR settings to independent adjustment of VR and VRH. The same applies to the LVD and LVDH. The ability to adjust set points varies in each product. The need to adjust set points should be considered along with the methods, skill, and tools required. Adjustment mechanisms vary from small "DIP" switches (miniature switches in a small package) and movement of jumpers to adjustments with a potentiometer. Jumpers are durable and the setting can be soldered into place to remove the environmental problems of DIP switches.

Set Points: These are the battery charge controller voltage points for PV array charging. The need to adjust set points should be considered along with the methods, skill, and tools required. Adjustment mechanisms vary from small "DIP" switches (miniature switches in a small package) and movement of jumpers to adjustments with a potentiometer.

Shunt (Interrupting): This algorithm terminates battery charging when the VR set point is reached by short-circuiting the PV array. This algorithm has been referred to as "pulse charging" due to the pulsing effect when reaching the finishing charge state. This should not be confused with pulse width modulation (PWM).

Shunt (Linear): This algorithm maintains the battery at a fixed voltage by using a control element in parallel with the battery. This control element turns on or closes when the VR set point is reached, shunting power away from the battery in a linear method (not on/off) and maintaining a constant voltage at the battery. A relatively simple controller design utilizes a Zener power diode, which is the limiting element.

Specific Gravity: The density of a liquid compared with water density. The specific gravity of the electrolyte is the weight of the electrolyte compared to the weight of an equal volume of pure water (the specific gravity of water is 1.0).

State of Charge: The amount of electrical energy stored in a battery at a given time expressed as a percentage of the energy when fully charged. A battery that has its entire capacity available is at a 100% state of charge; a battery with half its capacity removed is at a 50% state of charge.

Temperature Compensation: This is a very necessary feature if batteries will be operating in temperature exceeding 25°C with ±5°C swings. Typically, the VR is temperature compensated. The best temperature compensation is with an external probe mounted on the side of the battery midway or on the positive terminal post. However, compensation at the controller (internally) is better than nothing. Temperature compensation is particularly important for sealed batteries and batteries where water loss must be minimized.

Temperature Compensation Coefficients: Coefficients in charge controllers range from –3mV/°C/cell to –5mV/°C/cell, with –6mV/°C/cell recommended for flooded lead-antimony and lead-calcium batteries. Sealed gel and AGM type batteries require a more aggressive temperature compensation coefficient—typically, up to –6mV/°C/cell.

ELECTRICITY

AC: Alternating current (AC) is the standard form of electrical current supplied by the utility grid and by most fuel-powered generators. The polarity (and therefore the direction of current) alternates. In the United States, standard voltages for small water pumps are 115 and 230 V. Standards vary in different countries. See *inverter.*

Ampere (Amp, A, I): The unit of measure of electron flow rate or current through a circuit. Current that flows in a single direction is direct current (DC); current that changes direction is alternating current (AC).

Ampere-Hour: This is a unit of measure for energy capacity, obtained by multiplying the current in Amperes by the time in hours.

Amp-Hour, Ah, or Amp-Hr: This is an engineering unit to describe energy flow into and out of a PV cell or battery. The unit is not an exact measure of energy in that a voltage must be associated with the value (e.g., a battery that can deliver 5 A for 20 hours is 100 Ah of capacity).

Circuit: An electric circuit is the path of an electric current. A closed circuit has a complete path. An open circuit has a broken or disconnected path.

Circuit (Parallel): A circuit that provides more than one path for current flow. A parallel arrangement of batteries (usually of like voltage and capacity) would have all positive terminals connected to a conductor and all negative terminals connected to another conductor. If two 12 V batteries of 50 Ah capacity each are connected in parallel, the circuit voltage is 12 V, and the Ampere-hour capacity of the combination is 100 Ah.

Circuit (Series): This describes a circuit that has only one path for the current to flow. Batteries arranged in series are connected with the negative of the first to the positive of the second, negative of the second to thepositive of the third, etc. If two 12 V batteries of 50 Ah capacity each are connected in series, the circuit voltage is equal to the sum of the two battery voltages or 24 V, and the Ampere-hour capacity of the combination is 50 Ah.

Converter: This is a electronic device for DC power that steps up voltage and steps down current proportionally (or vice versa).

Coulomb: Charge, Q or q, in Coulombs (C). 1 C is a very large number of electrons: $1\ e = 1.6 \times 10^{-19}$ C, positive or negative

Current: This is rate of flow of electricity or the movement rate of electrons along a conductor. Measured in Amperes, commonly called Amps. Current $(I) = dq/dt,$ one number of charges moving past a point in 1 s; Ampere = Coulomb/second. An analogy for current is water flow rate in a water pipe.

Current (Alternating, AC): This is a current that varies periodically in magnitude and direction. A battery does not deliver alternating current (AC).

Current (Direct, DC): An electrical current flowing in an electrical circuit in one direction only. A battery delivers direct current (DC) and must be recharged with DC in the opposite direction of the discharge.

DC: Direct current (DC) is the type of power produced by PV panels and by storage batteries. The current flows in one direction and the polarity is fixed, defined as positive (+) and negative (–). Nominal PV system voltage ranges from 12 to 480 V.

Drop (Voltage): The net difference in the electrical potential (voltage) when measured across a resistance or impedance (Ohms) is the drop.

Efficiency: This is percentage of power that gets converted to useful work. For example, an electric pump that is 60% efficient converts 60% of the input energy into work—pumping water. The remaining 40% becomes waste heat.

Electron Volt: A unit of energy: $1\ eV = 1.6 * 10^{-19}$ J.

Energy: The product of power and time, measured in Watt-hours. 1,000 Wh = 1 kWh (kiloWatt-hour). Variation: The product of current and time is Ampere-hours (Ah). 1,000 W consumed for 1 hour = 1 kWh. Energy = V * Q. See *power*.

Ground: The reference potential of a circuit with respect to the Earth.

Inverter: An electronic device that converts low-voltage DC to high-voltage AC power. In solar-electric systems, an inverter may take the 12, 24, or 48 V DC and convert it to 110 or 220 V AC conventional household power.

Negative: Designating or pertaining to electrical potential (e.g., the negative battery terminal is the point from which electrons flow during discharge).

Ohm: A unit for measuring electrical resistance.

Ohm's Law: Expresses the relationship of Volts (V) and Amperes (A) in an electrical circuit with resistance (R). It can be expressed as follows: V = IR.

Positive: Designating or pertaining to a kind of electrical potential; opposite of negative (e.g., the positive battery terminal).

Power: The rate at which work is done (Joules/second). It is the product of voltage × current and measured in Watts. Power (P) = V * I; Watt = V * A; 1,000 W = 1 kW. An electric motor requires approximately 1 kW per horsepower (after typical efficiency losses).

Resistance (Electrical): The opposition to the free flow of current in a circuit. Measured in Ohms. Resistance (R) = V/I; Ohm = V/A.

Short Circuit: An unintended current bypass in an electric device or wiring, generally very low in resistance and thus causing a large current to flow.

System Load (kWh/d): The daily energy required to operate the energy-consuming devices (load) attached to the PV system. Depending on the system, the energy required may be either AC or DC energy.

System Load Control: The system component that controls when the system AC load is electrically disconnected from the system, usually to prevent damage to the battery bank. This component usually establishes the set points for V_{LVD}, V_{LVR}, and V_{HVD}.

Three-Phase Power AC: Three-phase power is AC that is carried by three wires in which the voltage in each two-wire combination is 120° ahead of or behind that in any other two-wire combination. Power delivery is smoother and more efficient than that of single-phase AC, and motors start more easily.

Transformer: An electrical device that steps up voltage and steps down current proportionally (or vice versa). Transformers work with AC only. For DC, see *converter*. Mechanical analogy: gears or belt drive.

Utility Grid: Commercial electric power distribution system, Frequency provided at 60 Hz in the United States and 50 Hz in Europe.

Voltage: The measurement of electrical potential. Electric potential (V) = energy/charge (V = E/Q, Volt = Joule/Coulomb). Analogy: pressure in a water pipe. Volts (V) = Amperes (I) × Ohms (R)

Voltage Drop: Loss of voltage (electrical pressure) caused by the resistance in wire and electrical devices. Proper wire sizing will minimize voltage drop, particularly over long distances. Voltage drop is determined by four factors: wire size, current (amperes), voltage, and length of wire. It is determined by consulting a wire sizing chart or formula available in various reference tests and is expressed as a percentage. Water analogy: friction loss in a pipe.

Voltage (Nominal): A way of naming a range of voltage to a standard. For example, a 12 V nominal system may operate in the range of 11–15 V. We call it 12 V for simplicity.

Watt: The unit for measuring electrical power (Joule/second)—that is, the rate of doing work in moving electrons by or against an electrical potential (Watts = Amperes × Volts).

Watt-Hour (Wh): The unit for measuring electrical energy over time, which equals Watts × hours.

PHOTOVOLTAICS

Array: The PV (solar to electric) system component composed of separate PV modules wired together in series and/or parallel. The PV modules are in turn composed of individual solar cells that are wired in series-connected strings within the module.

Array Maximum Power (P_{mp}): The maximum power available from the PV array at a given environmental operating condition, occurring at the maximum power point on the current voltage (I-V) curve.

Array Maximum-Power Voltage (V_{mp}): The voltage corresponding to the maximum power point on the array's current–voltage (I-V) curve.

Array Open-Circuit Voltage (V_{oc}): The voltage produced by the PV array in an open-circuit condition.

Array Power Rating at SRC (or STC) (P_{mp}): The maximum power available from the PV array at the standard reporting condition (SRC) specified by ASTM. The SRC commonly used by the PV industry (or standard test condition, STC) is for a solar irradiance of 1,000 W/m^2, a PV cell temperature of 25°C, and a standardized solar spectrum referred to as an air mass 1.5 spectrum (AM = 1.5).

Array Utilization: The percentage of daily DC energy available from the PV array that is actually used by the system. This value provides a gauge of how well the array's power-conditioning system tracks the maximum power point of the PV current–voltage (I-V) curve.

Band Gap: The gap between valence energy band and conduction energy band.

Daily Array-to-Load Energy Ratio (A:L): A ratio used to gauge the daily energy available from the PV array relative to the daily energy required by the load attached to the system. The ratio is both site dependent and system design dependent. For a resistive load, the daily average A:L ratio can be calculated for each month of the year. For a system with an AC load, the calculated A:L ratio will be lower than for a system with an equivalent DC load because of the energy losses associated with the inverter. For system design purposes, the A:L for the winter month with the lowest solar resource is typically used. For systems with a DC load, this design A:L ratio is typically in the range of 1.3–1.6. For systems with an AC load, the A:L ratio is typically in the range of 1.4–2.0.

Daily Array Efficiency (η_{PV}): The ratio of the daily energy available from the PV array at its maximum power point divided by the daily total solar insolation on the array; varies seasonally.

Daily Inverter Efficiency (η_{INV}): The ratio of the daily AC energy provided by the inverter divided by the total energy provided to the inverter from the battery and/or PV array.

Daily MPPT Efficiency (η_{MPPT}): The ratio of the daily energy actually provided by the PV array divided by the total daily energy available from the array if operated at its maximum power point.

Daily System Efficiency (η_{SYS}): The ratio of the daily AC energy provided by the inverter divided by the total daily energy available from the array if operated at its maximum power point.

Design Load (kWh/d): PV system design and optimization requires an accurate definition of the expected daily AC energy required from the system (load). The worst-case situation for a PV stand-alone system is typically the winter months, when the solar resource is minimal. Therefore, the design load is typically chosen as the daily AC energy expected on a typical winter day.

Design Month: The month chosen during system design to ensure that the PV system adequately meets the system load over the entire year. Typically, the design month is one of the winter months when the solar resource is lowest.

Diode: An electronic device that permits unidirectional current.

Electrical Inverter: The system component that converts direct current (DC) electrical energy from the PV array or battery to alternating current (AC) electrical energy required by the system load.

Equipment-Grounding Conductor: A conductor attached to metal surfaces of equipment that does not normally carry current, except during a fault condition. It is connected to earth ground and helps prevent electrical shocks and also helps overcurrent devices to operate properly.

Grounded: Term that indicates parts of an electrical system that are connected to an earth ground.

Grounded-Circuit Conductor: An electrical conductor that normally carries current in the system circuit that is connected to earth ground. Examples are the neutral conductor in AC wiring and the negative conductor in a PV array. Note that this conductor is distinct from the equipment-grounding conductor, which carries no current during normal operation.

Grounding Bond: In common usage, "bond" refers to the connection of the grounded conductor, the equipment-grounding conductor, and the grounding electrode conductor. Often a single common grounding point in the system.

Grounding Electrode: The ground rod or metallic device used to make physical contact with the Earth; it is typically a 5/8 in. diameter, 8 ft long copper rod.

Grounding-Electrode Conductor: The electrical conductor (wire) between the common single grounding point in a PV system and the grounding electrode.

Grounding-Electrode System: A wiring scheme with two or more grounding electrodes connected together. An example would be a home with an existing AC-grounding electrode when a new DC-grounding electrode is added for a PV system.

Heterojunction: The interface that occurs between two layers or regions of dissimilar crystalline semiconductors. These semiconducting materials have unequal band gaps as opposed to a homojunction, which is made from the same semiconductor material.

High-Voltage PV Array Disconnect (HVD): For voltage set point driven controllers, the HVD is the nominal voltage at which the charge controller disconnects the PV array from the battery bank to prevent overcharge. This is generally the highest voltage experienced by the battery bank in the system and is also referred to as the voltage regulation (VR) set point. The HVD is selected based on battery type, chemistry, and battery temperature.

Intrinsic Carrier: A semiconductor with valence band holes and conduction band electrons present in equal numbers

Load Current (A): The rated load current at which the charge controller can operate on a continuous basis.

Load Reconnect Voltage (LRV): The LRV is the nominal voltage at which the charge controller reconnects the load to the battery bank. The LVD subtracted from the LRV will give the load voltage disconnect hysteresis (LVDH).

Majority Carrier: The charge carrier that determines current. Majority carriers in a p-type material are holes and therefore its minority carriers are electrons. Majority carriers in an n-type material are electrons and its minority carriers are holes.

Nominal Load Current (A): The rated load current at which the charge controller can operate on a continuous basis.

Nominal System Voltage (V): The nominal voltage at which the system and charge controller operate—generally, the nominal battery bank voltage for the PV systems. Systems with higher load-power demands generally dictate a higher nominal voltage to reduce voltage drop and losses in the system and allow use of smaller wire sizes.

Photovoltaic: The phenomenon of converting light to electric power. Photo = light; Volt = electricity.

PV: The common abbreviation for photovoltaics.

PV Array: A group of PV modules (also called panels) arranged to produce the voltage and power desired.

PV Array—Direct: The use of electric power directly from a PV array, without storage batteries to store or stabilize it. Most solar water pumps work this way, utilizing a tank to store water.

PV Array Reconnect Voltage (RCV): The RCV is the nominal voltage at which the charge controller reconnects the PV array to the battery bank to resume charging. The HVD subtracted from the RCV will yield the regulation hysteresis (VRH).

PV Cell: The individual PV device. Most PV modules are made with around 36 or 72 silicon cells, each producing about 0.5 V.

PV Module: An assembly of PV cells framed into a weatherproof unit. Commonly called a PV panel. See *PV array.*

Short-Circuit Current (I_{sc}): The maximum PV array current at which the charge controller can operate for a short period, up to several minutes about 125% of array shout circuit current (I_{sc}). The duration and magnitude of the peak current is dependent on the PV array rating and the potential for higher than normal irradiance values due to reflection from the ground, snow, or edges of clouds. Relays and switches in the controller must be capable of handling this current.

Solar Tracker: A mounting rack for a PV array that automatically tilts to follow the daily path of the Sun through the sky. A tracking array will produce more energy through the course of the day than a fixed array (nontracking), particularly during the long days of summer.

Voltage, Open Circuit (V_{oc}): The voltage of a PV module or array with no load (when it is disconnected). The maximum PV open-circuit voltage that may be applied to a charge controller. A 12 V nominal PV module will produce about 20 V open circuit.

Voltage, Peak, or Maximum Power Point(V_{pp} or V_{mp}): The voltage at which a PV module or array transfers the greatest amount of power (Watts). A 12 V nominal PV module will typically have a peak power voltage of around 15–17 V. The solar array for a PV array-direct solar pump should reach this voltage in full-sun conditions or a multiple of this voltage.

Wave Function: Description of an electron using kinematic rather than spatial point descriptors. Similar to that used to describe sound and electromagnetic waves; however, whereas those need some material medium in order to propagate, the wave function describes the particle, although the function itself cannot be defined in terms of anything material. It can only be described by how it is related to physically observable effects.

SOLAR ENERGY CONCEPTS

Absolute Air Mass (AM_a): A dimensionless term used to describe the optical depth, or path length, that sunlight must traverse through the atmosphere before reaching the ground. When adjusted for the altitude or atmospheric pressure of a site, it is called absolute or pressure corrected. The reference value of 1.0 is for a site at sea level with the sun directly overhead at solar noon.

Daily (Peak) Sun-Hours: This is an alternative term used to quantify the daily solar insolation. In this case, the daily solar insolation in kiloWatt hours per square meter per day is divided by the standard solar irradiance of 1,000 W/m^2 to give units of hours per day.

Daily Solar Insolation (kWh/m^2): The cumulative daily solar irradiance in the plane of the PV array. This quantity is either measured directly or calculated from typical meteorological year (TMY) data for a specific geographic location. Sometimes expressed in sun-hours, where the daily insolation is divided by the standard solar irradiance of 1,000 W/m^2. That is, 5 kWh/m^2/d is referred to as 5 sun-hours.

Plane-of-Array Irradiance (I_{poa}, W/m^2): The total (global) solar irradiance in the plane of the PV array, measured using a pyranometer.

Solar Angle of Incidence (AOI, degrees): The angle between the direct beam from the Sun and a line perpendicular (normal) to the surface of the PV array.

Standard Reporting Condition (SRC or STC): The reference condition used by the PV industry for rating the power from PV modules, which has been standardized by organizations such as ASTM, IEEE, IEC, UL, and others. This condition has a solar irradiance of 1,000 W/m², PV cell temperature of 25°C, and a solar spectral distribution specified for an air mass equal to 1.5 (AM = 1.5). The condition is also commonly referred to as the standard test condition (STC).

SOLAR WATER-PUMPING

Booster Pump: A surface pump used to increase pressure in a water line or to pull from a storage tank and pressurize a water system. See *surface pump.*

Borehole (or Tube Well): Synonym for drilled well, especially outside North America.

Cable Splice: A joint in electrical cable. A submersible splice is protected by a water-tight seal.

Casing (Well): Plastic or steel tube that is permanently inserted in the well after drilling. Its size is specified according to its inside diameter.

Centrifugal Pump: A pumping mechanism that spins water in order to push it out by means of centrifugal force. See also *multistage centrifugal.*

Check Valve: A valve that allows water to flow one way but not the other.

Cut-in Pressure and Cut-out Pressure: Pressure (Pascal) points where automated pumping system will begin and end pumping. See *pressure switch.*

DC Motor (Brushless): High-technology motor used in more advanced solar submersibles. An electronic system is used to alternate the current precisely, causing the motor to spin. A submersible brushless motor is filled with water and requires no maintenance.

DC Motor (Brush Type): The traditional DC motor, in which small carbon blocks called brushes conduct current into the spinning portion of the motor. They are used in most solar surface pumps and in some low-power solar submersibles. The motor chamber must be filled with air and perfectly sealed from moisture. Brushes naturally wear down after years of use and must be replaced periodically.

DC Motor (Permanent Magnet): All DC solar pumps use this type of motor in some form. Because it is a variable speed motor by nature, reduced voltage (in low sun) produces proportionally reduced speed and causes no harm to the motor. Contrast: *induction motor.*

Diaphragm Pump: A type of pump in which water is drawn in and forced out of one or more chambers by a flexible diaphragm. Check valves let water into and out of each chamber.

Drawdown: Lowering of level of water in a well due to pumping.

Driller's Log: The document in which well characteristics are recorded by the well driller. In most states, drillers are required to register all water wells and to send a copy of the log to a state office. This supplies hydrological data and well performance test results to the well owner and the public.

Drop Pipe (Well): The pipe that carries water from a pump in a well up to the surface. It also supports the pump.

Float Switch: An electrical switch that responds to changes in water level. It may be used to prevent overflow of a tank by turning a pump off or to prevent a pump from running dry when the source level is low.

Float Valve: A valve that responds to changes in water level. It is used to prevent overflow of a tank by blocking the flow of water.

Foot Valve: A check valve placed in the water source below a surface pump. It prevents water from flowing back down the pipe and losing prime. See *check valve* and *priming.*

Friction Loss: The loss of pressure due to flow of water in a pipe. This is determined by four factors: pipe size (inside diameter), pipe material, flow rate, and length of pipe. It is determined by consulting a friction loss chart, available in an engineering reference book or from a pipe supplier. It is expressed in pounds per square inch or feet (equivalent additional feet of pumping). See Appendix B.

Gravity Flow: The use of natural gravity to produce pressure and water flow. A storage tank is elevated above the point of use so that water will flow with no further pumping required. A booster pump may be used to increase pressure. 2.31 vertical ft = 1 psi. 10 vertical m = 1 bar. See *pressure*.

Head: Total lift of water over a distance; this may also include friction (head) losses, measured in meters. In water distribution, a synonym is vertical drop or vertical lift. See *vertical lift* and *total dynamic head*.

Impeller: The round device that spins inside a centrifugal pump to push water upward; often staged in series in order to develop centrifugal force.

Induction Motor (AC): The type of electric motor used in conventional AC water pumps. It requires a high surge of current to start and a stable voltage supply, making it relatively expensive to run by solar power. See *inverter*.

Jet Pump: A surface-mounted centrifugal pump that uses an "ejector" (venturi) device to augment its suction capacity. In a deep well jet pump, the ejector is down in the well to assist the pump in overcoming the limitations of suction. (Some water is diverted back down the well, causing an increase in energy use.)

Linear Current Booster (LCB): An electronic device that varies the voltage and current of a PV array to match the needs of an array-direct pump, especially for a positive displacement pump. It allows the pump to start and to run under low sun conditions without stalling. Electrical analogy: variable transformer; mechanical analogy: automatic transmission. Also called *pump controller*.

Maximum Power Point Tracking (MPPT): An added refinement in some linear current boosters in which the input voltage tracks the variations of the output voltage of the PV array to draw the most possible solar power under varying conditions of temperature, solar intensity, and load.

Multistage Centrifugal: A centrifugal pump with more than one impeller and chamber, stacked in a sequence to produce higher pressure. Conventional AC deep-well submersible pumps and some solar submersibles work this way.

Open Discharge: The filling of a water vessel that is not sealed to hold pressure—for example, storage (holding) tank, pond, flood irrigation. Open system. Contrast: pressure tank.

Perforations (Well): Slits cut into the well casing to allow groundwater to enter. May be located at more than one level to coincide with water-bearing strata in the Earth.

Pitless Adapter: A special pipe fitting that fits on a well casing, below ground. It allows the pipe to pass horizontally through the casing so that no pipe is exposed above ground, where it could freeze. The pump may be installed and removed without further need to dig around the casing. This is done by using a 1 in. threaded pipe as a handle.

Positive Displacement Pump: Any mechanism that seals water in a chamber and then forces it out by reducing the volume of the chamber. Examples: piston, diaphragm, helical rotor, rotary vane. Used for low volume and high lift. Contrast with *centrifugal pump*. Synonyms: volumetric pump, force pump.

Pressure: The amount of force applied by water that is forced by a pump or by the gravity. Measured in pounds per square inch (psi) or bar (atmospheres). psi = vertical lift (or drop) in feet/2.31; 1 bar = 10 vertical m.

Pressure Switch: An electrical switch actuated by the pressure in a pressure tank. When the pressure drops to a low set point (cut-in), it turns a pump on. At a high point (cut-out), it turns the pump off.

Pressure Tank: A fully enclosed tank with an air space inside. As water is forced in, the air compresses. The stored water may be released after the pump has stopped. Most pressure tanks contain a rubber bladder to capture the air. If so, a synonym is "captive air tank."

Pressure Tank Precharge: The pressure of compressed air stored in a captive air pressure tank. A reading should be taken with an air pressure gauge (tire gauge) with water pressure at zero. The air pressure is then adjusted to about 3 psi lower than the cut-in pressure (see *pressure switch*). If precharge is not set properly, the tank will not work to full capacity, and the pump will cycle on and off more frequently.

Priming: The process of hand-filling the suction pipe and intake of a surface pump. Priming is generally necessary when a pump must be located above the water source. A self-priming pump is able to draw some air suction in order to prime itself, at least in theory. See *foot valve*.

Pulsation Damper: A device that absorbs and releases pulsations in flow produced by a piston or diaphragm pump. Consists of a chamber with air trapped within it or a length of flexible tube.

Pump Controller: An electronic device that controls or processes the power to a pump. It may perform any of the following functions: stopping and starting the pump, protection from overload, DC–AC conversion, voltage conversion, or power matching (see *linear current booster*). It may also have connections for low-water shutoff and full-tank shutoff devices and status indicators.

Pump Jack: A deep well piston pump. The piston and cylinder are submerged in the well water and actuated by a rod inside the drop pipe, powered by a motor at the surface. This is an old-fashioned system that is still used for extremely deep wells, including solar pumps as deep as 1,000 ft.

Recovery Rate (Well): Rate at which groundwater refills the casing after the level is drawn down. This is the term used to specify the production rate of the well.

Safety Rope (Pump): Rope used to secure the pump in case of pipe breakage.

Self-Priming Pump: Pump that automatically primes itself. See *priming*.

Static Water Level: Depth to the water surface in a well under static conditions (not being pumped). May be subject to seasonal changes or lowering due to depletion.

Submergence: Applied to submersible pumps: distance beneath the static water level at which a pump is set. Synonym: immersion level.

Submersible Cable: Electrical cable designed for in-well submersion. Conductor sizing is specified in square millimeters or (in the United States) American wire gauge (AWG), in which a higher number indicates smaller wire. It is connected to a pump by a cable splice.

Submersible Pump: A motor/pump combination designed to be placed entirely below the water surface.

Suction Lift: Applied to surface pumps: Vertical distance from the surface of the water in the source to a pump located above the surface. This distance is limited by physics to around 20 ft at sea level (subtract 1 ft per 1,000 ft altitude) and should be minimized for best results.

Surface Pump: A pump that is not submersible. It must be placed no more than about 20 ft above the surface of the water in the well. See *priming* (exception: see *jet pump*).

Total Dynamic Head: Vertical lift + friction loss in piping (see *vertical lift* and *friction loss*). Measured in meters.

Vane Pump (Rotary Vane): A positive displacement mechanism used in low-volume, high-lift surface pumps and booster pumps. Durable and efficient, but requires cleanly filtered water due to its mechanical precision.

Vertical Lift: The vertical distance that water is pumped (meters). This determines the pressure that the pump pushes against. Total vertical lift = vertical lift from surface of water source up to the discharge in the tank + (in a pressure system) discharge pressure. Synonym: static head. Note: Horizontal distance does *not* add to the vertical lift, except in terms of pipe friction loss, *nor* does the volume (weight) of water contained in pipe or tank. Submergence of the pump does *not* add to the vertical lift.

Wellhead: Top of the well, usually with some sort of physical cover.

Well Seal: Top plate of a well casing that provides a sanitary seal and support for the drop pipe and pump. Alternative: see *pitless adapter.*

Appendix A: World Insolation Data

ALGERIA

Biskra (34.85° N, 5.73°W, 125 m)	JAN	FEB	MAR	APR	MAY	JUN	JUL	AUG	SEP	OCT	NOV	DEC	YEAR
Latitude tilt -15° Fixed array	4.2	5.0	5.6	6.0	6.3	6.6	6.8	6.6	5.7	4.9	4.1	3.8	5.5
1-axis north-south tracking array	4.9	6.3	7.1	8.0	8.7	9.2	9.4	8.8	7.3	6.1	4.7	4.2	7.0
Latitude tilt Fixed array	4.9	5.5	5.8	5.8	5.9	6.1	6.3	6.3	5.7	5.2	4.6	4.4	5.6
1-axis north-south tracking array	5.6	7.0	7.4	7.9	8.3	8.6	8.8	8.6	7.5	6.6	5.4	4.9	7.2
Latitude tilt +15° Fixed array	5.2	5.7	5.7	5.4	5.3	5.3	5.5	5.7	5.5	5.3	4.9	4.8	5.4
1-axis north-south tracking array	6.0	7.2	7.3	7.4	7.4	7.4	7.7	7.9	7.2	6.7	5.7	5.3	6.9
Two axis tracking	6.1	7.2	7.4	8.0	8.8	9.3	9.4	8.9	7.5	6.7	5.7	5.4	7.4

ANGOLA

Luanda (8.82° N, 13.22° W, 42 m)	JAN	FEB	MAR	APR	MAY	JUN	JUL	AUG	SEP	OCT	NOV	DEC	YEAR
Latitude tilt -15° Fixed array	5.9	6.1	5.4	4.9	4.6	4.2	3.4	3.7	4.6	5.1	5.6	6.2	5.0
1-axis north-south tracking array	7.6	7.8	7.0	6.2	5.6	5.0	4.2	4.8	6.0	6.7	7.3	7.9	6.3
Latitude tilt Fixed array	5.6	5.9	5.5	5.2	5.1	4.8	3.7	4.0	4.7	5.0	5.3	5.7	5.0
1-axis north-south tracking array	7.2	7.7	7.2	6.7	6.3	5.8	4.8	5.2	6.2	6.6	6.9	7.4	6.5
Latitude tilt +15° Fixed array	4.9	5.4	5.3	5.3	5.4	5.1	3.9	4.0	4.6	4.7	4.8	5.0	4.9
1-axis north-south tracking array	6.3	7.0	6.9	6.8	6.7	6.3	5.1	5.4	6.1	6.1	6.1	6.3	6.3
Two axis tracking	7.7	7.8	7.2	6.8	6.7	6.3	5.1	5.4	6.2	6.7	7.3	8.0	6.8

ARGENTINA

Buenos Aires (34° 58'S, 58° 48' W, 25 m)	JAN	FEB	MAR	APR	MAY	JUN	JUL	AUG	SEP	OCT	NOV	DEC	YEAR
Latitude tilt -15° Fixed array	7.1	6.5	5.5	4.5	3.6	2.9	3.2	4.1	5.1	5.9	6.5	7.1	5.2
1-axis north-south tracking array	9.8	8.7	7.0	5.5	4.1	3.1	3.6	5.0	6.4	7.9	8.9	9.9	6.7
Latitude tilt Fixed array	6.6	6.2	5.5	4.8	4.0	3.4	3.7	4.5	5.2	5.7	6.0	6.5	5.2
1-axis north-south tracking array	9.2	8.5	7.2	6.0	4.6	3.7	4.1	5.5	6.7	7.8	8.5	9.2	6.8
Latitude tilt +15° Fixed array	5.8	5.6	5.2	4.8	4.3	3.7	4.0	4.6	5.1	5.3	5.3	5.7	4.9
1-axis north-south tracking array	8.1	7.7	6.9	6.0	4.9	4.0	4.4	5.7	6.5	7.2	7.4	7.9	6.4
Two axis tracking	9.9	8.7	7.2	6.1	4.9	4.0	4.5	5.7	6.7	7.9	8.9	9.9	7.0
Corrientes (27.47° S; 58.82° W, 50 m)	JAN	FEB	MAR	APR	MAY	JUN	JUL	AUG	SEP	OCT	NOV	DEC	YEAR
Latitude tilt -15° Fixed array	5.4	5.2	4.1	4.1	3.8	3.9	4.2	5.3	4.5	4.9	5.1	5.1	4.6
1-axis north-south tracking array	6.8	6.8	5.5	5.4	4.9	4.8	5.3	6.7	5.9	6.4	6.6	6.6	6.0
Latitude tilt Fixed array	5.1	5.1	4.1	4.3	4.2	4.4	4.7	5.7	4.6	4.8	4.9	4.8	4.7

	JAN	FEB	MAR	APR	MAY	JUN	JUL	AUG	SEP	OCT	NOV	DEC	YEAR
1-axis north-south tracking array	6.5	6.6	5.6	5.8	5.5	5.6	6.0	7.3	6.1	6.3	6.3	6.1	6.1
Latitude tilt +15° Fixed array	4.5	4.7	4.0	4.4	4.4	4.7	5.0	6.0	4.5	4.5	4.4	4.2	4.6
1-axis north-south tracking array	5.6	6.0	5.4	5.8	5.8	6.0	6.4	7.5	6.0	5.9	5.5	5.2	5.9
Two axis tracking	6.9	6.8	5.6	5.9	5.8	6.1	6.5	7.6	6.1	6.4	6.6	6.6	6.4

San Carlos de Bariloche (41° 09.04'S, 71° 18.0'W , 845 m)

	JAN	FEB	MAR	APR	MAY	JUN	JUL	AUG	SEP	OCT	NOV	DEC	YEAR
Latitude tilt -15° Fixed array	7.0	6.8	5.5	4.0	2.8	1.8	2.4	3.4	4.8	5.9	7.0	6.6	4.8
1-axis north-south tracking array	9.6	9.0	7.1	4.7	2.9	1.8	2.4	3.8	6.1	7.9	9.5	9.2	6.2
Latitude tilt Fixed array	6.8	6.8	5.9	4.3	3.2	2.1	2.7	3.7	5.0	5.9	6.8	6.4	4.9
1-axis north-south tracking array	9.0	8.7	7.2	5.1	3.3	2.1	2.8	4.2	6.4	7.8	9.0	8.6	6.2
Latitude tilt +15° Fixed array	6.5	6.6	5.5	4.4	3.4	2.3	3.0	3.9	5.0	5.6	6.5	6.1	4.9
1-axis north-south tracking array	7.5	7.9	6.9	5.1	3.5	2.3	3.0	4.3	6.3	7.2	7.9	7.3	5.8
Two axis tracking	9.6	9.0	7.3	5.2	3.5	2.3	3.0	4.4	6.4	8.0	9.5	9.3	6.4

Santiago del Estero (27° 48'0"S, 64°16'48" W, 198 m)

	JAN	FEB	MAR	APR	MAY	JUN	JUL	AUG	SEP	OCT	NOV	DEC	YEAR
Latitude tilt -15° Fixed array	6.4	6.0	4.4	3.6	3.2	3.8	4.6	5.2	6.2	6.5	6.5	6.5	5.2
1-axis north-south tracking array	8.7	8.0	5.5	4.2	3.6	4.4	5.5	6.5	8.2	8.8	8.8	8.8	6.6
Latitude tilt Fixed array	6.0	5.8	4.7	4.1	3.7	4.3	5.0	5.3	6.1	6.1	6.0	6.0	5.2
1-axis north-south tracking array	8.2	7.9	5.9	4.8	4.2	5.1	6.1	6.9	8.1	8.4	8.2	8.2	6.8
Latitude tilt +15° Fixed array	5.3	5.3	4.8	4.3	4.0	4.6	5.1	5.2	5.6	5.4	5.2	5.2	5.0
1-axis north-south tracking array	7.1	7.1	6.0	5.1	4.5	5.5	6.3	6.7	7.5	7.4	7.1	7.1	6.4
Two axis tracking	8.7	8.0	6.0	5.1	4.6	5.5	6.3	6.9	8.2	8.8	8.9	8.9	7.0

Patagones (40° 46' 41"S, 62° 58' 49" 'W, 45 m)

	JAN	FEB	MAR	APR	MAY	JUN	JUL	AUG	SEP	OCT	NOV	DEC	YEAR
Latitude tilt -15° Fixed array	6.9	6.5	5.5	4.4	3.4	3.0	2.8	4.1	4.9	5.6	6.5	6.9	5.0
1-axis north-south tracking array	9.4	8.6	7.0	5.4	3.6	3.0	2.9	4.7	6.2	7.5	8.9	9.6	6.4
Latitude tilt Fixed array	6.7	6.4	5.6	4.8	3.9	3.5	3.3	4.5	5.1	5.5	6.4	6.7	5.2
1-axis north-south tracking array	8.9	8.4	7.2	5.9	4.2	3.5	3.4	5.3	6.5	7.5	8.5	9.0	6.5
Latitude tilt +15° Fixed array	6.4	6.3	5.5	5.0	4.1	3.8	3.5	4.7	5.1	5.3	6.1	6.3	5.2
1-axis north-south tracking array	7.8	7.6	6.9	5.9	4.4	3.8	3.6	5.4	6.4	6.9	7.5	7.7	6.2
Two axis tracking	9.5	8.6	7.2	6.0	4.4	3.9	3.6	5.4	6.6	7.6	9.0	9.7	7.0

AUSTRALIA

Darwin (12.43° N, 30.87° W, 30 m)	JAN	FEB	MAR	APR	MAY	JUN	JUL	AUG	SEP	OCT	NOV	DEC	YEAR
Latitude tilt -15° Fixed array	5.2	5.3	5.6	5.1	5.1	5.0	5.3	6.1	6.4	6.5	6.2	5.7	5.6
1-axis north-south tracking array	6.8	7.0	7.2	6.3	6.2	5.8	6.2	7.5	8.1	8.4	8.1	7.4	7.1

	JAN	FEB	MAR	APR	MAY	JUN	JUL	AUG	SEP	OCT	NOV	DEC	YEAR
Latitude tilt Fixed array	4.9	5.2	5.6	5.4	5.8	5.7	6.0	6.7	6.6	6.4	5.9	5.3	5.8
1-axis north-south tracking array	6.5	6.9	7.3	6.8	6.9	6.7	7.1	8.2	8.4	8.3	7.7	6.9	7.3
Latitude tilt +15° Fixed array	4.4	4.8	5.4	5.4	6.1	6.2	6.5	7.0	6.5	5.9	5.2	4.7	5.7
1-axis north-south tracking array	5.6	6.2	7.0	6.9	7.3	7.2	7.5	8.4	8.2	7.7	6.7	5.9	7.1
Two axis tracking	6.9	7.0	7.4	6.9	7.3	7.3	7.6	8.4	8.4	8.4	8.1	7.5	7.6

Melbourne (37.82° N, 44.97° W, 35 m)

	JAN	FEB	MAR	APR	MAY	JUN	JUL	AUG	SEP	OCT	NOV	DEC	YEAR
Latitude tilt -15° Fixed array	7.2	6.4	4.0	4.1	3.5	3.1	3.3	3.7	4.6	5.4	5.4	5.9	4.7
1-axis north-south tracking array	10.0	8.6	5.4	5.1	3.9	3.3	3.6	4.4	5.9	7.3	7.6	8.5	6.2
Latitude tilt Fixed array	6.6	6.1	3.9	4.4	4.0	3.7	3.8	4.1	4.7	5.2	5.0	5.5	4.7
1-axis north-south tracking array	9.4	8.4	5.5	5.5	4.5	3.9	4.2	4.9	6.2	7.2	7.3	7.9	6.2
Latitude tilt +15° Fixed array	5.8	5.5	3.7	4.5	4.2	4.0	4.1	4.2	4.6	4.8	4.5	4.8	4.5
1-axis north-south tracking array	8.2	7.7	5.3	5.6	4.7	4.2	4.5	5.0	6.0	6.7	6.4	6.8	5.9
Two axis tracking	10.0	8.7	5.5	5.6	4.8	4.3	4.5	5.0	6.2	7.3	7.6	8.5	6.5

BRAZIL

Porto Nacional (10° 70'S, 48° 42' W, 200 m)

	JAN	FEB	MAR	APR	MAY	JUN	JUL	AUG	SEP	OCT	NOV	DEC	YEAR
Latitude tilt -15° Fixed array	5.7	5.3	5.1	5.0	5.0	4.9	5.1	5.7	5.7	5.6	5.4	5.4	5.3
1-axis north-south tracking array	7.4	6.9	6.6	6.3	6.1	5.8	6.0	7.0	7.2	7.3	7.1	7.1	6.7
Latitude tilt Fixed array	5.3	5.1	5.1	5.3	5.6	5.6	5.7	6.2	5.8	5.5	5.1	5.1	5.5
1-axis north-south tracking array	7.0	6.8	6.8	6.8	6.9	6.7	6.9	7.7	7.5	7.3	6.8	6.6	7.0
Latitude tilt +15° Fixed array	4.7	4.7	4.9	5.4	6.0	6.2	6.2	6.5	5.7	5.1	4.6	4.5	5.4
1-axis north-south tracking array	6.1	6.1	6.5	6.9	7.2	7.2	7.4	7.9	7.4	6.7	5.9	5.7	6.8
Two axis tracking	7.4	6.9	6.8	6.9	7.3	7.3	7.4	7.9	7.6	7.4	7.1	7.2	7.3

Sao Paulo (23.6° S, 46.6° W, 60 m)

	JAN	FEB	MAR	APR	MAY	JUN	JUL	AUG	SEP	OCT	NOV	DEC	YEAR
Latitude tilt -15° Fixed array	5.3	5.3	4.6	4.3	3.5	3.5	3.4	3.8	4.7	5.2	5.4	5.3	4.5
1-axis north-south tracking array	7.3	7.1	6.1	5.4	4.2	4.1	3.9	4.7	6.0	6.9	7.3	7.3	5.8
Latitude tilt Fixed array	5.0	5.1	4.6	4.5	3.9	4.0	3.8	4.1	4.8	5.1	5.1	5.0	4.6
1-axis north-south tracking array	6.8	7.0	6.3	5.8	4.7	4.8	4.5	5.2	6.3	6.9	7.0	6.8	6.0
Latitude tilt +15° Fixed array	4.5	4.7	4.4	4.6	4.1	4.4	4.1	4.2	4.6	4.7	4.6	4.4	4.4
1-axis north-south tracking array	6.0	6.3	6.0	5.9	5.0	5.1	4.8	5.3	6.1	6.4	6.1	5.8	5.7
Two axis tracking	7.3	7.1	6.3	5.9	5.0	5.2	4.8	5.3	6.3	7.0	7.3	7.3	6.2

Praia, Cabo Verde (14.90° S, 23.52° W , 27 m)

	JAN	FEB	MAR	APR	MAY	JUN	JUL	AUG	SEP	OCT	NOV	DEC	YEAR
Latitude tilt -15° Fixed array	5.5	6.2	7.3	7.8	7.5	7.2	6.2	5.7	5.9	5.9	5.1	4.5	6.2

	JAN	FEB	MAR	APR	MAY	JUN	JUL	AUG	SEP	OCT	NOV	DEC	YEAR
1-axis north-south tracking array	6.3	7.5	9.1	9.9	9.6	9.3	8.1	7.5	7.6	7.2	6.1	5.3	7.8
Latitude tilt Fixed array	6.2	6.8	7.6	7.6	7.0	6.7	5.8	5.5	6.0	6.3	5.7	5.2	6.4
1-axis north-south tracking array	7.3	8.2	9.5	9.9	9.1	8.7	7.6	7.3	7.8	7.8	6.9	6.1	8.0
Latitude tilt +15° Fixed array	6.7	7.0	7.5	7.1	6.2	5.8	5.1	5.1	5.8	6.4	6.1	5.6	6.2
1-axis north-south tracking array	7.8	8.4	9.3	9.2	8.1	7.5	6.7	6.7	7.5	7.9	7.3	6.6	7.7
Two axis tracking	7.8	8.5	9.5	10.0	9.6	9.4	8.1	7.5	7.8	7.9	7.3	6.7	8.3

CHINA

Shanghai (31.21° N, 21.47° W, 3 m)	JAN	FEB	MAR	APR	MAY	JUN	JUL	AUG	SEP	OCT	NOV	DEC	YEAR
Latitude tilt -15° Fixed array	3.4	3.1	4.3	4.9	5.3	4.7	5.8	6.0	5.2	4.4	3.5	3.1	4.5
1-axis north-south tracking array	3.7	3.6	5.5	6.6	7.4	6.6	8.0	8.0	6.7	5.4	3.9	3.4	5.7
Latitude tilt Fixed array	3.8	3.3	4.4	4.7	5.0	4.3	5.4	5.7	5.2	4.7	3.9	3.6	4.5
1-axis north-south tracking array	4.3	3.9	5.8	6.5	7.0	6.2	7.5	7.8	6.9	5.8	4.5	3.9	5.9
Latitude tilt +15° Fixed array	4.1	3.3	4.2	4.3	4.5	3.8	4.7	5.2	5.0	4.7	4.1	3.8	4.3
1-axis north-south tracking array	4.6	4.0	5.7	6.0	6.2	5.3	6.5	7.1	6.6	5.9	4.7	4.2	5.6
Two axis tracking	4.6	4.0	5.8	6.6	7.4	6.7	8.1	8.1	6.9	5.9	4.7	4.3	6.1

ECUADOR

Quito (0° 28'S, 78° 53'W, 2850 m)	JAN	FEB	MAR	APR	MAY	JUN	JUL	AUG	SEP	OCT	NOV	DEC	YEAR
Latitude tilt -15° Fixed array	5.4	5.2	4.1	4.1	3.8	3.9	4.2	5.3	4.5	4.9	5.1	5.1	4.6
1-axis north-south tracking array	6.8	6.8	5.5	5.4	4.9	4.8	5.3	6.7	5.9	6.4	6.6	6.6	6.0
Latitude tilt Fixed array	5.1	5.1	4.1	4.3	4.2	4.4	4.7	5.7	4.6	4.8	4.9	4.8	4.7
1-axis north-south tracking array	6.5	6.6	5.6	5.8	5.5	5.6	6.0	7.3	6.1	6.3	6.3	6.1	6.1
Latitude tilt +15° Fixed array	4.5	4.7	4.0	4.4	4.4	4.7	5.0	6.0	4.5	4.5	4.4	4.2	4.6
1-axis north-south tracking array	5.6	6.0	5.4	5.8	5.8	6.0	6.4	7.5	6.0	5.9	5.5	5.2	5.9
Two axis tracking	6.9	6.8	5.6	5.9	5.8	6.1	6.5	7.6	6.1	6.4	6.6	6.6	6.4

FRANCE

Paris - St. Maur (48.82° N, 2.50° W, 50 m)	JAN	FEB	MAR	APR	MAY	JUN	JUL	AUG	SEP	OCT	NOV	DEC	YEAR
Latitude tilt -15° Fixed array	1.8	2.5	3.8	4.3	5.0	5.4	5.1	4.6	4.0	2.7	1.7	1.6	3.5
1-axis north-south tracking array	1.8	2.5	4.6	6.0	7.4	8.0	7.7	6.6	5.0	3.0	1.7	1.6	4.7
Latitude tilt Fixed array	2.1	2.8	3.9	4.3	4.8	5.1	4.9	4.5	4.0	3.0	2.0	1.8	3.6
1-axis north-south tracking array	2.1	2.8	4.8	6.0	7.1	7.5	7.2	6.5	5.2	3.3	2.0	1.8	4.7
Latitude tilt +15° Fixed array	2.2	2.9	3.9	4.0	4.4	4.6	4.5	4.2	3.9	3.0	2.1	2.0	3.5

	JAN	FEB	MAR	APR	MAY	JUN	JUL	AUG	SEP	OCT	NOV	DEC	YEAR
1-axis north-south tracking array	2.2	2.9	4.7	5.5	6.2	6.5	6.3	5.9	5.0	3.3	2.1	2.0	4.4
Two axis tracking	2.2	2.9	4.8	6.1	7.4	8.1	7.7	6.6	5.2	3.3	2.1	2.0	4.9

GUYANA

Georgetown (7.8° N, 58.1° W, 1 m)	JAN	FEB	MAR	APR	MAY	JUN	JUL	AUG	SEP	OCT	NOV	DEC	YEAR
Latitude tilt -15° Fixed array	4.3	4.8	5.0	5.3	4.9	4.8	5.3	5.6	5.6	5.1	4.5	4.0	4.9
1-axis north-south tracking array	5.2	6.0	6.5	6.9	6.4	6.3	6.9	7.2	7.2	6.4	5.6	4.8	6.3
Latitude tilt Fixed array	4.7	5.1	5.2	5.2	4.7	4.5	5.0	5.4	5.7	5.4	5.0	4.5	5.0
1-axis north-south tracking array	5.9	6.6	6.8	6.8	6.1	5.9	6.5	7.0	7.4	7.0	6.3	5.6	6.5
Latitude tilt +15° Fixed array	5.0	5.3	5.1	4.8	4.2	4.0	4.4	5.0	5.5	5.5	5.3	4.8	4.9
1-axis north-south tracking array	6.3	6.8	6.6	6.3	5.4	5.1	5.6	6.4	7.1	7.0	6.7	6.0	6.3
Two axis tracking	6.4	6.8	6.8	6.9	6.4	6.4	6.9	7.2	7.4	7.1	6.7	6.1	6.8

KENYA

Nairobi (1.30°‑, 36.75°W, 1800 m)	JAN	FEB	MAR	APR	MAY	JUN	JUL	AUG	SEP	OCT	NOV	DEC	YEAR
Latitude tilt -15° Fixed array	6.9	7.1	6.4	5.3	4.4	4.1	3.5	4.0	5.3	5.8	5.9	6.5	5.4
1-axis north-south tracking array	8.6	9.0	8.2	6.8	5.5	5.1	4.4	5.2	6.8	7.4	7.5	8.1	6.9
Latitude tilt Fixed array	6.5	6.9	6.5	5.7	4.9	4.7	3.8	4.3	5.4	5.7	5.6	6.0	5.5
1-axis north-south tracking array	8.1	8.7	8.4	7.3	6.2	5.9	5.0	5.7	7.1	7.4	7.1	7.5	7.0
Latitude tilt +15° Fixed array	5.7	6.3	6.3	5.8	5.1	5.0	4.0	4.4	5.3	5.3	5.0	5.2	5.3
1-axis north-south tracking array	7.0	7.9	8.0	7.4	6.6	6.3	5.3	5.8	6.9	6.8	6.3	6.4	6.7
Two axis tracking	8.6	9.0	8.4	7.4	6.6	6.4	5.4	5.8	7.1	7.5	7.5	8.2	7.3

INDIA

New Delhi (28.58°N, 77.20°W, 210 m)	JAN	FEB	MAR	APR	MAY	JUN	JUL	AUG	SEP	OCT	NOV	DEC	YEAR
Latitude tilt -15° Fixed array	5.0	6.4	7.1	7.1	7.4	6.8	4.5	5.5	5.7	6.1	5.6	4.9	6.0
1-axis north-south tracking array	6.4	8.1	8.6	9.2	9.8	9.2	6.3	7.4	7.2	7.3	7.5	6.1	7.8
Latitude tilt Fixed array	5.8	7.0	7.3	6.9	6.9	6.2	4.2	5.3	5.7	6.6	6.4	5.7	6.2
1-axis north-south tracking array	7.4	9.0	9.0	9.2	9.4	8.5	5.9	7.3	7.4	8.0	8.6	7.1	8.1
Latitude tilt +15° Fixed array	6.3	7.3	7.2	6.4	6.1	5.4	3.8	4.8	5.5	6.7	6.9	6.3	6.0
1-axis north-south tracking array	7.9	9.2	8.8	8.5	8.3	7.3	5.2	6.6	7.1	8.1	9.1	7.7	7.8
Two axis tracking	7.9	9.2	9.1	9.3	9.9	9.2	6.3	7.5	7.5	8.1	9.1	7.8	8.4

JAPAN

Tokyo (35.68° N, 39.77° W, 5 m)	JAN	FEB	MAR	APR	MAY	JUN	JUL	AUG	SEP	OCT	NOV	DEC	YEAR

MEXICO

	JAN	FEB	MAR	APR	MAY	JUN	JUL	AUG	SEP	OCT	NOV	DEC	YEAR
Latitude tilt −15° Fixed array	3.0	3.2	3.4	3.6	3.8	3.3	3.7	3.8	3.0	2.6	2.6	2.7	3.2
1-axis north-south tracking array	3.1	3.6	4.5	5.2	5.6	5.0	5.5	5.5	4.3	3.0	2.8	2.8	4.2
Latitude tilt Fixed array	3.3	3.5	3.5	3.5	3.6	3.1	3.4	3.6	3.0	2.7	2.9	3.1	3.3
1-axis north-south tracking array	3.6	4.0	4.7	5.2	5.3	4.7	5.2	5.4	4.4	3.2	3.2	3.2	4.4
Latitude tilt +15° Fixed array	3.6	3.5	3.4	3.2	3.2	2.8	3.1	3.3	2.8	2.7	3.1	3.3	3.2
1-axis north-south tracking array	3.9	4.1	4.6	4.8	4.7	4.0	4.5	4.9	4.2	3.3	3.4	3.5	4.2
Two axis tracking	3.9	4.2	4.8	5.3	5.6	5.1	5.5	5.5	4.4	3.3	3.4	3.5	4.5

Chihuahua City (28° 38'N, 106°05' W, 1500 m)

	JAN	FEB	MAR	APR	MAY	JUN	JUL	AUG	SEP	OCT	NOV	DEC	YEAR
Latitude tilt −15° Fixed array	5.1	5.8	6.6	7.1	7.4	7.0	7.0	6.8	6.7	6.3	5.2	4.5	6.3
1-axis north-south tracking array	6.4	7.2	8.1	9.2	9.8	9.5	9.3	8.9	8.4	7.5	6.8	5.5	8.1
Latitude tilt Fixed array	5.8	6.4	6.8	6.9	6.9	6.4	6.4	6.5	6.8	6.8	6.0	5.3	6.4
1-axis north-south tracking array	7.4	7.9	8.5	9.1	9.3	8.8	8.8	8.7	8.7	8.2	7.8	6.4	8.3
Latitude tilt +15° Fixed array	6.3	6.7	6.7	6.4	6.1	5.6	5.6	5.9	6.5	7.0	6.4	5.7	6.2
1-axis north-south tracking array	7.9	8.2	8.3	8.4	8.2	7.6	7.6	7.9	8.3	8.3	8.2	6.9	8.0
Two axis tracking	8.0	8.2	8.5	9.2	9.8	9.6	9.4	9.0	8.7	8.3	8.2	7.0	8.7

Guaymas (27.925° N, 110.892°W, 80 m)

	JAN	FEB	MAR	APR	MAY	JUN	JUL	AUG	SEP	OCT	NOV	DEC	YEAR
Latitude tilt −15° Fixed array	5.1	5.8	6.6	7.1	7.4	7.4	7.0	6.8	6.7	6.7	5.2	4.5	6.4
1-axis north-south tracking array	6.4	7.2	8.1	9.2	9.8	9.9	9.3	8.9	8.4	8.2	6.5	5.6	8.1
Latitude tilt Fixed array	5.8	6.4	6.8	6.9	6.9	6.7	6.4	6.5	6.8	7.3	6.0	5.3	6.5
1-axis north-south tracking array	7.4	7.9	8.5	9.1	9.3	9.2	8.8	8.7	8.7	9.6	7.6	6.7	8.5
Latitude tilt +15° Fixed array	6.3	6.7	6.7	6.4	6.1	5.8	5.6	5.9	6.5	7.4	6.4	5.7	6.3
1-axis north-south tracking array	7.9	8.2	8.3	8.4	8.2	7.9	7.6	7.9	8.3	9.0	8.4	7.7	8.2
Two axis tracking	8.0	8.2	8.5	9.2	9.8	10.0	9.4	9.0	8.7	8.8	8.2	7.0	8.7

Mexico City DF (19° 33'N, 99° 18'W, 2270 m)

	JAN	FEB	MAR	APR	MAY	JUN	JUL	AUG	SEP	OCT	NOV	DEC	YEAR
Latitude tilt −15° Fixed array	4.3	6.2	7.7	6.2	5.9	4.9	4.9	5.4	5.0	4.5	4.5	4.5	5.4
1-axis north-south tracking array	5.1	7.4	9.5	8.1	7.8	6.7	6.6	7.2	6.5	5.7	5.3	5.5	6.8
Latitude tilt Fixed array	4.9	6.9	8.0	6.1	5.6	4.6	4.6	5.2	5.0	4.8	5.1	5.2	5.5
1-axis north-south tracking array	5.9	8.2	10.0	8.0	7.5	6.2	6.2	7.0	6.7	6.2	6.0	6.5	7.0
Latitude tilt +15° Fixed array	5.2	7.1	7.9	5.6	5.0	4.1	4.1	4.8	4.8	4.9	5.4	5.7	5.4
1-axis north-south tracking array	6.2	8.4	9.7	7.4	6.6	5.3	5.4	6.4	6.4	6.2	6.4	7.0	6.8
Two axis tracking	6.3	8.4	10.0	8.1	7.9	6.7	6.7	7.2	6.7	6.3	6.4	7.1	7.3

Navojoa (27° 6'N, 109°25' W, 35 m)

	JAN	FEB	MAR	APR	MAY	JUN	JUL	AUG	SEP	OCT	NOV	DEC	YEAR
Latitude tilt -15° Fixed array	4.3	5.0	5.6	6.1	6.3	6.3	5.7	5.8	5.4	5.4	4.4	3.7	5.3
1-axis north-south tracking array	5.1	6.0	7.0	8.0	8.4	8.5	7.7	7.7	7.0	6.5	5.3	4.3	6.8
Latitude tilt Fixed array	4.9	5.4	5.8	5.9	5.9	5.8	5.3	5.5	5.5	5.8	4.9	4.3	5.4
1-axis north-south tracking array	5.9	6.6	7.4	7.9	8.0	7.9	7.2	7.5	7.2	7.1	6.0	5.0	7.0
Latitude tilt +15° Fixed array	5.2	5.6	5.6	5.5	5.2	5.0	4.6	5.0	5.2	5.9	5.2	4.6	5.2
1-axis north-south tracking array	6.3	6.8	7.2	7.3	7.0	6.7	6.2	6.8	6.9	7.1	6.3	5.4	6.7
Two axis tracking	6.3	6.8	7.4	8.0	8.4	8.5	7.7	7.7	7.2	7.2	6.3	5.5	7.3

Puerto Vallarta (20.5126° 0'N, 105.2500° W, 5 m)

	JAN	FEB	MAR	APR	MAY	JUN	JUL	AUG	SEP	OCT	NOV	DEC	YEAR
Latitude tilt -15° Fixed array	4.6	5.2	5.8	5.9	6.1	5.9	6.1	6.0	5.5	5.2	4.6	4.1	5.4
1-axis north-south tracking array	5.5	6.3	7.3	7.8	8.1	7.9	8.0	7.9	7.1	6.4	5.4	4.9	6.9
Latitude tilt Fixed array	5.2	5.7	6.0	5.8	5.7	5.5	5.6	5.8	5.5	5.6	5.2	4.7	5.5
1-axis north-south tracking array	6.3	6.9	7.7	7.7	7.6	7.3	7.5	7.7	7.2	6.9	6.1	5.7	7.1
Latitude tilt +15° Fixed array	5.6	5.9	5.9	5.4	5.1	4.8	5.0	5.2	5.3	5.6	5.5	5.1	5.4
1-axis north-south tracking array	6.7	7.1	7.5	7.1	6.7	6.3	6.5	7.0	6.9	7.0	6.5	6.1	6.8
Two axis tracking	6.7	7.1	7.7	7.8	8.1	8.0	8.1	7.9	7.3	7.0	6.5	6.2	7.4

Tacubaya (19.40'N, 99.10° W, 2,300 meters)

	JAN	FEB	MAR	APR	MAY	JUN	JUL	AUG	SEP	OCT	NOV	DEC	YEAR
Latitude tilt -15° Fixed array	4.7	5.5	6.2	6.0	5.7	5.5	4.9	5.1	4.5	4.5	4.6	4.5	5.1
1-axis north-south tracking array	5.5	6.6	7.8	7.9	7.5	7.3	6.6	6.8	5.9	5.7	5.4	5.5	6.5
Latitude tilt Fixed array	5.4	6.0	6.4	5.9	5.3	5.1	4.5	4.9	4.5	4.8	5.2	5.2	5.3
1-axis north-south tracking array	6.4	7.3	8.1	7.8	7.2	6.8	6.2	6.6	6.1	6.1	6.2	6.5	6.8
Latitude tilt +15° Fixed array	5.8	6.2	6.3	5.5	4.8	4.5	4.1	4.5	4.3	4.9	5.5	5.7	5.2
1-axis north-south tracking array	6.8	7.5	8.0	7.2	6.3	5.9	5.4	6.0	5.8	6.2	6.5	7.0	6.5
Two axis tracking	6.8	7.5	8.2	7.9	7.6	7.4	6.6	6.8	6.1	6.2	6.5	7.1	7.1

Todos Santos (23.0° N, 110.0° W, 145 m)

	JAN	FEB	MAR	APR	MAY	JUN	JUL	AUG	SEP	OCT	NOV	DEC	YEAR
Latitude tilt -15° Fixed array	4.4	5.0	5.7	6.1	6.7	6.7	6.2	6.4	5.8	5.5	4.6	3.9	5.6
1-axis north-south tracking array	5.3	6.0	7.1	7.9	8.9	8.9	8.3	8.5	7.4	6.6	5.6	4.5	7.1
Latitude tilt Fixed array	5.0	5.5	5.8	5.9	6.3	6.1	5.8	6.2	5.8	5.8	5.2	4.5	5.7
1-axis north-south tracking array	6.1	6.6	7.5	7.9	8.5	8.2	7.8	8.2	7.6	7.2	6.4	5.3	7.3
Latitude tilt +15° Fixed array	5.4	5.6	5.7	5.5	5.6	5.3	5.1	5.6	5.6	5.9	5.5	4.8	5.5
1-axis north-south tracking array	6.5	6.8	7.3	7.3	7.4	7.0	6.8	7.5	7.2	7.2	6.7	5.7	7.0
Two axis tracking	6.6	6.8	7.5	8.0	8.9	9.0	8.4	8.5	7.6	7.3	6.7	5.8	7.6

Tuxtla Gutierrez (17.50° N, 93.0° W, 525 m)	JAN	FEB	MAR	APR	MAY	JUN	JUL	AUG	SEP	OCT	NOV	DEC	YEAR
Latitude tilt -15° Fixed array	3.9	4.5	4.8	4.7	4.7	4.4	4.7	4.6	4.0	4.1	4.0	3.7	4.4
1-axis north-south tracking array	4.7	5.8	6.2	6.2	6.4	6.0	6.4	6.2	5.4	5.2	4.8	4.3	5.6
Latitude tilt Fixed array	4.4	5.1	4.9	4.6	4.5	4.1	4.5	4.5	4.1	4.3	4.4	4.2	4.5
1-axis north-south tracking array	5.4	6.4	6.5	6.2	6.1	5.6	6.0	6.1	5.6	5.6	5.5	5.1	5.8
Latitude tilt +15° Fixed array	4.7	5.3	4.8	4.3	4.1	3.7	4.0	4.1	3.9	4.3	4.7	4.5	4.4
1-axis north-south tracking array	5.8	6.6	6.3	5.7	5.4	4.8	5.2	5.5	5.3	5.7	5.8	5.5	5.6
Two axis tracking	5.8	6.6	6.5	6.3	6.4	6.0	6.4	6.2	5.6	5.7	5.8	5.5	6.1

Veracruz (19.20°N, 96.13W, 1 m)	JAN	FEB	MAR	APR	MAY	JUN	JUL	AUG	SEP	OCT	NOV	DEC	YEAR
Latitude tilt -15° Fixed array	4.0	5.2	5.4	5.8	6.9	7.1	5.8	6.5	6.1	5.9	5.1	5.0	5.5
1-axis north-south tracking array	4.7	6.3	7.1	7.6	9.0	9.2	7.6	8.4	7.8	7.2	5.9	6.4	7.3
Latitude tilt Fixed array	4.5	5.7	6.1	5.8	6.4	6.5	5.4	6.2	6.2	6.4	5.8	5.9	5.9
1-axis north-south tracking array	5.4	7.0	7.8	7.2	8.5	8.6	7.2	8.2	8.0	7.8	6.8	7.5	7.5
Latitude tilt +15° Fixed array	4.8	5.9	6.2	5.7	5.7	5.6	4.7	5.6	6.0	6.5	6.2	6.4	5.8
1-axis north-south tracking array	5.8	7.2	7.8	7.2	7.5	7.4	6.2	7.4	7.7	7.9	7.1	8.0	7.3
Two axis tracking	5.8	7.2	8.2	7.9	9.0	9.3	7.7	8.4	8.0	8.0	7.2	8.1	7.9

MONGOLIA

Ulan-Bator (47.86°N, 6.75°W, 1330 m).	JAN	FEB	MAR	APR	MAY	JUN	JUL	AUG	SEP	OCT	NOV	DEC	YEAR
Latitude tilt -15° Fixed array	4.1	5.0	5.8	5.6	6.7	6.1	5.7	5.6	5.0	4.5	3.4	3.2	5.1
1-axis north-south tracking array	4.1	5.7	7.8	7.6	9.4	8.9	8.4	7.7	6.6	5.3	3.5	3.2	6.5
Latitude tilt Fixed array	4.8	5.6	6.1	5.6	6.3	5.7	5.4	5.4	5.1	4.9	4.0	3.9	5.2
1-axis north-south tracking array	4.8	6.3	8.2	7.6	9.0	8.3	7.9	7.6	6.8	5.8	4.0	3.9	6.7
Latitude tilt +15° Fixed array	5.3	6.0	6.1	5.3	5.8	5.2	5.0	5.1	5.0	5.1	4.4	4.3	4.8
1-axis north-south tracking array	5.3	6.5	8.1	7.0	7.9	7.1	6.9	6.9	6.5	5.9	4.4	4.3	6.4
Two axis tracking	5.3	6.5	8.2	4.7	9.4	9.0	8.4	7.8	6.8	5.9	4.4	4.3	7.0

PUERTO RICO

San Juan (18° 26'N, 66° 00' W, 20 m)	JAN	FEB	MAR	APR	MAY	JUN	JUL	AUG	SEP	OCT	NOV	DEC	YEAR
Latitude tilt -15° Fixed array	4.3	4.9	5.8	6.1	5.6	5.9	5.9	5.9	5.4	4.9	4.5	4.1	5.3
1-axis north-south tracking array	5.6	6.3	7.5	7.9	6.8	7.5	7.5	7.5	6.9	6.3	5.7	5.3	6.7
Latitude tilt Fixed array	5.0	5.4	6.0	6.1	5.3	5.5	5.6	5.7	5.5	5.3	5.1	4.8	5.4
1-axis north-south tracking array	6.0	6.7	7.7	7.8	6.6	7.2	7.2	7.4	7.0	6.6	6.2	5.8	6.9
Latitude tilt +15° Fixed array	5.4	5.6	6.0	5.8	4.8	4.9	5.0	5.3	5.4	5.4	5.4	5.2	5.4

	JAN	FEB	MAR	APR	MAY	JUN	JUL	AUG	SEP	OCT	NOV	DEC	YEAR
1-axis north-south tracking array	6.3	6.8	7.7	7.6	6.3	6.8	6.8	7.0	6.9	6.7	6.4	6.1	6.8
Two axis tracking	6.4	6.8	7.8	7.9	6.9	7.7	7.6	7.5	7.0	6.7	6.5	6.2	7.1

SWEDEN

Stockholm (59.35°, 17.95° W, 45 m)

	JAN	FEB	MAR	APR	MAY	JUN	JUL	AUG	SEP	OCT	NOV	DEC	YEAR
Latitude tilt -15° Fixed array	1.4	2.5	3.9	4.1	5.2	5.5	5.3	4.6	3.5	2.1	1.1	1.1	3.3
1-axis north-south tracking array	1.4	2.5	4.6	5.8	8.2	8.9	8.5	6.8	4.4	2.2	1.1	1.1	4.6
Latitude tilt Fixed array	1.7	2.8	4.0	4.1	4.9	5.1	5.0	4.4	3.5	2.3	1.3	1.2	3.4
1-axis north-south tracking array	1.7	2.8	4.9	5.8	7.8	8.3	8.0	6.6	4.5	2.4	1.3	1.2	4.6
Latitude tilt +15° Fixed array	1.8	2.9	4.0	3.8	4.5	4.7	4.6	4.1	3.4	2.3	1.3	1.4	3.2
1-axis north-south tracking array	1.8	2.9	4.7	5.3	6.8	7.1	7.0	6.0	4.3	2.4	1.3	1.4	4.3
Two axis tracking	1.8	2.9	4.9	5.9	8.2	9.0	8.6	6.8	4.5	2.4	1.3	1.4	4.8

THAILAND

Bangkok (13°45'N, 100°30'E, 20 m)

	JAN	FEB	MAR	APR	MAY	JUN	JUL	AUG	SEP	OCT	NOV	DEC	YEAR
Latitude tilt -15° Fixed array	5.0	5.6	5.2	5.6	5.6	5.3	4.5	4.7	4.3	4.4	5.2	5.0	5.0
1-axis north-south tracking array	5.8	6.9	6.7	7.3	7.4	7.0	6.1	6.2	5.7	5.5	6.2	5.8	6.4
Latitude tilt Fixed array	5.6	6.1	5.4	5.5	5.3	4.9	4.3	4.5	4.4	4.6	5.8	5.7	5.2
1-axis north-south tracking array	6.7	7.5	7.0	7.3	7.0	6.5	5.7	6.1	5.9	6.0	7.0	6.7	6.6
Latitude tilt +15° Fixed array	6.0	6.3	5.3	5.1	4.8	4.4	3.8	4.2	4.2	4.7	6.2	6.3	5.1
1-axis north-south tracking array	7.1	7.7	6.9	6.8	6.2	5.6	5.0	5.6	5.6	6.1	7.4	7.2	6.4
Two axis tracking	7.2	7.7	7.0	7.4	7.4	7.1	6.1	6.3	5.9	6.1	7.4	7.3	6.9

UNITED STATES of AMERICA

Birmingham, Alabama (33° 34'N, 86° 45' W, 190 m)

	JAN	FEB	MAR	APR	MAY	JUN	JUL	AUG	SEP	OCT	NOV	DEC	YEAR
Latitude tilt -15° Fixed array	2.9	3.6	4.6	5.5	6.1	5.8	5.6	5.4	5.2	4.7	3.5	2.7	4.6
1-axis north-south tracking array	3.6	4.6	6.2	7.3	7.9	7.3	7.0	6.8	6.7	6.0	4.4	3.5	6.0
Latitude tilt Fixed array	3.3	4.0	4.8	5.4	5.8	5.4	5.3	5.3	5.3	5.1	3.9	3.2	4.7
1-axis north-south tracking array	3.9	4.8	6.3	7.3	7.7	7.0	6.8	6.7	6.8	6.3	4.8	3.8	6.0
Latitude tilt +15° Fixed array	3.5	4.1	4.7	5.1	5.2	4.7	4.7	4.9	5.2	5.3	4.2	3.4	4.6
1-axis north-south tracking array	4.1	4.9	6.3	7.1	7.3	6.6	6.4	6.4	6.6	6.4	5.0	4.0	5.9
Two axis tracking	4.2	4.9	6.3	7.4	8.0	7.4	7.1	6.8	6.8	6.4	5.0	4.1	6.2

Fairbanks, Alaska (64° 49'M, 147° 52'W, 140 m)

	JAN	FEB	MAR	APR	MAY	JUN	JUL	AUG	SEP	OCT	NOV	DEC	YEAR
Latitude tilt -15° Fixed array	0.3	2.3	4.9	5.8	5.8	5.4	5.3	4.1	3.6	2.5	0.8	0.0	3.4

	JAN	FEB	MAR	APR	MAY	JUN	JUL	AUG	SEP	OCT	NOV	DEC	YEAR
1-axis north-south tracking array	0.3	2.6	6.4	8.5	9.0	8.4	7.9	5.7	4.8	2.9	0.9	0.0	4.8
Latitude tilt Fixed array	0.4	2.5	5.2	5.7	5.4	5.0	4.9	3.9	3.7	2.7	1.0	0.0	3.4
1-axis north-south tracking array	0.4	2.8	6.6	8.5	8.8	8.2	7.7	5.6	4.9	3.1	1.0	0.0	4.8
Latitude tilt +15° Fixed array	0.4	2.6	5.1	5.3	4.8	4.3	4.3	3.6	3.5	2.7	1.0	0.0	3.1
1-axis north-south tracking array	0.4	2.9	6.6	8.2	8.4	7.8	7.3	5.4	4.8	3.1	1.0	0.0	4.7
Two axis tracking	0.4	2.9	6.6	8.5	9.1	8.7	8.1	5.8	4.9	3.2	1.0	0.0	4.9

Phoenix, Arizona (33° 26'N, 112° 01'W, 340 m)

	JAN	FEB	MAR	APR	MAY	JUN	JUL	AUG	SEP	OCT	NOV	DEC	YEAR
Latitude tilt -15° Fixed array	4.5	5.7	6.9	7.9	8.5	8.2	7.6	7.5	7.3	6.2	5.2	4.2	6.6
1-axis north-south tracking array	5.9	7.6	9.5	11.2	12.5	11.9	10.0	10.3	9.9	8.3	6.7	5.4	9.1
Latitude tilt Fixed array	5.3	6.4	7.3	7.8	8.0	8.5	7.1	7.3	7.6	6.8	6.1	4.9	6.9
1-axis north-south tracking array	6.5	8.2	9.9	11.2	12.2	11.5	9.7	10.1	10.1	8.8	7.4	6.0	9.3
Latitude tilt +15° Fixed array	5.8	6.8	7.3	7.4	7.1	6.5	6.3	6.7	7.4	7.1	6.6	5.4	6.7
1-axis north-south tracking array	6.9	8.4	9.8	10.9	11.6	10.8	9.1	9.7	10.0	8.9	7.8	6.4	9.2
Two axis tracking	7.0	8.4	9.9	11.3	12.7	12.2	10.1	10.3	10.1	8.9	7.8	6.5	9.6

Daggett, California (32° 52'N, 116° 47'W, 585 m)

	JAN	FEB	MAR	APR	MAY	JUN	JUL	AUG	SEP	OCT	NOV	DEC	YEAR
Latitude tilt -15° Fixed array	4.4	5.5	6.6	7.8	8.3	8.3	8.2	7.9	7.3	6.2	4.8	4.0	6.6
1-axis north-south tracking array	5.5	7.2	9.1	11.2	12.1	12.5	11.9	11.0	10.2	8.2	6.1	4.9	9.2
Latitude tilt Fixed array	5.1	6.1	7.0	7.7	7.8	7.7	7.6	7.7	7.6	6.8	5.5	4.7	6.8
1-axis north-south tracking array	6.1	7.7	9.4	11.2	11.8	12.0	11.6	10.9	10.4	8.7	6.7	5.5	9.3
Latitude tilt +15° Fixed array	5.5	6.5	7.0	7.3	6.9	6.5	6.6	7.0	7.4	7.1	6.0	5.1	6.6
1-axis north-south tracking array	6.4	7.9	9.4	10.9	11.3	11.4	10.9	10.5	10.3	8.9	7.1	5.8	9.2
Two axis tracking	6.5	7.9	9.4	11.3	12.3	12.8	12.2	11.1	10.4	8.8	78.1	5.9	9.6

Fresno, California (36° 46'N, 119° 43' W, 100 m)

	JAN	FEB	MAR	APR	MAY	JUN	JUL	AUG	SEP	OCT	NOV	DEC	YEAR
Latitude tilt -15° Fixed array	2.9	4.3	6.0	1.3	7.4	8.3	8.4	8.0	7.5	6.0	4.1	2.5	6.1
1-axis north-south tracking array	3.4	5.4	8.3	10.3	11.7	12.3	12.6	11.5	10.6	7.9	5.2	2.9	8.5
Latitude tilt Fixed array	3.3	4.7	6.4	7.3	7.5	7.6	7.8	7.8	7.8	6.7	4.7	2.8	6.2
1-axis north-south tracking array	3.8	5.8	8.5	10.3	11.4	11.9	12.2	11.3	10.8	8.4	5.7	3.1	8.6
Latitude tilt +15° Fixed array	3.5	4.9	6.3	6.9	6.6	6.5	6.8	7.2	7.7	6.9	5.0	3.0	5.9
1-axis north-south tracking array	3.9	5.9	8.5	10.0	10.9	11.2	11.6	10.9	10.7	8.5	5.9	3.3	8.4
Two axis tracking	4.0	5.9	8.5	10.4	11.9	12.6	12.8	11.5	10.8	8.5	6.0	3.3	8.9

Sacramento, California (38° 31'N, 121° 30'W, 8 m)

	JAN	FEB	MAR	APR	MAY	JUN	JUL	AUG	SEP	OCT	NOV	DEC	YEAR
Latitude tilt -15° Fixed array	2.8	3.8	5.6	7.0	7.6	8.1	8.4	8.1	7.3	5.6	3.9	2.5	5.9
1-axis north-south tracking array	3.3	4.9	7.6	10.0	11.3	12.3	12.4	11.6	10.2	7.3	4.9	2.9	8.2

	JAN	FEB	MAR	APR	MAY	JUN	JUL	AUG	SEP	OCT	NOV	DEC	YEAR
Latitude tilt Fixed array	3.2	4.2	5.9	7.0	7.2	7.5	7.8	7.9	7.6	6.1	4.5	2.8	6.0
1-axis north-south tracking array	3.7	5.2	7.8	10.0	11.0	11.9	12.1	11.4	10.4	7.7	5.4	3.2	8.3
Latitude tilt +15° Fixed array	3.4	4.4	5.9	6.6	6.4	6.4	6.8	7.2	7.4	6.3	4.8	3.0	5.7
1-axis north-south tracking array	3.8	5.3	7.8	9.7	10.5	11.2	11.4	11.0	10.2	7.8	5.6	3.4	8.2
Two axis tracking	3.9	5.3	7.8	10.0	11.4	12.6	12.7	11.6	10.4	7.8	5.6	3.4	8.6

San Diego, California (32° 44'N, 117° 10' W, 9 m)

	JAN	FEB	MAR	APR	MAY	JUN	JUL	AUG	SEP	OCT	NOV	DEC	YEAR
Latitude tilt -15° Fixed array	4.1	5.1	5.9	6.5	6.3	6.2	6.9	6.9	5.9	5.3	4.4	3.9	5.6
1-axis north-south tracking array	5.3	6.6	7.9	8.6	8.1	7.8	8.8	8.9	7.5	6.8	5.6	4.9	7.2
Latitude tilt Fixed array	4.7	5.7	6.2	6.4	6.0	5.8	6.4	6.8	6.1	5.8	5.1	4.5	5.8
1-axis north-south tracking array	5.8	7.0	8.2	8.6	7.9	7.5	8.5	8.8	7.6	7.2	6.1	5.5	7.4
Latitude tilt +15° Fixed array	5.1	5.9	6.2	6.1	5.4	5.1	5.7	6.3	5.9	6.0	5.4	5.0	5.7
1-axis north-south tracking array	6.1	7.2	8.2	8.3	7.5	7.0	8.0	8.4	7.5	7.3	6.4	5.8	7.3
Two axis tracking	6.2	7.2	8.2	8.6	8.2	7.9	8.9	8.9	7.6	7.3	6.4	5.9	7.6

Denver, Colorado (39° 45'N, 104° 52' W, 1625 m)

	JAN	FEB	MAR	APR	MAY	JUN	JUL	AUG	SEP	OCT	NOV	DEC	YEAR
Latitude tilt -15° Fixed array	4.3	4.9	6.4	6.7	7.1	7.2	7.3	6.8	6.8	5.9	4.4	4.1	6.0
1-axis north-south tracking array	5.5	6.6	8.9	9.4	10.1	10.3	10.3	9.4	9.4	7.9	5.6	5.1	8.2
Latitude tilt Fixed array	5.1	5.5	6.8	6.7	6.7	6.7	6.8	6.7	7.0	6.5	5.1	4.8	6.2
1-axis north-south tracking array	6.1	7.0	9.2	9.4	9.9	9.9	10.0	9.3	9.6	8.4	6.2	5.7	8.4
Latitude tilt +15° Fixed array	5.5	5.8	6.8	6.2	6.0	5.8	6.0	6.1	6.9	6.8	5.4	5.3	6.1
1-axis north-south tracking array	6.4	7.2	9.2	9.1	9.4	9.3	9.5	8.9	9.5	8.6	6.5	6.1	8.3
Two axis tracking	6.5	7.2	9.2	9.4	10.2	10.5	10.5	9.4	9.6	8.5	6.5	6.2	8.7

Washington D.C. (38° 57'N, 77° 27'W, 90 m)

	JAN	FEB	MAR	APR	MAY	JUN	JUL	AUG	SEP	OCT	NOV	DEC	YEAR
Latitude tilt -15° Fixed array	2.8	3.5	4.5	5.1	5.5	5.9	5.4	5.7	4.8	4.0	3.1	2.3	4.4
1-axis north-south tracking array	3.4	4.3	5.8	6.8	7.3	7.6	6.8	7.3	6.0	5.0	3.8	2.7	5.6
Latitude tilt Fixed array	3.2	3.8	4.7	5.1	5.2	5.5	5.1	5.5	4.9	4.4	3.5	2.6	4.5
1-axis north-south tracking array	3.8	4.6	6.0	6.8	7.1	7.4	6.6	7.2	6.1	5.3	4.2	3.0	5.7
Latitude tilt +15° Fixed array	3.5	3.9	4.6	4.8	4.7	4.8	4.5	5.1	4.7	4.5	3.7	2.8	4.3
1-axis north-south tracking array	4.0	4.7	5.9	6.5	6.7	6.9	6.2	6.9	6.0	5.3	4.3	3.1	5.6
Two axis tracking	4.0	4.7	6.0	6.8	7.4	7.8	6.9	7.4	6.1	5.3	4.3	3.2	5.8

Miami, Florida (25° 48' N 80° 16' W, 2 m)

	JAN	FEB	MAR	APR	MAY	JUN	JUL	AUG	SEP	OCT	NOV	DEC	YEAR
Latitude tilt -15° Fixed array	3.8	4.7	5.4	6.2	5.6	5.2	5.4	5.4	4.9	4.4	4.3	3.8	4.9
1-axis north-south tracking array	4.8	6.0	7.2	8.0	7.1	6.2	6.7	6.5	6.0	5.7	5.5	4.9	6.2
Latitude tilt Fixed array	4.3	5.2	5.7	6.1	5.3	4.9	5.1	5.3	5.0	4.8	4.9	4.4	5.1

	JAN	FEB	MAR	APR	MAY	JUN	JUL	AUG	SEP	OCT	NOV	DEC	YEAR
1-axis north-south tracking array	5.2	6.4	7.4	8.0	7.0	6.0	6.5	6.4	6.1	5.9	6.0	5.4	6.4
Latitude tilt +15° Fixed array	4.6	5.5	5.7	5.8	4.8	4.3	4.6	4.9	4.9	4.9	5.2	4.8	5.0
1-axis north-south tracking array	5.4	6.6	7.4	7.7	6.6	5.6	6.1	6.1	6.0	6.0	6.2	5.7	6.3
Two axis tracking	5.5	6.6	7.4	8.0	7.2	6.4	6.8	6.5	6.1	6.0	6.3	5.8	6.6
Orlando, Florida (28° 33' N, 81° 20' W, 35 m)													
Latitude tilt -15° Fixed array	3.8	4.6	5.4	6.3	6.3	5.6	5.6	5.4	4.9	4.7	4.3	3.5	5.0
1-axis north-south tracking array	4.9	5.8	7.4	8.6	8.3	6.9	6.8	6.6	6.0	5.9	5.5	4.4	6.4
Latitude tilt Fixed array	4.4	5.1	5.7	6.2	6.0	5.3	5.3	5.2	5.0	5.0	4.9	4.1	5.2
1-axis north-south tracking array	5.3	6.2	7.6	8.5	8.1	6.6	6.6	6.5	6.1	6.2	6.0	4.8	6.5
Latitude tilt +15° Fixed array	4.7	5.3	5.7	5.8	5.4	4.7	4.7	4.8	4.8	5.2	5.2	4.4	5.1
1-axis north-south tracking array	5.6	6.3	7.6	8.3	7.6	6.2	6.2	6.2	5.9	6.2	6.2	5.1	6.5
Two axis tracking	5.6	6.3	7.6	8.6	8.4	7.0	6.9	6.6	6.1	6.2	6.3	5.2	6.7
Atlanta, Georgia (33° 39' N, 84° 36' W, 315 m)													
Latitude tilt -15° Fixed array	2.9	3.6	4.8	5.6	6.3	5.8	5.9	5.8	4.7	4.8	3.7	3.0	4.7
1-axis north-south tracking array	3.7	4.6	6.5	7.6	8.3	7.5	7.5	7.5	5.9	6.2	4.8	3.8	6.1
Latitude tilt Fixed array	3.3	4.0	5.0	5.5	6.0	5.4	5.5	5.7	4.8	5.3	4.2	3.4	4.8
1-axis north-south tracking array	4.0	4.8	6.6	7.5	8.1	7.2	7.3	7.3	6.0	6.5	5.2	4.2	6.2
Latitude tilt +15° Fixed array	3.5	4.1	4.9	5.2	5.3	4.8	4.9	5.3	4.7	5.4	4.5	3.7	4.7
1-axis north-south tracking array	4.2	4.9	6.6	7.3	7.7	6.7	6.9	7.0	5.8	6.6	5.4	4.4	6.1
Two axis tracking	4.2	5.0	6.6	7.6	8.4	7.6	7.6	7.5	6.0	6.6	5.4	4.5	6.4
Honolulu, Hawaii (21° 20' N, 157° 16' W, 5 m)													
Latitude tilt -15° Fixed array	4.0	4.6	5.3	5.7	6.2	6.1	6.3	6.3	5.9	5.0	4.2	3.8	5.3
1-axis north-south tracking array	5.0	5.8	6.8	7.4	7.9	7.6	8.2	8.3	7.7	6.5	5.4	4.8	6.8
Latitude tilt Fixed array	4.5	5.1	5.5	5.7	5.9	5.7	5.9	6.2	6.1	5.4	4.8	4.4	5.4
1-axis north-south tracking array	5.4	6.2	7.0	7.4	7.7	7.3	7.9	8.1	7.9	6.8	5.8	5.3	6.9
Latitude tilt +15° Fixed array	4.8	5.3	5.5	5.4	5.3	5.1	5.3	5.7	6.0	5.6	5.1	4.8	5.3
1-axis north-south tracking array	5.6	6.3	6.9	7.1	7.3	6.9	7.4	7.8	7.7	6.8	6.0	5.6	6.8
Two axis tracking	5.7	6.3	7.0	7.4	8.0	7.8	8.3	8.3	7.9	6.9	6.1	5.6	7.1
New Orleans, Louisiana (29° 59' N, 90° 15' W, 3 m)													
Latitude tilt -15° Fixed array	3.1	4.0	5.0	5.6	6.2	5.8	5.6	5.6	5.2	5.0	3.7	3.1	4.8
1-axis north-south tracking array	3.9	4.9	6.7	7.4	8.1	7.4	6.9	7.0	6.6	6.5	4.7	3.9	6.2
Latitude tilt Fixed array	3.6	4.4	5.2	5.6	5.9	5.5	5.2	5.5	5.4	5.5	4.2	3.6	5.0
1-axis north-south tracking array	4.2	5.2	6.8	7.4	7.9	7.2	6.7	7.0	6.7	6.8	5.1	4.2	6.3

	JAN	FEB	MAR	APR	MAY	JUN	JUL	AUG	SEP	OCT	NOV	DEC	YEAR
Latitude tilt +15° Fixed array	3.8	4.5	5.2	5.3	5.3	4.8	4.7	5.1	5.3	5.7	4.4	3.9	4.8
1-axis north-south tracking array	4.4	5.3	6.8	7.1	7.5	6.7	6.3	6.7	6.6	6.9	5.3	4.5	6.2
Two axis tracking	4.5	5.3	6.8	7.4	8.2	7.6	7.0	7.1	6.7	6.9	5.3	4.5	6.5
Caribou, Maine (46° 52'N, 68° 01'W; 190 m)													
Latitude tilt -15° Fixed array	2.3	3.8	5.1	5.3	5.2	5.6	5.5	5.1	4.1	3.2	1.9	1.8	4.1
1-axis north-south tracking array	2.7	4.7	6.6	7.0	7.3	7.7	7.6	6.9	5.1	4.0	2.2	2.1	5.3
Latitude tilt Fixed array	2.7	4.2	5.3	5.2	5.0	5.2	5.2	4.9	4.2	3.4	2.1	2.1	4.1
1-axis north-south tracking array	3.0	5.0	6.8	7.0	7.1	7.5	7.4	6.8	5.2	4.2	2.4	2.3	5.4
Latitude tilt +15° Fixed array	2.8	4.3	5.3	4.9	4.4	4.5	4.6	4.5	4.1	3.5	2.2	2.2	3.9
1-axis north-south tracking array	3.2	5.1	6.8	6.8	6.8	7.0	7.0	6.5	5.1	4.2	2.5	2.4	5.3
Two axis tracking	3.2	5.1	6.8	7.1	7.4	7.9	7.8	6.9	5.2	4.2	2.5	2.5	5.5
Boston, Massachusetts (41° 40'N, 71° 10'W, 10 m)													
Latitude tilt -15° Fixed array	2.2	3.1	4.1	4.5	5.2	5.9	5.5	5.0	4.9	3.6	2.4	2.0	4.0
1-axis north-south tracking array	2.6	3.9	5.3	5.8	7.3	8.3	7.1	6.4	6.5	4.5	2.9	2.3	5.2
Latitude tilt Fixed array	2.5	3.4	4.3	4.4	4.9	5.4	5.1	4.8	5.0	3.9	2.7	2.3	4.1
1-axis north-south tracking array	2.9	4.2	5.4	5.8	7.1	8.0	6.8	6.3	6.6	4.7	3.1	2.6	5.3
Latitude tilt +15° Fixed array	2.7	3.6	4.2	4.2	4.4	4.8	4.6	4.4	4.9	3.9	2.9	2.5	3.9
1-axis north-south tracking array	3.0	4.3	5.3	5.6	6.7	7.6	6.4	6.0	6.5	4.7	3.3	2.8	5.2
Two axis tracking	3.0	4.3	5.4	5.8	7.4	8.5	7.2	6.4	6.6	4.7	3.3	2.8	5.5
Detroit, Michigan (42° 25'N, 83° 01'W, 190 m)													
Latitude tilt -15° Fixed array	2.0	3.1	3.8	4.9	5.6	5.9	6.0	5.3	4.9	3.8	2.2	1.6	4.1
1-axis north-south tracking array	2.4	3.9	5.0	6.4	7.4	7.8	7.9	6.8	6.2	4.7	2.5	1.8	5.2
Latitude tilt Fixed array	2.3	3.4	3.9	4.8	5.3	5.5	5.6	5.2	5.0	4.1	2.4	1.8	4.1
1-axis north-south tracking array	2.6	4.2	5.1	6.4	7.2	7.5	7.6	6.7	6.3	4.9	2.7	1.9	5.3
Latitude tilt +15° Fixed array	2.4	3.5	3.8	4.5	4.8	4.8	5.0	4.8	4.8	4.1	2.5	1.9	3.9
1-axis north-south tracking array	2.7	4.2	5.0	6.2	6.8	7.1	7.2	6.4	6.1	5.0	2.8	2.0	5.1
Two axis tracking	2.7	4.2	5.1	6.5	7.4	7.9	8.0	6.8	6.3	5.0	2.8	2.1	5.4
Columbia, Missouri (38° 49'N, 92°13'W, 270 m)													
Latitude tilt -15° Fixed array	2.2	2.9	4.0	5.1	6.2	6.0	6.4	5.9	4.8	3.8	2.5	1.8	4.3
1-axis north-south tracking array	2.6	3.2	4.8	6.9	8.5	8.9	9.1	8.2	6.3	4.6	2.8	1.9	5.7
Latitude tilt Fixed array	2.4	3.1	4.1	5.0	5.8	5.5	6.0	5.7	4.9	4.1	2.7	1.9	4.3
1-axis north-south tracking array	2.5	3.4	4.9	6.9	8.3	8.6	8.9	8.1	6.3	4.8	3.0	2.0	5.7
Latitude tilt +15° Fixed array	2.4	3.1	4.1	4.7	5.2	4.8	5.3	5.3	4.8	4.1	2.8	2.0	4.1

	JAN	FEB	MAR	APR	MAY	JUN	JUL	AUG	SEP	OCT	NOV	DEC	YEAR
1-axis north-south tracking array	2.6	3.5	4.9	6.7	7.9	8.2	8.4	7.8	6.2	4.8	3.1	2.1	5.5
Two axis tracking	2.6	3.5	5.0	6.9	8.6	9.2	9.3	8.3	6.3	4.8	3.1	2.1	5.8
Great Falls, Montana (47° 29'N, 111° 22'W, 1115 m)													
Latitude tilt -15° Fixed array	2.5	3.7	5.2	5.6	6.0	6.6	7.6	6.9	5.6	4.5	2.9	2.3	5.0
1-axis north-south tracking array	3.0	4.6	6.9	7.5	8.0	9.0	11.3	9.8	7.6	5.9	3.5	2.7	6.7
Latitude tilt Fixed array	2.9	4.1	5.5	5.6	5.7	6.1	7.1	6.7	5.8	5.0	3.3	2.6	5.0
1-axis north-south tracking array	3.3	4.9	7.2	7.5	7.8	8.7	10.9	9.7	7.7	6.2	3.8	3.0	6.7
Latitude tilt +15° Fixed array	3.1	4.2	5.4	5.2	5.1	5.4	6.2	6.1	5.6	5.1	3.5	2.8	4.8
1-axis north-south tracking array	3.4	5.0	7.1	7.3	7.4	8.2	10.4	9.3	7.6	6.3	4.0	3.1	6.6
Two axis tracking	3.5	5.0	7.2	7.6	8.0	9.2	11.4	9.8	7.7	6.3	4.0	3.2	6.9
Omaha, Nebraska (41° 25'N, 26° 5'W, 320 m)													
Latitude tilt -15° Fixed array	3.4	4.3	4.7	5.4	6.4	6.7	6.5	6.4	5.3	4.5	3.4	2.8	5.0
1-axis north-south tracking array	4.2	5.7	6.3	7.1	9.3	9.6	9.0	8.9	7.1	5.8	4.1	3.3	6.7
Latitude tilt Fixed array	4.0	4.8	5.0	5.3	6.0	6.2	6.1	6.2	5.4	4.9	3.8	3.2	5.1
1-axis north-south tracking array	4.7	6.1	6.5	7.1	9.1	9.2	8.7	8.8	7.2	6.2	4.5	3.7	6.8
Latitude tilt +15° Fixed array	4.3	5.0	5.0	5.0	5.4	5.4	5.4	5.7	5.2	5.1	4.1	3.5	4.9
1-axis north-south tracking array	4.9	6.2	6.4	6.9	8.7	8.7	8.3	8.5	7.1	6.3	4.7	3.9	6.7
Two axis tracking	5.0	6.2	6.5	7.1	9.4	9.8	9.1	9.0	7.2	6.3	4.7	4.0	7.0
Elko, Nevada (40o 50'N, 115o47' W, 1550 meters)													
Latitude tilt -15° Fixed array	3.7	5.3	5.8	6.6	7.3	7.9	8.2	8.1	7.5	5.9	4.0	3.3	6.1
1-axis north-south tracking array	4.6	7.0	7.9	9.4	10.9	11.8	12.0	11.5	10.5	7.8	5.0	4.1	8.6
Latitude tilt Fixed array	4.3	5.9	6.1	6.6	6.9	7.3	7.6	7.9	7.7	6.5	4.5	3.9	6.3
1-axis north-south tracking array	5.1	7.4	8.1	9.4	10.7	11.4	11.7	11.4	10.7	8.3	5.5	4.6	8.7
Latitude tilt +15° Fixed array	4.7	6.2	6.1	6.1	6.2	6.3	6.6	7.2	7.6	6.7	4.9	4.3	6.1
1-axis north-south tracking array	5.4	7.7	8.1	9.1	10.2	10.8	11.0	10.9	10.6	8.4	5.7	4.9	8.6
Two axis tracking	5.4	7.7	8.1	9.4	11.1	12.2	12.3	11.6	10.8	8.4	5.7	4.9	9.0
Las Vegas, Nevada (36o 05'N, 115o10' W, 665 meters)													
Latitude tilt -15° Fixed array	4.8	6.1	7.3	8.3	8.4	8.5	7.9	7.9	7.6	6.3	5.1	4.1	6.9
1-axis north-south tracking array	6.2	8.2	10.2	11.9	12.4	12.5	11.3	11.1	10.7	8.5	6.6	5.2	9.6
Latitude tilt Fixed array	5.6	6.9	7.7	8.2	7.9	7.8	7.4	7.7	7.9	6.9	5.9	4.9	7.1
1-axis north-south tracking array	6.9	8.8	10.6	11.9	12.1	12.0	10.9	10.9	11.0	8.9	7.2	5.8	9.8
Latitude tilt +15° Fixed array	6.2	7.3	7.7	7.7	7.0	6.7	6.4	7.1	7.8	7.2	6.3	5.4	6.9
1-axis north-south tracking array	7.3	9.1	10.6	11.6	11.5	11.3	10.3	10.5	10.8	9.1	7.6	6.2	9.7

	JAN	FEB	MAR	APR	MAY	JUN	JUL	AUG	SEP	OCT	NOV	DEC	YEAR
Two axis tracking	7.4	9.1	10.6	12.0	12.6	12.8	11.5	11.1	10.9	9.1	7.6	6.3	10.1
Albuquerque, New Mexico (35° 03'N, 106° 37'W, 1620 m)													
Latitude tilt -15° Fixed array	4.5	5.6	6.5	7.9	8.2	8.1	7.8	7.6	7.1	6.5	5.3	4.4	6.6
1-axis north-south tracking array	5.9	7.4	8.8	10.9	11.6	11.9	10.5	10.3	9.8	8.6	7.1	5.7	9.0
Latitude tilt Fixed array	5.3	6.3	6.9	7.8	7.8	7.4	7.3	7.4	7.4	7.1	6.2	5.3	6.8
1-axis north-south tracking array	6.5	8.0	9.1	10.9	11.3	11.5	10.1	10.2	10.0	9.1	7.7	6.4	9.2
Latitude tilt +15° Fixed array	5.7	6.7	6.9	7.4	6.9	6.4	6.4	6.8	7.2	7.4	6.7	5.8	6.7
1-axis north-south tracking array	6.8	8.2	9.1	10.5	10.8	10.8	9.5	9.8	9.9	9.3	8.1	6.8	9.1
Two axis tracking	6.9	8.2	9.2	10.9	11.7	12.2	10.6	10.4	10.0	9.3	8.2	6.9	9.5
Syracuse, New York (43° 07'N, 76° 07'W, 125 m)													
Latitude tilt -15° Fixed array	1.8	2.3	3.4	4.7	5.1	5.4	5.6	5.3	4.3	3.1	1.8	1.3	3.7
1-axis north-south tracking array	2.0	2.8	4.3	6.3	6.5	7.0	7.1	6.6	5.4	3.7	2.0	1.5	4.6
Latitude tilt Fixed array	1.9	2.4	3.5	4.6	4.8	5.0	5.2	5.1	4.4	3.3	1.9	1.5	3.7
1-axis north-south tracking array	2.2	2.9	4.4	6.3	6.4	6.8	6.9	6.5	5.5	3.9	2.1	1.6	4.6
Latitude tilt +15° Fixed array	2.0	2.5	3.4	4.3	4.3	4.4	4.7	4.7	4.3	3.3	2.0	1.5	3.5
1-axis north-south tracking array	2.2	3.0	4.3	6.0	6.0	6.4	6.5	6.2	5.4	3.9	2.2	1.7	4.5
Two axis tracking	2.3	3.0	4.4	6.3	6.6	7.2	7.2	6.6	5.5	3.9	2.2	1.7	4.7
Raleigh-Durham, North Carolina (35° 52'N, 78° 45' W, 135 m)													
Latitude tilt -15° Fixed array	2.8	3.9	4.7	5.6	5.6	5.8	5.8	5.5	4.8	4.3	3.7	2.8	4.6
1-axis north-south tracking array	3.5	5.0	6.3	7.7	7.2	7.3	7.2	7.0	6.0	5.4	4.7	3.4	5.9
Latitude tilt Fixed array	3.2	4.3	4.9	5.6	5.3	5.4	5.4	5.4	4.9	4.6	4.2	3.2	4.7
1-axis north-south tracking array	3.8	5.3	6.5	7.7	7.0	7.0	6.9	6.9	6.1	5.6	5.1	3.7	6.0
Latitude tilt +15° Fixed array	3.4	4.4	4.9	5.2	4.8	4.8	4.8	5.0	4.8	4.8	4.5	3.5	4.6
1-axis north-south tracking array	4.0	5.4	6.4	7.4	6.6	6.6	6.5	6.6	5.9	5.7	5.3	3.9	5.9
Two axis tracking	4.0	5.4	6.5	7.7	7.2	7.4	7.3	7.0	6.1	5.7	5.3	4.0	6.1
Bismarck, North Dakota (46° 46'N, 100° 45' W, 500 m)													
Latitude tilt -15° Fixed array	2.8	4.1	4.9	5.4	6.1	6.5	7.1	6.7	5.4	4.2	2.8	2.6	4.9
1-axis north-south tracking array	3.3	5.3	6.5	7.3	8.7	9.1	9.8	9.5	7.4	5.3	3.4	3.1	6.6
Latitude tilt Fixed array	3.2	4.6	5.1	5.3	5.8	6.0	6.6	6.5	5.6	4.6	3.2	3.0	5.0
1-axis north-south tracking array	3.7	5.6	6.7	7.2	8.5	8.8	9.5	9.4	7.5	5.6	3.7	3.5	6.7
Latitude tilt +15° Fixed array	3.4	4.8	5.1	5.0	5.2	5.3	5.8	6.0	5.5	4.7	3.4	3.3	4.8
1-axis north-south tracking array	3.9	5.8	6.7	7.0	8.1	8.3	9.0	9.1	7.4	5.7	3.9	3.7	6.5

	JAN	FEB	MAR	APR	MAY	JUN	JUL	AUG	SEP	OCT	NOV	DEC	YEAR
Two axis tracking	3.9	5.8	6.7	7.3	8.8	9.3	10.0	9.6	7.5	5.7	3.1	3.7	6.9
Oklahoma City, Oklahoma (35° 24'N, 97 ° 36' W, 395 m)													
Latitude tilt -15° Fixed array	3.7	4.1	5.1	6.1	5.9	6.5	6.6	6.7	5.7	5.1	4.1	3.1	5.2
1-axis north-south tracking array	4.9	5.4	6.8	8.4	7.9	8.8	8.8	8.9	7.6	6.6	5.4	3.9	7.0
Latitude tilt Fixed array	4.3	4.6	5.3	6.0	5.6	6.1	6.2	6.5	5.9	5.5	4.7	3.6	5.4
1-axis north-south tracking array	5.4	5.7	7.0	8.4	7.7	8.5	8.5	8.8	7.7	7.0	5.8	4.3	7.1
Latitude tilt +15° Fixed array	4.7	4.8	5.3	5.7	5.0	5.3	5.5	6.0	5.7	5.7	5.0	3.9	5.2
1-axis north-south tracking array	5.7	5.8	7.0	8.1	7.3	8.0	8.0	8.5	7.6	7.1	6.1	4.5	7.0
Two axis tracking	5.7	5.9	7.1	8.4	8.0	9.0	8.9	9.0	7.7	7.1	6.2	4.6	7.3
Medford, Oregon (42° 22'N, 122° 52' W, 400 m)													
Latitude tilt -15° Fixed array	1.8	3.2	4.6	5.6	6.6	7.2	8.0	7.4	6.3	4.3	2.3	1.3	4.9
1-axis north-south tracking array	2.0	3.8	5.8	7.4	9.0	10.2	11.7	10.3	8.4	5.3	2.7	1.4	6.5
Latitude tilt Fixed array	2.0	3.4	4.8	5.5	6.3	6.7	7.5	7.2	6.5	4.6	2.6	1.4	4.9
1-axis north-south tracking array	2.1	4.0	6.0	7.3	8.7	9.9	11.3	10.2	8.6	5.6	2.9	1.5	6.5
Latitude tilt +15° Fixed array	2.1	3.5	4.7	5.2	5.6	5.8	6.6	6.7	6.3	4.7	2.7	1.5	4.6
1-axis north-south tracking array	2.2	4.1	5.9	7.1	8.3	9.3	10.7	9.8	8.4	5.7	3.0	1.6	6.3
Two axis tracking	2.2	4.1	6.0	7.4	9.0	10.4	11.8	10.4	8.5	5.7	3.0	1.6	6.7
Pittsburgh, Pennsylvania (40° 30'N, 80°13' W, 303 m)													
Latitude tilt -15° Fixed array	2.0	2.4	3.3	4.7	5.2	5.4	5.6	5.2	4.5	3.6	2.2	1.4	3.8
1-axis north-south tracking array	2.4	2.8	4.1	6.1	6.7	6.9	7.0	6.4	5.6	4.4	2.6	1.6	4.7
Latitude tilt Fixed array	2.3	2.6	3.4	4.6	4.9	5.1	5.3	5.1	4.6	3.8	2.4	1.6	3.8
1-axis north-south tracking array	2.6	3.0	4.2	6.0	6.5	6.6	6.8	6.3	5.7	4.6	2.8	1.7	4.7
Latitude tilt +15° Fixed array	2.4	2.6	3.3	4.3	4.4	4.5	4.7	4.7	4.4	3.9	2.5	1.7	3.6
1-axis north-south tracking array	2.7	3.0	4.1	5.8	6.2	6.3	6.4	6.1	5.5	4.6	2.8	1.8	4.6
Two axis tracking	2.7	3.0	4.2	6.1	6.8	7.0	7.1	6.5	5.7	4.6	2.8	1.8	4.9
Nashville, Tennessee (36° 07'N, 86°41' W, 180 m)													
Latitude tilt -15° Fixed array	2.4	3.2	4.3	5.3	5.5	6.3	6.0	5.8	5.1	4.2	3.0	2.2	4.5
1-axis north-south tracking array	2.8	3.9	5.6	6.7	6.9	8.0	7.7	7.1	6.6	5.3	3.6	2.6	5.6
Latitude tilt Fixed array	2.7	3.5	4.5	5.3	5.3	5.9	5.7	5.7	5.2	4.6	3.3	2.6	4.5
1-axis north-south tracking array	3.1	4.2	5.7	6.7	6.7	7.7	7.4	7.0	6.7	5.6	3.9	2.9	5.6
Latitude tilt +15° Fixed array	2.9	3.6	4.4	5.0	4.8	5.2	5.1	5.3	5.0	4.7	3.5	2.8	4.4
1-axis north-south tracking array	3.2	4.2	5.7	6.4	6.3	7.3	7.0	6.7	6.6	5.7	4.1	3.1	5.5

	JAN	FEB	MAR	APR	MAY	JUN	JUL	AUG	SEP	OCT	NOV	DEC	YEAR
Two axis tracking	3.3	4.2	5.7	6.7	6.9	8.2	7.8	7.1	6.7	5.7	4.1	3.1	5.8
Austin, Texas (30° 18'N, 97° 42' W, 190 m)													
Latitude tilt -15° Fixed array	3.6	4.1	5.2	5.1	5.8	6.4	6.6	6.2	5.5	4.9	4.0	3.6	5.1
1-axis north–south tracking array	4.7	5.4	7.1	6.6	7.5	8.4	8.8	8.1	7.1	6.4	5.3	4.7	6.7
Latitude tilt Fixed array	4.1	4.6	5.5	5.0	5.5	5.9	6.2	6.0	5.6	5.4	4.6	4.2	5.2
1-axis north–south tracking array	5.1	5.7	7.2	6.6	7.3	8.0	8.5	8.0	7.2	6.8	5.7	5.2	6.8
Latitude tilt +15° Fixed array	4.5	4.8	5.4	4.8	5.0	5.2	5.5	5.6	5.5	5.5	4.9	4.6	5.1
1-axis north–south tracking array	5.3	5.9	7.2	6.4	6.9	7.5	8.0	7.6	7.1	6.9	5.9	5.5	6.7
Two axis tracking	5.4	5.9	7.3	6.7	7.5	8.5	8.9	8.1	7.2	6.9	6.0	5.6	7.0
Brownsville, Texas (25° 54' N, 97° 26'W, 5 m)													
Latitude tilt -15° Fixed array	3.3	4.1	5.0	6.2	6.0	6.4	6.8	6.7	5.7	4.8	3.7	3.2	5.2
1-axis north–south tracking array	4.1	5.2	6.6	7.8	7.4	8.6	9.2	9.1	7.2	6.2	4.8	3.9	6.7
Latitude tilt Fixed array	3.7	4.5	5.2	6.1	5.7	5.9	6.4	6.6	5.8	5.2	4.2	3.6	5.3
1-axis north–south tracking array	4.4	5.5	6.8	7.8	7.2	8.3	8.9	9.0	7.3	6.5	5.2	4.3	6.8
Latitude tilt +15° Fixed array	4.0	4.7	5.2	5.8	5.2	5.2	5.6	6.1	5.7	5.3	4.5	3.9	5.1
1-axis north–south tracking array	4.6	5.6	6.7	7.6	6.8	7.8	8.4	8.6	7.2	6.6	5.4	4.5	6.6
Two axis tracking	4.7	5.6	6.8	7.9	7.4	8.8	9.4	9.1	7.3	6.7	5.4	4.6	7.0
El Paso, Texas (31° 45'N, 106° 20'W, 1200 m)													
Latitude tilt -15° Fixed array	4.7	6.1	6.9	7.8	8.3	8.2	7.7	7.4	6.8	6.5	5.2	4.6	6.7
1-axis north–south tracking array	6.3	8.1	9.5	11.0	11.8	11.5	10.6	10.0	9.3	8.9	7.0	6.0	9.2
Latitude tilt Fixed array	5.5	6.9	7.4	7.8	7.8	7.6	7.2	7.2	7.0	7.2	6.1	5.4	6.9
1-axis north–south tracking array	6.9	8.7	9.8	11.0	11.5	11.1	10.2	9.9	9.5	9.4	7.7	6.7	9.4
Latitude tilt +15° Fixed array	6.0	7.2	7.4	7.3	6.9	6.5	6.3	6.6	6.9	7.5	6.6	5.6	6.8
1-axis north–south tracking array	7.3	9.0	9.8	10.7	10.9	10.4	9.6	9.5	9.3	9.6	8.0	7.2	9.3
Two axis tracking	7.4	9.0	9.8	11.0	1190.0	11.8	10.8	10.1	9.5	9.6	8.1	7.3	9.7
Fort Worth, Texas (32°50'N 90° 20'W, 225 m)													
Latitude tilt -15° Fixed array	3.3	4.1	5.3	5.1	5.9	6.7	6.9	6064.0	5.9	4.9	4.0	3.3	5.2
1-axis north–south tracking array	4.1	5.3	7.0	6.8	7.7	9.1	9.6	9.0	8.0	6.3	5.2	4.2	6.9
Latitude tilt Fixed array	3.8	4.6	5.6	5.1	5.6	6.2	6.4	6.5	6.1	5.3	4.6	3.9	5.3
1-axis north–south tracking array	4.5	5.6	7.2	6.8	7.5	8.7	9.3	8.8	8.1	6.7	5.6	4.6	697.0
Latitude tilt +15° Fixed array	4.1	4.8	5.6	4.9	5.0	5.4	5.7	6.0	6.0	5.5	4.9	4.2	5.2
1-axis north–south tracking array	4.7	5.8	7.2	6.6	7.1	8.2	8.8	8.5	8.0	6.8	5.9	4.9	6.9

	JAN	FEB	MAR	APR	MAY	JUN	JUL	AUG	SEP	OCT	NOV	DEC	YEAR
Two axis tracking	4.8	4.8	7.2	6.8	7.8	9.3	9.8	9.0	8.1	6.8	5.9	5.0	7.2

Bryce Canyon, Utah (37° 42'N, 112° 09'W, 2310 m)

	JAN	FEB	MAR	APR	MAY	JUN	JUL	AUG	SEP	OCT	NOV	DEC	YEAR
Latitude tilt -15° Fixed array	4.5	5.5	6.6	7.7	7.7	7.9	7.5	7.3	7.3	6.2	4.9	4.3	6.5
1-axis north-south tracking array	5.8	7.5	9.2	10.9	11.7	12.0	10.9	10.4	10.4	8.3	6.5	5.3	9.1
Latitude tilt Fixed array	5.3	6.2	7.0	7.6	7.3	7.3	7.0	7.1	7.6	6.8	5.7	5.1	6.7
1-axis north-south tracking array	6.4	8.0	9.5	10.9	11.4	11.6	10.6	10.3	10.6	8.8	7.1	6.0	9.3
Latitude tilt +15° Fixed array	5.8	6.5	7.0	7.2	6.5	6.3	6.1	6.5	7.4	7.1	6.1	5.6	6.5
1-axis north-south tracking array	6.8	8.2	9.5	10.6	10.9	11.0	10.0	9.9	10.4	8.9	7.4	6.4	9.2
Two axis tracking	6.8	8.2	9.5	11.0	11.8	12.3	11.1	10.4	10.6	8.9	7.5	6.5	9.6

Seattle, Washington (47° 27'N, 122° 18'W, 120 m)

	JAN	FEB	MAR	APR	MAY	JUN	JUL	AUG	SEP	OCT	NOV	DEC	YEAR
Latitude tilt -15° Fixed array	1.3	2.1	3.7	4.6	5.4	5.5	6.5	5.9	4.6	2.9	1.5	1.0	3.8
1-axis north-south tracking array	1.4	2.4	4.7	6.1	7.2	7.3	8.9	8.0	5.9	3.5	1.7	1.2	4.9
Latitude tilt Fixed array	1.4	2.2	3.9	4.6	5.1	5.1	6.1	5.8	4.7	3.1	1.7	1.2	3.7
1-axis north-south tracking array	1.5	2.5	4.8	6.0	7.0	7.0	8.7	7.9	6.0	3.6	1.8	1.3	4.9
Latitude tilt +15° Fixed array	1.4	2.3	3.8	4.3	4.6	4.5	5.4	5.3	4.5	3.1	1.7	1.2	3.5
1-axis north-south tracking array	1.6	2.5	4.8	5.8	6.6	6.6	8.2	7.5	5.9	3.6	1.9	1.3	4.7
Two axis tracking	1.6	2.5	4.9	6.1	7.2	7.4	9.1	8.0	6.0	3.7	1.9	1.3	5.0

Madison, Wisconsin (43° 08' N, 89° 20' W, 260 meters)

	JAN	FEB	MAR	APR	MAY	JUN	JUL	AUG	SEP	OCT	NOV	DEC	YEAR
Latitude tilt -15° Fixed array	2.7	3.7	5.0	4.9	5.5	5.8	6.1	6.0	5.2	3.8	2.6	2.0	4.4
1-axis north-south tracking array	3.2	4.6	6.4	6.4	7.5	7.7	8.0	8.0	6.7	4.7	3.1	2.2	5.7
Latitude tilt Fixed array	3.1	4.1	5.3	4.9	5.2	5.4	5.7	5.9	5.3	4.1	2.9	2.3	4.5
1-axis north-south tracking array	3.6	4.9	6.6	6.4	7.3	7.5	7.7	7.9	6.8	4.9	3.4	2.4	5.8
Latitude tilt +15° Fixed array	3.4	4.2	5.2	4.6	4.6	4.8	5.1	5.4	5.2	4.2	3.1	2.4	4.4
1-axis north-south tracking array	3.8	5.0	6.6	6.2	6.9	7.0	7.3	7.5	6.7	5.0	3.5	2.6	5.7
Two axis tracking	3.8	5.0	6.7	6.5	7.6	7.9	8.1	8.0	6.8	5.0	3.6	2.6	6.0

VENEZUELA

Caracas (10° 50'N; 66° 88'W, 860 m)	JAN	FEB	MAR	APR	MAY	JUN	JUL	AUG	SEP	OCT	NOV	DEC	YEAR
Latitude tilt -15° Fixed array	5.0	6.0	6.1	6.0	5.0	5.2	5.6	5.8	5.7	4.9	4.5	4.7	5.4
1-axis north-south tracking array	6.0	7.3	7.8	7.7	6.6	6.8	7.2	7.6	7.3	6.2	5.5	5.6	6.8
Latitude tilt Fixed array	5.6	6.5	6.3	5.9	4.8	4.9	5.2	5.7	5.8	5.2	5.0	5.4	5.5
1-axis north-south tracking array	6.8	8.0	8.1	7.7	6.3	6.4	6.8	7.4	7.5	6.7	6.2	6.4	7.0
Latitude tilt +15° Fixed array	6.1	6.7	6.2	5.5	4.3	4.3	4.6	5.2	5.6	5.3	5.3	5.8	5.4

1-axis north-south tracking array	7.3	8.2	7.9	7.1	5.6	5.5	5.9	6.7	7.2	6.8	6.6	6.9	6.8
Two axis tracking	7.3	8.2	8.1	7.8	6.6	6.9	7.3	7.6	7.5	6.8	6.6	7.0	7.3

* Insolation data compiled from Sandia National Laboratories, New Mexico State University, and Universidad Autónoma de Ciudad Juárez.

Appendix B: Friction Loss Factors

TABLE B.1
Friction Loss Factors in Rigid PVC Pipe

Flow (L/s)	Pipe size (inches)								
	0.5	**0.75**	**1**	**1.25**	**1.5**	**2**	**2.5**	**3**	**4**
0.10	4.20	1	0.25	0.08					
0.15	8.80	2.20	0.53	0.17	0.07				
0.20	15	3.70	0.90	0.28	0.12				
0.25	22	5.50	1.35	0.44	0.18				
0.30	31	7.80	1.90	0.60	0.25				
0.35	41	10	2.45	0.80	0.34				
0.40	53	13	3.10	1	0.43				
0.45	66	16.30	4	1.25	0.54	0.13			
0.50		19	4.80	1.50	0.65	0.16			
0.55		23.50	5.60	1.80	0.78	0.19			
0.60		27.50	6.60	2.10	0.90	0.22			
0.65		32	7.80	1.40	1.04	0.25			
0.70		36	8.70	2.70	1.19	0.28			
0.75		41	9.90	3.10	1.32	0.33	0.10		
0.80		45	11	3.50	1.05	0.37	0.12		
0.85		52	12.50	4	1.70	0.41	0.14		
0.90		57	14	4.50	1.90	0.45	0.15		
0.95		0.63	15	4.90	2.10	0.50	0.17		
1			16.50	5.40	2.25	0.55	0.18	0.08	
1.05			18	5.80	2.50	0.60	0.20	0.09	
1.10			19.50	6.30	2.70	0.67	0.22	0.10	
1.15			21.50	6.90	2.95	0.71	0.24	0.10	
1.20			23	7.30	3.20	0.78	0.26	0.11	
1.30			26.50	8.60	3.75	0.90	0.29	0.13	
1.40			30	10	4.25	1	0.34	0.15	
1.50			35	11.20	4.90	1.15	0.39	0.17	
1.60			39	120.50	5.50	1.30	0.43	0.19	
1.70			44	14.20	6.05	1.45	0.49	0.21	
1.80			49	15.90	6.90	1.60	0.54	0.24	
1.90			55	17.40	7.50	1.80	0.60	0.26	
2			60	19	8	2	0.66	0.28	
2.20				22.50	9.70	2.35	0.79	0.34	
2.40				26.80	11.50	2.75	0.90	0.40	
2.60				31	13.30	3.20	1.05	0.45	
2.80				35.10	15.20	3.70	1.20	0.52	
3				40	17	4.20	1.36	0.60	
3.20				45	19.30	4.70	1.52	0.68	
3.40				50	21.90	5.25	1.70	0.75	
3.60				56	24	5.80	1.90	0.84	0.20
3.80				62	26	6.30	2.10	0.90	0.22
4				69	29	7	2.30	1	0.24

TABLE B.1 (continued)
Friction Loss Factors in Rigid PVC Pipe

				Pipe size (inches)					
Flow (L/s)	0.5	0.75	1	1.25	1.5	2	2.5	3	4
4.50					36	8.80	2.80	1.20	0.30
5					44	10.50	3.50	1.50	0.37
5.50					62	12.50	4.20	1.75	0.44
6						14.70	4.90	2.10	0.52
6.50						17	5.60	2.40	0.60
7						19.50	6.50	2.80	0.70

Notes: Approximate factors in meters per 100 m (percentages). New rigid PVC pipe. L/s = liters per second.

TABLE B.2
Friction Loss Factors in Galvanized Steel Pipe

				Pipe size (inches)					
Flow (L/s)	0.5	0.75	1	1.25	1.5	2	2.5	3	4
0.10	5.90	1.58	0.38	0.12					
0.15	12.25	3.40	0.82	0.26					
0.20	21.45	5.65	1.40	0.44	0.19				
0.25	31.65	8.50	2.10	0.68	0.28				
0.30	44.91	11.90	2.90	0.92	0.40				
0.35	58.20	15.80	3.80	1.20	0.52				
0.40	75.50	19.90	4.80	1.55	0.67				
0.45	91.90	25	6	1.93	0.84				
0.50		30	7.30	2.35	1	0.25			
0.55		36	8.70	2.75	1.20	0.30			
0.60		42	10.20	3.25	1.40	0.35			
0.65		48	11.90	3.80	1.63	0.40			
0.70		55	13.6	4.35	1.82	0.46			
0.75		63	15.40	4.90	2.15	0.52	0.17		
0.80			17.40	5.55	2.40	0.59	0.19		
0.85			19.40	6.15	2.65	0.68	0.21		
0.90			21.80	6.90	2.90	0.74	0.23		
0.95			24	7.50	3.25	0.82	0.28		
1			26.20	8.20	3.60	0.80	0.28	0.12	
1.05			28.50	9	3.90	0.97	0.31	0.13	
1.10			31	9.80	4.20	1.05	0.34	0.15	
1.15			34.60	10.60	4.80	1.15	0.37	0.16	
1.20			36	11.50	5	1.25	0.39	0.17	
1.30			42.50	13.30	5.70	1.45	0.45	0.20	
1.40			48	15.30	6.60	1.65	0.52	0.23	
1.50			55	17.50	7.65	1.90	0.59	0.26	
1.60			62	19.50	8.45	2.10	0.67	0.29	
1.70			69	22	9.50	2.35	0.75	0.33	
1.80				24.20	10.50	2.60	0.82	0.36	
1.90				24.50	11.70	2.85	0.90	0.40	

TABLE B.2 (continued)
Friction Loss Factors in Galvanized Steel Pipe

				Pipe size (inches)					
Flow (L/s)	**0.5**	**0.75**	**1**	**1.25**	**1.5**	**2**	**2.5**	**3**	**4**
2				29.50	12.80	3.20	1	0.44	
2.20				35	15.30	3.80	1.20	0.52	
2.40				42	17.90	4.45	1.40	0.61	
2.60				48.50	20.50	5.15	1.60	0.71	0.17
2.80				55	24	5.95	1.85	0.82	0.20
3				62.50	26.70	6.70	2.10	0.92	0.22
3.20					30	7.60	2.35	1.02	0.25
3.40					34	8.40	2.65	1.15	0.28
3.60					38	9.40	2.95	1.28	0.32
3.80					41	10.30	3.25	1.42	0.35
4					45	11.20	3.55	1.55	0.38
4.50					56	14	4.45	1.95	0.46
5						17	5.45	2.25	0.56
5.50						20	6.50	2.80	0.68
6						24	7.50	3.35	0.80
6.50						28	8.85	3.90	0.92
7						32	10	4.45	1.05

Notes: Approximate factors in meters per 100 (percentages). New pipe. L/s = liters per second.

Appendix C: Present Value Factors

TABLE C.1
Present Value Factor (PVF) of a Payment with Interest

Year	1%	2%	3%	4%	5%	6%	7%	8%	9%	10%	11%	12%	13%	14%	15%
1	0.9901	0.9804	0.9709	0.9615	0.9524	0.9434	0.9346	0.9259	0.9174	0.9091	0.9009	0.8929	0.8850	0.8772	0.8696
2	0.9803	0.9612	0.9426	0.9246	0.9070	0.8900	0.8734	0.8573	0.8417	0.8264	0.8116	0.7972	0.7831	0.7695	0.7561
3	0.9706	0.9423	0.8890	0.8890	0.8638	0.8396	0.8163	0.7938	0.7722	0.7513	0.7312	0.7118	0.6931	0.6750	0.6575
4	0.9610	0.9238	0.8548	0.8548	0.8227	0.7921	0.7629	0.7350	0.7084	0.6830	0.6587	0.6355	0.6133	0.5921	0.5718
5	0.9515	0.9057	0.8219	0.8219	0.7835	0.7473	0.7130	0.6806	0.6499	0.6209	0.5935	0.5674	0.5428	0.5194	0.4972
6	0.9420	0.8880	0.7903	0.7903	0.7462	0.7050	0.6663	0.6302	0.5963	0.5645	0.5346	0.5066	0.4803	0.4556	0.4323
7	0.9327	0.8706	0.7599	0.7599	0.7107	0.6651	0.6227	0.5835	0.5470	0.5132	0.4817	0.4523	0.4251	0.3996	0.3759
8	0.9235	0.8535	0.7307	0.7307	0.6768	0.6274	0.5820	0.5403	0.5019	0.4665	0.4339	0.4039	0.3762	0.3506	0.3269
9	0.9143	0.8368	0.7026	0.7026	0.6446	0.5919	0.5439	0.5002	0.4604	0.4241	0.3909	0.3606	0.3329	0.3075	0.2843
10	0.9053	0.8203	0.6756	0.6756	0.6139	0.5584	0.5083	0.4632	0.4224	0.3855	0.3522	0.3220	0.2946	0.2697	0.2472
11	0.8963	0.8043	0.6496	0.6496	0.5847	0.5268	0.4751	0.4289	0.3875	0.3505	0.3173	0.2875	0.2607	0.2366	0.2149
12	0.8874	0.7885	0.6246	0.6246	0.5568	0.4970	0.4440	0.3971	0.3555	0.3186	0.2858	0.2567	0.2307	0.2076	0.1869
13	0.8787	0.7730	0.6006	0.6006	0.5303	0.4688	0.4150	0.3677	0.3262	0.2897	0.2575	0.2292	0.2042	0.1821	0.1625
14	0.8700	0.7579	0.5775	0.5775	0.5051	0.4423	0.3878	0.3405	0.2992	0.2633	0.2320	0.2046	0.1807	0.1597	0.1413
15	0.8613	0.7430	0.5553	0.5553	0.4810	0.4173	0.3624	0.3152	0.2745	0.2394	0.2090	0.1827	0.1599	0.1401	0.1229
16	0.8528	0.7284	0.5339	0.5339	0.4581	0.3936	0.3387	0.2919	0.2519	0.2176	0.1883	0.1631	0.1415	0.1229	0.1069
17	0.8444	0.7142	0.5134	0.5134	0.4363	0.3714	0.3166	0.2703	0.2311	0.1978	0.1696	0.1456	0.1252	0.1078	0.0929
18	0.8360	0.7002	0.4936	0.4936	0.4155	0.3503	0.2959	0.2502	0.2120	0.1799	0.1528	0.1300	0.1108	0.0946	0.0808
19	0.8277	0.6864	0.4746	0.4746	0.3957	0.3305	0.2765	0.2317	0.1945	0.1635	0.1377	0.1161	0.0981	0.0829	0.0703
20	0.8195	0.6730	0.4564	0.4564	0.3769	0.3118	0.2584	0.2145	0.1784	0.1486	0.1240	0.1037	0.0868	0.0728	0.0611
21	0.8114	0.6598	0.4388	0.4388	0.3589	0.2942	0.2415	0.1987	0.1637	0.1351	0.1117	0.0926	0.0768	0.0638	0.0531
22	0.8034	0.6468	0.4220	0.4220	0.3418	0.2775	0.2257	0.1839	0.1502	0.1228	0.1007	0.0826	0.0680	0.0560	0.0462
23	0.7954	0.6342	0.4057	0.4057	0.3256	0.2618	0.2109	0.1703	0.1378	0.1117	0.0907	0.0738	0.0601	0.0491	0.0402
24	0.7876	0.6217	0.3901	0.3901	0.3101	0.2470	0.1971	0.1577	0.1264	0.1015	0.0817	0.0659	0.0532	0.0431	0.0349
25	0.7798	0.6095	0.3751	0.3751	0.2953	0.2330	0.1842	0.1460	0.1160	0.0923	0.0736	0.0588	0.0471	0.0378	0.0304

TABLE C.2
PVFA: Present Value Factor of Fixed Annual Payments

Year	1%	2%	3%	4%	5%	6%	7%	8%	9%	10%	11%	12%	13%	14%	15%
1	0.9901	0.9804	0.9709	0.9615	0.9524	0.9434	0.9346	0.9259	0.9174	0.9091	0.9009	0.8929	0.8850	0.8772	0.8696
2	1.9704	1.9416	1.9135	1.8861	1.8594	1.8334	1.8080	1.7833	1.7591	1.7355	1.7125	1.6901	1.6681	1.6467	1.6257
3	2.9410	2.8839	2.8286	2.7751	2.7232	2.6730	2.6243	2.5771	2.5313	2.4869	2.4437	2.4018	2.3612	2.3216	2.2832
4	3.9020	3.8077	3.7171	3.6299	3.5460	3.4651	3.3872	3.3121	3.2397	3.1699	3.1024	3.0373	2.9745	2.9137	2.8550
5	4.8534	4.7135	4.5797	4.4518	4.3295	4.2124	4.1002	3.9927	3.8897	3.7908	3.6959	3.6048	3.5172	3.4331	3.3522
6	5.7955	5.6014	5.4172	5.2421	5.0757	4.9173	4.7665	4.6229	4.4859	4.3553	4.2305	4.1114	3.9975	3.8887	3.7845
7	6.7282	6.4720	6.2303	6.0021	5.7864	5.5824	5.3893	5.2064	5.0330	4.8684	4.7122	4.5638	4.4226	4.2883	4.1604
8	7.6517	7.3255	7.0197	6.7327	6.4632	6.2098	5.9713	5.7466	5.5348	5.3349	5.1461	4.9676	4.7988	4.6389	4.4873
9	8.5660	8.1622	7.7861	7.4353	7.1078	6.8017	6.5152	6.2469	5.9952	5.7590	5.5370	5.3282	5.1317	4.9464	4.7716
10	9.4713	8.9826	8.5302	8.1109	7.7217	7.3601	7.0236	6.7101	6.4177	6.1446	5.8892	5.6502	5.4262	5.2161	5.0188
11	10.3676	9.7868	9.2526	8.7605	8.3064	7.8869	7.4987	7.1390	6.8052	6.4951	6.2065	5.9377	5.6869	5.4527	5.2337
12	11.2551	10.5753	9.9540	9.3851	8.8633	8.3838	7.9427	7.5361	7.1607	6.8137	6.4924	6.1944	5.9176	5.6603	5.4206
13	12.1337	11.3484	10.6350	9.9856	9.3936	8.8527	8.3577	7.9038	7.4869	7.1034	6.7499	6.4235	6.1218	5.8424	5.5831
14	13.0037	12.1062	11.2961	10.5631	9.8986	9.2950	8.7455	8.2442	7.7862	7.3667	6.9819	6.6282	6.3025	6.0021	5.7245
15	13.8651	12.8493	11.9379	11.1184	10.3797	9.7122	9.1079	8.5595	8.0607	7.6061	7.1909	6.8109	6.4624	6.1422	5.8474
16	14.7179	13.5777	12.5611	11.6523	10.8378	10.1059	9.4466	8.8514	8.3126	7.8237	7.3792	6.9740	6.6039	6.2651	5.9542
17	15.5623	14.2919	13.1661	12.1657	11.2741	10.4773	9.7632	9.1216	8.5436	8.0216	7.5488	7.1196	6.7291	6.3729	6.0472
18	16.3983	14.9920	13.7535	12.6593	11.6896	10.8276	10.0591	9.3719	8.7556	8.2014	7.7016	7.2497	6.8399	6.4674	6.1280
19	17.2260	15.6785	14.3238	13.1339	12.0853	11.1581	10.3356	9.6036	8.9501	8.3649	7.8393	7.3658	6.9380	6.5504	6.1982
20	18.0456	16.3514	14.8775	13.5903	12.4622	11.4699	10.5940	9.8181	9.1285	8.5136	7.9633	7.4694	7.0248	6.6231	6.2593
21	18.8570	17.0112	15.4150	14.0292	12.8212	11.7641	10.8355	10.0168	9.2922	8.6487	8.0751	7.5620	7.1016	6.6870	6.3125
22	19.6604	17.6580	15.9369	14.4511	13.1630	12.0416	11.0612	10.2007	9.4424	8.7715	8.1757	7.6446	7.1695	6.7429	6.3587
23	20.4558	18.2922	16.4436	14.8568	13.4886	12.3034	11.2722	10.3711	9.5802	8.8832	8.2664	7.7184	7.2297	6.7921	6.3988
24	21.2434	18.9139	16.9355	15.2470	13.7986	12.5504	11.4693	10.5288	9.7066	8.9847	8.3481	7.7843	7.2829	6.8351	6.4338
25	22.0232	19.5235	17.4131	15.6221	14.0939	12.7834	11.6536	10.6748	9.8226	9.0770	8.4217	7.8431	7.3300	6.8729	6.4641

Appendix D: Table of Approximate PV Pumping-System Costs

TABLE D-1
Approximate Installed Costs of PV Pumping Systems

Insolation (Peak Solar Hours)						Approximate System Cost(*) (US Dollars)							
3	4	5	6	7	8								
20,000	26,700	33,300	40,000	46,700	53,300	$8,300	$9,600	$11,400	$13,600	$16,300	$16,500	**	**
13,500	18,000	22,500	27,000	31,500	36,000	$8,200	$8,900	$9,300	$12,400	$13,400	$13,500	$17,200	**
10,000	13,300	16,700	20,000	23,300	26,700	$7,000	$8,400	$8,300	$10,300	$10,600	$12,400	$16,500	$17,800
6,500	8,700	10,800	13,000	15,200	17,300	$6,700	$7,000	$8,100	$8,800	$9,800	$11,600	$13,500	$16,400
5,000	6,700	8,400	10,000	11,700	13,300	$6,500	$6,700	$7,100	$8,100	$8,700	$10,500	$12,800	$14,500
4,000	5,300	6,600	8,000	9,300	10,700	$6,100	$6,300	$6,800	$7,900	$8,000	$9,400	$11,800	$12,700
2,500	3,300	4,200	5,000	5,800	6,700	$3,600	$3,700	$5,200	$6,500	$7,200	$8,700	$10,500	$11,300
2,000	2,700	3,400	4,000	4,800	5,400	$2,800	$3,300	$4,300	$5,600	$6,500	$8,500	$10,300	$10,800
1,500	2,000	2,500	3,000	3,500	4,000	$2,600	$2,800	$3,900	$4,400	$4,700	$5,500	$7,000	$9,800
1,000	1,300	1,700	2,000	2,300	2,700	$2,100	$2,400	$3,200	$3,500	$3,600	$4,100	$5,000	$6,200
500	700	800	1,000	1,200	1,300	$1,600	$1,800	$2,300	$2,500	$2,600	$3,000	$3,400	$3,300
Yield (liters / day)						5	10	15	20	30	40	50	60
						Total Dynamic Head (meters)							

Notes:

(*) Approximate costs, which vary by country. Includes professional installation. Excludes cost of guarantees, import taxes, and VAT, which can vary widely.

(**) Larger solar pumps for this size of application are not common.

Index